Shaping Science

Shaping Science

Organizations, Decisions, and Culture on NASA's Teams

JANET VERTESI

The University of Chicago Press
Chicago and London

The University of Chicago Press, Chicago 60637
The University of Chicago Press, Ltd., London

Published 2020
Paperback edition 2023
Printed in the United States of America

32 31 30 29 28 27 26 25 24 23   1 2 3 4 5

ISBN-13: 978-0-226-69108-4 (cloth)
ISBN-13: 978-0-226-82955-5 (paper)
ISBN-13: 978-0-226-69111-4 (e-book)
DOI: https://doi.org/10.7208/chicago/9780226691114.001.0001

Library of Congress Cataloging-in-Publication Data

Names: Vertesi, Janet, author.
Title: Shaping science : organizations, decisions, and culture on NASA's teams / Janet Vertesi.
Description: Chicago : University of Chicago Press, 2020. | Includes bibliographical references and index.
Identifiers: LCCN 2020014489 | ISBN 9780226691084 (cloth) | ISBN 9780226691114 (ebook)
Subjects: LCSH: Spacelabs, Inc. | United States. National Aeronautics and Space Administration. | Space robotics—Research—Social aspects—United States. | Research teams. | Research—Management—Social aspects. | Knowledge management—Social aspects. | Organizational sociology.
Classification: LCC Q180.55.G77 V47 2020 | DDC 629.430973—dc23
LC record available at https://lccn.loc.gov/2020014489

♾ This paper meets the requirements of ANSI/NISO Z39.48-1992 (Permanence of Paper).

Every mission is kind of like a living organism. It has a personality and it has a style, and that personality and style is sort of gained at the beginning of the mission, and it never changes; even though the people who migrate through it change, you change out the people and you still have the same mission personality.

JAMES, planetary scientist

CONTENTS

PREFACE

This is a book about how organizations matter to what we know and how we know it: specifically, how a team's daily work, interactions, hierarchies, leadership, and decision-making affect the very discoveries we make about the world. I will argue that these aspects of everyday teamwork are not ancillary or hostile to science but actually play a central role in how research collaborations work, ultimately shaping their scientific findings and their project outcomes. To witness this in action, readers will take a deep dive into a special kind of collaboration, NASA's robotic spacecraft teams.

Spaceflight may seem a counterintuitive place to witness knowledge-making organizations in action. Public accounts of space exploration typically focus on the robots on the frontiers of the solar system bravely exploring new worlds, on the technical achievements of rocketry, or on the virtues of public versus private spaceflight. But the robots we send into space are not autonomous: their commands are issued from millions of miles away by a very human team of scientists and engineers on Earth. This group is large, distributed, interdisciplinary, and in some cases, international. They undergo extensive training and must dedicate countless hours and resources to build up the specialist laboratories, techniques, knowledge, and assistants necessary to produce results. They also gather together to contribute their best thinking, expertise, and resources to solve essential questions about the world. They are the quintessential example of a scientific and technical team, working across interdisciplinary boundaries to solve complex problems while operating at a high standard of excellence.

Ultimately, however, scientists and engineers alike are only human. And the way that they organize their collaborations looks a lot like how we organize our companies, our nonprofit groups, our governments, or any other group of people with a common venture that is too large to undertake alone. Replace "mission" in the epigraph quote with "company," "team," or "university" and it likely resonates with a lot of readers. The stories I will tell here could just as easily be a comparison of contrasting organizations such as Apple and Microsoft, the Commonwealth and the United Nations, or a community animal shelter and the ASPCA. The difference is that what results from spacecraft collaboration is not some kind of product, service, or set of laws. What results is knowledge about the world—or in this case, other worlds.

This book builds on the social studies of science to argue that *the organization of a scientific or technical collaboration shapes that collaboration's findings*. While the idea that science is organizationally shaped is this book's chief contribution, the idea that science is subject to social forces is hardly new. For more than forty years, social scientists have ventured into scientific laboratories to observe where knowledge comes from—especially scientific knowledge, which philosophers call "justified true belief." For instance, sociologist Karin Knorr-Cetina compares CERN with a biology laboratory to show how disciplinary differences affect knowledge work, while anthropologist Sharon Traweek describes how the "beam time" of the Stanford Linear Accelerator sweeps physicists' lives up in its thrall. Meanwhile, sociologist Diane Vaughan's study of the space shuttle reveals how institutional norms and work practices can routinize and normalize risk, historian Peter Galison describes how theorists and experimentalists come up with hybrid languages to communicate across divides, and Harry Collins details the decades of community work and tacit knowledge necessary to discover gravitational waves. Such studies and many others show how producing robust matters of fact requires particular tools, texts, and people as well as particular groups. Spacecraft are no different. Because robotic spacecraft teams often share the same people, institutions, and tools, *and yet are organized differently*, examining these teams in detail reveals how our knowledge of the world is interlaced with organizational concerns.

This is the second book based on my extended ethnography of the planetary science community, each one examining an aspect of everyday life on NASA's robotic spacecraft teams. Keen to avoid the pitfalls of the "sequel" effect, I've envisioned this project as something of a song cycle: an approach in classical music where a series of unique pieces are linked thematically, allowing the composer to illuminate different aspects of that theme with each tune. While *Seeing like a Rover* examined image work in detail, this book takes the reader inside two missions—to Saturn and to Mars—to demonstrate the relationship between how we work and what we know. The two missions are based in the same interdisciplinary field of planetary science, share team members, and occur at the same time: their explorations began and ended within months of each other. In addition to their considerable scientific findings, the teams are also model organisms (so to speak) for the critical role that organizations play in our social, technical, and natural worlds. In my view, this insight is no less important than discovering past water on Mars or new subtleties of Saturn's rings and deserves to be added to these missions' considerable list of contributions to human understanding.

The ethnographic studies I draw together here occurred in a relatively short sequence. Within months of leaving the Paris team, I was embedded with Helen, introduced to its new project scientist by a mutual friend, and gained badged access to the mission's home laboratory. I stayed at Helen for a year and a half, until I moved to the East Coast for a new academic position. From there, I kept tabs on the mission for five more years, joining their science team meetings intermittently and hiring other ethnographers to access different sides of the mission as my professional commitments intensified.

I did not come to planetary science expecting to study its people, institutions, cultures, robots, and findings for more than a decade. Like any laboratory ethnographer, I was drawn to specific elements of the field site as I encountered it. But the longer I stayed, the more I noticed. The paths of the individuals I followed and cared about began to expand, diverge, and converge again. As a sociologist, I found myself surrounded with new and emerging questions, unable to look away. Thus the characters and the robots return with more to teach us and more to say. To the many people and institutions that repeatedly opened their doors and their arms to me over the years, I offer my endless gratitude and my heartfelt thanks.

Tour guides never miss the opportunity to enter this building, walking visitors up to a mezzanine deck for a panoramic view and photos. I would often linger by these internal windows for a few extra minutes before heading to the offices upstairs where my own work took place. The screens marked calls from distant probes I had only read about, phoning home from across the solar system, alongside pings from probes I eventually came to know well. The room is a nerve center for planetary exploration, reaching across the country and past our planet to connect Earth and space. Even when no one is in the room, the scene is impressive. Watching the board, you feel at once small in the solar system yet also part of a grand and enormous adventure that far outsizes you and envelops you in its cause. *You are on a mission.*

It is not for nothing that spacecraft teams are called "missions," a word that orients their members toward a common purpose, uniting individuals across boundaries in a common site of focus, energy, and negotiation. Yet the work of exploring other planets is just that: work. A large team of scientists and engineers behind the scenes meets regularly to decide what their vehicle is going to do, produce commands for spacecraft activity, conduct analysis, and write papers about the data that their robots return from space. Like most contemporary organizations, the group's expertise is complex, but resources are limited. Each team member may have a different idea about where their spacecraft should go and what data it should collect, but there is only one spacecraft, and it can only do a few things at a time.

For this reason, serving on a spacecraft team is much like being on a very large bus with several hundred people. Everyone has a different idea about where that bus should go, what they would like to visit, and what they should take pictures of—but there is only one steering wheel. To solve this problem, each team sets up its own collaboration rules and roles for decision-making and develops a local culture to guide how observations will be allocated among its various scientists (i.e., for telescope time, see McCray 2000; Traweek 1988; on culture as problem solving, see Van Maanen and Barley 1985a). These organizational techniques vary from mission to mission, but they exert an important and enduring effect on the way that we understand the planets in our solar system and our collaborators as well.

In this book, I will argue that *the organization of scientific teams affects the construction of knowledge claims*—that is, different forms of organizing and enacting a scientific collaboration play into how decisions are made, how instruments are built and operated, and how objects in the solar system are crafted and known. Scientists typically dismiss organizational issues as a question of red tape, drama, or bureaucratic headaches in the way of their work. But if organizations affect outcomes, shaping scientific and technical practices, personnel development, and even results, then scientists and technical experts in research collaborations across the board must take note of this essential missing variable. Drawing on organizational sociology and science studies, I introduce organizations as a crucial factor in scientific knowledge, apart from scientific discipline or field, and I develop new ways of understanding laboratory forms, research technologies, and scientific objects wherever we may find them. This suggests that science teams can take a proactive approach to organizing their collaboration, rather than waiting for a strong leader or simply keeping their heads in the sand.

I will develop this argument through a comparative ethnographic case study of two long-term collaborative projects within the same field: a planetary science mission to Saturn and a mission to Mars. Both are outstanding teams involving top experts, impressive discoveries, impactful publications, and immense technical feats. They also overlap in terms of discipline, personnel, participating institutions, and time period of investigations, but a combination of funding shifts and business trends in the mid-1990s shaped them into two different organizational forms. As such, both address the essential question of contemporary collaboration: How to integrate large groups of people with a wide range of expertise in order to solve complex problems (Girard and Stark 2003; Heckscher and Adler 2006; Heckscher and Donnellon 1994; de Vaan, Vedres, and Stark 2015; Vedres and Stark 2010)? But like two different banks, start-ups, or nonprofit groups (Michel and Wortham 2009; Polletta 2002; Turco 2016), they make different organizational choices to solve this problem. In the coming pages, I will show their everyday practices of making decisions and prioritizing scientific observations. And I will show how the scientific knowledge, data, and even individuals that emerge from these organizations are shaped by these organizational processes, producing different visions of the planets as a result.

A Framework: Organized Science

Organizational sociologists have studied NASA before: most famously Diane Vaughan, whose study of the tragic *Challenger* accident detailed technical experts' extraordinary efforts to process tremendous amounts of information, reduce uncertainty, manage complexity, make decisions, and produce discoveries. As Vaughan noted, the institutions and organizations involved have "powerful and continuous effects" on the production of technoscientific knowledge in the way they gathered and processed information (Vaughan 1999, 931). A few years after her classic book about the organizational communication—and miscommunication—that lead to the fateful decision to launch, she posed a provocative question to sociologists of science: "How do the organizations for which you work . . . affect your own knowledge processes" (Vaughan 1999, 934–35)?

This question resonates in the sociology of scientific knowledge, where scholars are keen to understand the relationship between science and society. In an early, famous formulation of this connection, sociologist Steven Shapin and historian Simon Schaffer describe the early days of the Royal Society in London, arguably the first scientific institution of its kind. Competing visions were afoot as to how to establish the society and proceed with its knowledge-making enterprise. Comparing Thomas Hobbes's and Robert Boyle's contrasting politics in Restoration England and their corresponding ideas for their nascent community, Shapin and Schaffer argue that "solutions to the problem of knowledge are solutions to the problem of social order" (1985, 332)—that is, scientific institutions must first decide *how* they will know what they know, addressing the very practical question of how to set up their community, before they can decide *what* they know.

Since these foundational writings, scholars have attended to the many ways in which cultural narratives, institutions, or everyday interactions impact scientific work (Garfinkel, Lynch, and Livingston 1981; Jasanoff 2004; Martin 1991). Early laboratory ethnographers made note of the power differences between PIs and graduate students, the circulation of datasets, or how people interacted with their instruments and objects of study (Knorr-Cetina and Amann 1990; Latour and Woolgar 1979; Lynch 1985b; Traweek 1988). Yet the importance of laboratory organization soon faded against the importance of laboratory outcomes: the moral dilemmas of biological citizenship, the circulation of test objects, the rise of the scientific persona, the clash of experts and policy makers, the management of databases, and so on. The relationship between how we organize ourselves and how we produce knowledge about the world remains an open question.

Attention to what happens when research circulates in the world is essential for good citizenry and for responsible science. But the organizational layers that infuse such knowledge in the first place deserve our investigation as well. If how science is *organized* affects how science is *done* and what a collaboration can and cannot prove, then those of us who care deeply about scientific outcomes and their surrounding debates must also care deeply about how such knowledge work is organized at its point of inception. And if scientific work is concerned with the construction and stabilization of objects and subjects—that is, the natural world and the people who explore it—then how *organizations* participate in this process must be understood. At stake is nothing less than knowledge itself.

This book therefore returns to Shapin and Schaffer's claim, inverting it to ask: *Do different solutions to the problem of social order present different solutions to the problem of knowledge?* In other words, does the way that we order ourselves, our communities, and our machines in some way determine, prefigure, or help constitute the kinds of things we know about the world? And if so, *how*? To answer these questions in detail, I turn to the organizational—or "meso"—level, a perspective that allows us to witness the interweaving of science and society on the ground. I introduce the analytic framework of *organized science*: a multipart framework aimed at unpacking how organizations contribute to the constitution of knowledge and demonstrating the role of organizational dynamics in the production and circulation of scientific facts. The principles of *organized science* are threefold:

1. **Science teams are organizations.** Each scientific collaboration, research team, or knowledge-making community is an organization, with its own routines, rituals, practices, and interactional norms that members use to conduct their scientific and technical work. Each collaboration also has its own rules for decision-making, patterns of authority, status-based interactions, and lines of accountability. Taken together, this comprises their organizational order. Due to the expansive character of contemporary scientific teams, these organizations need not be confined to specific institutions, geographic areas, or disciplines. While certain types of laboratories may organize themselves in the same way, with common forms shared across a scientific field, the epistemic cultures associated with scientific collaborations arise from their local social organization and not, primarily, from their discipline.
2. **Scientific organizations shape scientific outcomes.** A scientific collaboration's data, publications about its findings, and the careers of individuals in the organization are produced and shaped through the team's organizational practices. This includes the results of research, the patterns of conversation, and the knowledge that emerges from such collaborations—all bear the indelible stamp of the organizations that produce them. Even the off-the-shelf or homegrown technologies that collaborations use to facilitate their work are used in local, organizationally acceptable ways. Hence research findings, instrumentation, and data do not present some sort of external constraint upon or justification of scientific outcomes: they are themselves the outcomes of organizational processes. In other words, *scientific collaborations that are organized differently*—with different interactional norms and authority structures—*produce different kinds of scientific knowledge*. This is the core tenet of *organized science*. This does not mean that facts are entirely foreordained by the groups that produce them. But because

facts are produced and enacted in an organizational context, this gives them a texture and contour that is isomorphic with the organization from which they emerge.

3. **Scientific outcomes impact scientific organizations.** A knowledge collaboration produces streams of data, scientific papers, technological advancements, and even generations of personnel. How do these elements participate in stabilizing, authorizing, or reproducing the organization? I demonstrate that in scientific collaborations, such outcomes feed back into the organization as a form of looping effect. Observing this iterative process in action over many years reveals how collaborations produce a robust, trustworthy constellation of social, technical, and scientific objects. Not only do these iterative loops produce a scientific organization's outcomes; they also make a collaboration's local culture, its collaborators, and its artifacts appear natural and necessary to the team's scientific work.

Scientists—even social scientists—reading this book may find these principles to be a dubious claim. Surely the natural world is simply out there, waiting there for us to discover true, binding, and universal facts. But time and again, the history and sociology of science and technology show that our knowledge of the natural world is predicated upon our communities and commitments, our cultures and our interactions. Our instruments, questions, and observations are keenly attuned to particular phenomena of interest so that we may examine them in detail (Hanson 1958). This means that we miss out on elements of the natural world that we haven't yet thought to examine or built the right apparatus to explore. We also build our databases, encyclopedias, and catalogs around the kinds of things (and people) that we know, leaving other categories and experiences out (Bowker and Star 1999). Even the power of the scientific method, with its emphasis on hypothesis, test, and experimental replication, is hampered by issues such as who has prior experience, tacit knowledge, or "magic hands" in operating a finicky instrument (Collins 1985, 2004). Because we cannot know what we don't know how to know, decisions about how knowledge work should proceed are central to long-term scientific collaborations. This means addressing questions such as which organizational model guarantees good governance and good science and how individuals should interact with each other and their instruments to apprehend and share scientific truths. Careful consideration about how organizations shape science should matter to everyone who has ever worked on a research team, in industry or academia; to those who aim to lead such teams; and to those who fund or support them.

Answering such questions requires bringing together empirical investigation of scientific collaborations with the literatures from the sociology of science and technology on the one hand, and organizational sociology on the other. Science studies brings the tradition of laboratory ethnography and a sensibility to the construction of scientific objects and subjects. This serves us well in disentangling the ways in which local scientific practices shape understanding of the world, produce particular relationships to our instruments, and craft scientific selves. Meanwhile, organizational ethnography offers ways to link the local myths, rituals, and interactional norms of an organizational milieu to larger-scale institutional phenomena, such as how organizations produce particular kinds of value or how they reproduce themselves. The structure of scientific collaborations plays a key role as well, no less important than posited in the early days of the sociology of science. This requires attention to a collaboration's locus of decision-making and its status hierarchies, as well as to the way that different forms produce legitimacy, authority, and solidarity. I will discuss these issues below.

Answering these questions also requires an on-the-ground investigation of how organizations matter to scientific work. NASA provides a useful site from which to examine these claims: not because of its totalizing or bureaucratic nature as an organization itself but because it sponsors robust, long-term, interdisciplinary, and interinstitutional collaborations in the form of spacecraft teams. These teams are selected to build and operate a spacecraft in order to explore some aspect of the solar system. They unite scientists and engineers at universities, research centers, and agency affiliates, working together in a concerted effort yet under different organizational forms. In this way, teams that share the same personnel, resources, institutional constraints, and disciplinary affiliations may yet still diverge in the kinds of knowledge they produce. Detailed ethnographic investigation of two such teams demonstrates the role of the collaboration's organization in producing scientific facts. In other words, when choosing one model of organizing their team over another, more is at stake for researchers than just how people prefer to work together.

The first half of this book, then, will examine the organizational elements of two contrasting scientific teams, introducing literature and thinking from organizational sociology, organizational behavior, team science, and sociotechnical organizations. The second half turns to the implications of these organizational choices for knowledge: which questions are asked and answered, what data are acquired and circulated, whose careers rise and fall. Ultimately, the two stories I tell here offer new ways of thinking about how to set up, study, and evaluate a knowledge-making endeavor—whether a scientific collaboration, a technical team, a research unit, or a community organization—with an eye to the role of organizations in the production of facts, artifacts, and individuals.

Organizational Orders

Weaving together the literatures of science studies and organizational sociology offers us fresh ways to analyze organizations as a meaningful fact of laboratory life. Early work in the sociology of scientific knowledge demonstrated that we know what we know scientifically by virtue of being embedded in a community with refined practices and skills as much as by virtue of training in the scientific method or access to special instruments. Since then, attention in science studies has focused on how laboratory cultures impact scientific reports. For instance, Bruno Latour and Steve Woolgar, in their seminal *Laboratory Life,* portray life in the Salk laboratory, while Sharon Traweek describes Japanese versus American styles of building and managing particle accelerators at KEK and at Stanford (1979; 1988). Amid these foundational studies, Karin Knorr-Cetina formulates the concept of a discipline's "epistemic culture," demonstrating that high energy physics collaborations such as the ones at CERN possess a specific way of producing knowledge that differs from, for instance, that of a microbiology laboratory.

Since then, social scientists have deployed the framework of *culture* to discuss the intangible, tacit "ways of doing things around here," including the "microcultures" or "ideocultures" of group work in scientific collaborations. For instance, sociologist of science Harry Collins identifies different "evidential cultures" associated with institutions in gravity wave science that dictates the data collected and circulated among colleagues (1998); sociologist Gary Allen Fine shows how one national weather office's watch-and-wait culture versus another's rush-to-report culture produces different forecasts for their respective regions (2010); anthropologist Hugh Gusterson investigates peace-oriented scientists' narratives that justify their working at a nuclear weapons research lab; and sociologist David Peterson describes the "bench building" that distinguishes between natural and social scientists (2015).

If culture is a key explanatory resource for the social construction of knowledge, structure is often left out of the picture. Organizational structures have not gone unexamined, however. Sociologist Robert Merton, working from the logical positivist tradition, highlighted the importance of lineages between scientists and their students as well as the independence of disciplines and fields (1965; 1968). He investigated these structural considerations as separate from—although no less important than—cultural considerations such as the cultural preconditions for science (1938), or his famous "norms" of science: communalism, disinterestedness, and organized skepticism among them (1942; contested in Mitroff 1974). Post-Mertonian studies of science have made great strides toward understanding the structures of scientific work, such as by characterizing the different configurations that collaborations adopt across fields (Shrum, Genuth, and Chompalov 2007). Following the rise of poststructuralism in the academy, however, structuralist elements such as hierarchies and organizational forms are largely absent from many contemporary accounts of science—including schools of thought in constructivism, actor-networks, and agential realism—or are reconfigured to address the question of practice or self-discipline.

The two collaborations I discuss here do have different cultures, but they are also organized along contrasting lines, which distribute authority and accountability differently among team members. This matters a great deal for their work, as I will show. The consequences of these choices are clearly visible in these field sites as well. For instance, solidarity effects can be traced to each group's organization of tasks in a manner predicted by Émile Durkheim's writings on the division of labor (1893). Although Durkheim famously associated flattened hierarchies and low differentiation of expertise with early stages of development in society, these social forms are now immensely popular ways of organizing in the advanced technology sector. And scientists involved in disputes appeal to charismatic or bureaucratic authority, both forms of "legitimate domination" according to Max Weber (1968). Thinking organizationally, *culture* alone cannot do all the explanatory work for understanding how organizations shape science: structure must play a role as well.

Fortunately, both the conditions of scientific collaboration as I encountered them and contemporary techniques for studying organizations address precisely this problem. Like other funding organizations in the late twentieth and early twenty-first centuries that privileged "collaboratories," these NASA teams span institutional and disciplinary boundaries, with many members themselves trained in the "interdiscipline" of planetary science. The same period saw the rise of "flat" organizations in the sciences and engineering, especially popular in the technology start-up sector as a paragon of innovation, to encourage cross talk between experts. While these changes might raise challenges for a classic structuralist, as ready distinctions between fields (and field sites) are less visible, they also offer an advantage. Sociologists who study such organizations—now known as "postbureaucracies" or "heterarchies"—show that these groups distribute decision-making among members in a participatory fashion, reserving a more limited role for hierarchical authority. Because interactions among team members produce the decision-making authority of the organization, such interactions now carry structural weight and importance. Where it was once useful to identify "formal" versus "informal" aspects of the organization to catalog how members of an organization evade, outwit, or ritualize its authority structures (Blau 1980; Lipsky 2010), in "flat" companies, these elements are now relatively indistinguishable in practice. For instance, an employee in a traditional firm might know when it is better to talk to the secretary instead of the CEO to get something done, but flat team members legitimate a course of action among themselves and know which lines of authority to appeal to as they do so. As such, contemporary ethnographers of organizations are less interested in mapping formal versus informal interactions to structural versus cultural mechanisms, focusing instead on characterizing where and how decisions are made and who can speak within or for a team (Freeland and Sivan 2018; Heckscher 1994; Kunda 2006; Turco 2016).

Rather than addressing structure as a lofty, idealized element invisible in practice, then, or something imposed by a fleet of managers wielding an impersonal organization chart (or by sociologists wielding a hefty volume of Weber), structural issues like authority, status, decision-making, and legitimation are observable and reportable, performed daily in routine tasks and interactions by individuals in the ranks. In a postbureaucracy, these actions make sense to members of the organization as cultural elements, such as shared forms of talk and group sense-making, and are at the same time part of structuring the organization. For the analyst, this means that we can witness on-the-ground practices of ordering people and machines.[1] Further, decisions about which observations to take not only reveal each team's organizational structure but also produce the datasets that fuel scientific discoveries. I will therefore pay special attention to how these groups make decisions, including to whom and what they appeal in order to do so.

Early social investigations of scientific work were, for the most part, situated in hierarchical scientific and engineering laboratories or the bureaucratic environments of National Labs. It was perhaps easier in those cases to hold *culture* to account because these structures did not exhibit much variety and because culture is what ethnographers investigate. But sociologist Michael Burawoy (1979, 6) asks that we pay attention to how the theories we develop about organizations, labor, and workplace dynamics often obscure the historically specific conditions that give rise to those very dynamics.

Thus examining contemporary, lateral collaborations reveals certain base assumptions in our theorizing that may not hold true of labor under postmodern conditions. Studies of ritual company meetings reveal corporate control enacted through shared forms of talk among employees (Kunda 2006). Studies of hospitals and technology companies show how organizational change or cohesion is visible in how people talk and work among the ranks (Kellogg 2009; Powell and Colyvas 2013). In cases where hierarchical structures are unclear, studies of civil rights organizations (Polletta 2002), participatory festivals (Chen 2009), and technology start-ups (Turco 2016) describe how those who reject hierarchy struggle to find the vocabulary and effective practices that support their local collaborative goals. Even the poststructuralist's interest in tensions of structure/agency, human/nonhuman relations, or power/knowledge is grappled with daily in empirical cases of contemporary flat organizations. Removing the assumption of hierarchy and bureaucracy may therefore reveal the organizational premises of science all along, given both the variety of forms that ensue and the tools to examine those forms in action.

In this book, then, I embrace the insights that come from considering elements such as authority, hierarchy, division of labor, and legitimacy but do so by setting aside the tired structure-culture binary. I focus instead on decision-making practices that members accept as legitimate (or decry as illegitimate). I examine those observable microinteractions, rituals, and narratives that elucidate the organization's local order (Barley 1996; Garfinkel 1967; Powell and Colyvas 2013). And I allow for the local, artful fusion of multiple appeals to authority—hierarchies, collectives, charisma—as is common in these organizations. Inspired by both microinteractional approaches to organizations and science studies, then, I treat structure and culture not as a binary, or as a situation wherein one is held stable and the other becomes an independent variable, but as integrated components of an *organizational order*—or, *pace* Donna Haraway, as a hybrid "structure-culture." This blends what we once conceptualized as discrete organizational elements, attending instead to members' local ways of managing authority, accountability, and decision-making on the ground.[2]

This approach offers several contributions to science studies and to organizational sociology. The first is to highlight the role of social structures in constructivist accounts of scientific work. In this way, we can address how groups of scientists confront and solve problems and allow for institutional heterogeneity and comparability across disciplines and sites, without giving up the benefits of a microsociological lens on group interactions. This also allows us to trace which elements of the collaboration scientists appeal to for decision-making and when and to examine how scientists assume a role in a collaborative team, assert that they are doing their jobs well or poorly, and recognize others associated with their group. Even shifting the locus of variation from *disciplinary* or *institutional* cultures to *organizational* settings allows for new axes of comparison among scientific collaborations. In particular, it allows us to explain why groups with different collaborative logics and architectures might *do science differently*—even if they involve the same disciplines, institutions, and people.

The second is to locate another site of resonance between the sociology of knowledge and studies of ideology and institutions. Studies of nuclear power in postwar France, cybernetics in socialist Chile, and computer networking in the Soviet Union reveal how ideological commitments inspire national research agendas, configuring arrangements of people and machines in support of a political program in what historian Gabrielle Hecht calls "technopolitics" (Hecht 1998; Medina 2014; Peters 2016). Largely due to connections to statecraft, these studies engage directly with questions of social organization in technical collaboration. So too do studies of scientific institutions such as national laboratories or transnational committees, which frequently assume hierarchy and bureaucratic oversight (Doing 2004; Gusterson 1996; Law 1994; O'Reilly 2017; Thorpe and Shapin 2000; Vaughan 1996). There is of course a time, a place, and a national agenda to the spacecraft teams I will describe as well, explaining their commitments to decentralized decision-making, individual expertise, and strong leadership at the same time. But when it comes to the politics of their organizational choices, there is no smoking gun. When asked, team leaders equate their organizational choices with simply doing "good science," and they work hard to buffer themselves from what they see as transnational politics or NASA's institutional concerns that threaten their group work. I therefore follow my actors' lead and analyze the politics of their organizations as such politics become perspicuous for them, as in chapter 5. Elsewhere, I focus on the role that their organizational choices play in group solidarity, human-robot relations, and the formation of scientific questions. This demonstrates the power of an organizational lens across scientific sites, even beyond ones of national significance.

Contemporary work in science studies confirms that social practices not only produce *knowledge*; they craft *objects* as well. For this reason, this book also develops an analysis of the role of objects in organizational work. Anthropologists have already convincingly demonstrated that the planets are infused with cultural meanings through scientific practice. Lisa Messeri describes how planetary scientists transform distant worlds into knowable places, while Stefan Helmreich shows how studies of the deep ocean or hydrothermal vents are ways of studying life on Earth and in space (Helmreich 2009; Helmreich, Roosth, and Friedner 2015; Messeri 2016). Others have also shown how robots of various forms are also imbued with cultural sensibilities and interactional expectations (Alač 2009; Šabanović 2010; Suchman 2011) and how modes of "organizational ordering" narrate agency within locally meaningful assemblages of people and objects (Law 1994).

These insights are of special importance here. In this frame of view, objects are not immaterial social constructs, nor are they external constraints upon action. They are brought into focus—even into being—through actions that engage both humans and the material world, and they are made legible through human interaction and narrative (Barad 2007; Haraway 2007; Law and Mol 2002; Mol 2002; Woolgar and Lezaun 2013). In other words, they are both material and meaningful at the same time. As a naïve example, assembling a few pieces of wood into a dining table can transform a dinner party, as an obdurate object such as a table takes on cultural meanings, expectations, and interactional scripts for a certain kind of meal, conversation, and even utensils. When objects are infused with social meaning, the material world that emerges from local practices anchors our social relations and vice versa.

This approach is useful when it comes to analyzing the spacecraft—and planets—involved in planetary science as they participate in their mission's social relations. People learn how to use and interact with their spacecraft as part of their mission's organizational order. The opportunities they perceive for its investigations and the limitations they place on its movement are dictated by the pressures of a hostile planetary environment and aging machinery as much as by the way that their Earth-bound team has learned to care for their craft, make decisions about its future activities, and allocate responsibility for its management. For this reason, I will not resort to technology as the explanation for particular kinds of behaviors but will describe the role these technologies play in local organizational orders (as in Leonardi 2012b; Leonardi and Barley 2010; Orlikowski 2010). Instead of explaining how the spacecraft requires or constrains certain human actions, I will turn this statement around to show how *an organization's ways of enacting the spacecraft* actually perform and even anchor social relations on Earth. Instead of describing how the planets require particular kinds of actions or interactions on the part of the spacecraft or the team, I will show how spacecraft actions and team interactions construct a particular vision of the planet, encouraging certain topics for investigation that are isomorphic with each team's organizational order. And I will describe how these activities are organizationally situated as well as constitutive.

This requires thinking about how the analytical objects that a collaboration *produces* are the result of local organizational orders as well. To take an example well known in science studies, the hospital that enacts a disease differently in each of its units (Mol 2002) speaks as much to *organizational distinctions* between a hospital's various departments and patterns of their collaboration and communication as it does to a disciplinary difference in how a radiologist versus an internist might understand the same disease. In my own site, there is an *organizational* "ontological choreography" (Thompson 2005) afoot that builds different pictures of Mars and Saturn, enacting these planets along organizational contours. Because the organizational, the epistemic, and the ontological are intertwined in the practices and interactions of scientific work, this approach offers new insight into how a collaboration's local fault lines and interactional norms reproduce themselves as perceived natural qualities of objects in the world. It also demonstrates how power is produced alongside facts and artifacts (Winner 1993), injecting organizational topology into the networks, actions, and interactions that produce scientific knowledge, objects, and communities.

Why These Missions?

Sociological comparisons must establish commonalities as well as an axis of variation to show social mechanisms at work in the world. These two missions set good examples. They are both large, multiyear, multinational collaborative spacecraft teams with a shared center of operations located at a contract facility, which I call Spacelabs. They both placed a remote spacecraft in the solar system at or on their planet of interest in the same year, losing contact with these probes within months of each other more than a decade later. And both missions are comprised of a large team on Earth responsible for spacecraft activities, in addition to the probes themselves.

There are differences too. The Paris mission drives two vehicles upon the surface of Mars; Helen is a large orbital probe that flew through the Saturn system in a "tour," sweeping past its moons, its rings, and through its plasma field regularly in a dynamic dance. The twin Paris robots feature a half-dozen onboard instruments and, at time of fieldwork, approximately 150 scientists and engineers; Helen has twelve instruments and around 300 scientists (although this is a low estimate[3]); these were differences in complexity that, for their operators at least, place them on "opposite ends of the spectrum" (Cheng et al. 2008). Americans worked alongside citizens of several European nations on all instrument teams,[4] although Helen involves the European Space Agency (ESA) and the Italian Space Agency (ASI) as partners, while Paris has a few international investigators and instruments but was not founded based on international collaboration. I will describe the importance of the above issues for the different teams later in the book. However, central to this study is the difference in each team's sociotechnical organization. Each mission is oriented toward a different model of knowledge integration, responding to a problem facing many twenty-first-century organizations—namely, how to organize a diverse group of experts in such a way as to capitalize on their wide-ranging expertise.

All planetary science missions face this challenge, and each spacecraft team approaches the problem differently. All must achieve some balance between autonomy and depth of expertise on the one hand and connectivity, interdependence, or breadth on the other. Like many scientific collaborations, some produce teams in which shared questions trump disciplinary or institutional affiliations, while others allow many different groups to pursue their own goals, bringing findings together in a process of patchwork (Balakrishnan et al. 2011). To establish interdisciplinary ties, some projects employ singular bridge-builders between domains (Burt 2004), others establish overlapping teams (de Vaan, Vedres, and Stark 2015; Vedres and Stark 2010), while others still rely on a visionary leader to bring everyone together across fault lines (Chen 2009; Isaacson 2011; Thorpe and Shapin 2000). Helen and Paris both stake out positions on this continuum between independence versus interdependence associated with the integration of expertise using contrasting organizational forms.

The Paris mission solves this problem with a flat hierarchy, a style of management popular in technology start-ups at the turn of the twenty-first century. A single principal investigator leads a team of scientific experts unified under a single umbrella. The scientists may choose to use a particular spacecraft instrument for the majority of their work but are not officially placed into separate camps; instead, they work collectively as a unit. In principle, any scientist or engineer can request any observation from any instrument at any time. Their spacecraft too features an integrated systems approach, with instruments producing combinable datasets that are frequently viewed together. This approach holds across the team, sharing expertise and flattening social relations between scientists and engineers (figure 1).

On the other side of the solar system, Helen solves the problem of knowledge integration with a matrix organization. This is a team structure made popular in aerospace industries in the 1960s and 1970s. Its popularity in the 1980s and 1990s coincided with the rise of Total Quality Management, a response to competition from the Japanese automobile industry,

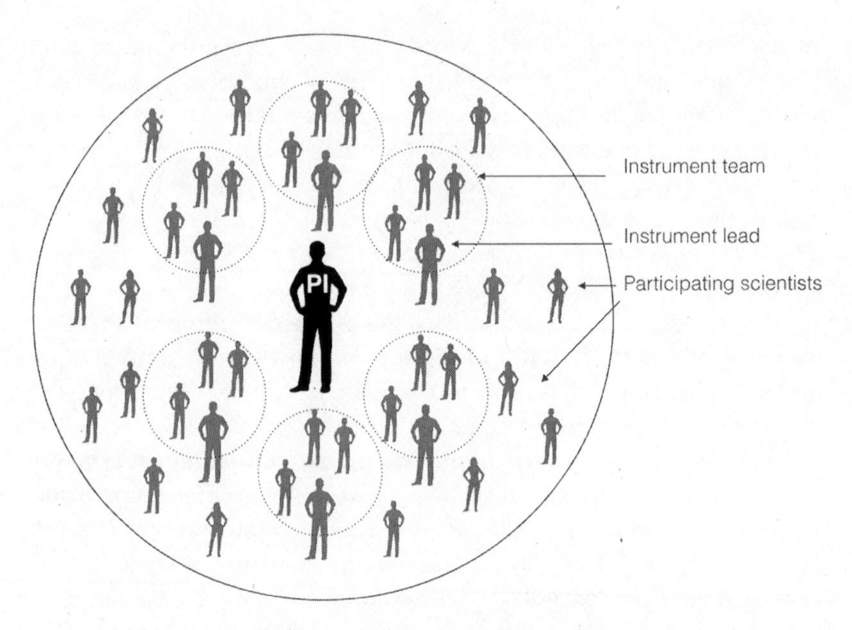

1. Diagram of Paris's organizational form

which placed responsibility for action lower in the corporate hierarchy than ever before (Appelbaum, Nadeau, and Cyr 2008; Burns 1989; Cleland 1981; Davis and Lawrence 1977; Ford and Randolph 1992; Galbraith 2009; Morrill 1991). In a matrix organization, workers are sorted into groups based on their substantive expertise, and each of these groups has leaders and managers much like a traditional bureaucratic hierarchy. But individuals are also grouped across these bureaucratic lines into project teams, each with their own project manager. In this way, matrix organizations are said to combine both vertical and lateral lines of accountability.

Adopting this accountability structure, Helen features twelve instrument teams, each with a distinct group of scientists who work as specialists with their team's instrument. Scientists also belong secondarily to a crosscutting working group corresponding to the part of the Saturn system that they study. At the top of the instrument teams' hierarchies is a roster of principal investigators and team leaders, while leading the crosscutting groups are individual interdisciplinary scientists (IDS), a NASA-selected role that allows them to combine more than one instrumental specialty (figure 2). A project scientist and his or her deputy collect recommendations from the scientists on the mission to give advice to the engineers; the latter are organized under a project manager and deputy who oversee operations within a hierarchical reporting structure.

The matrix and the collective solve for certain problems but share others with similar types of organizations. One concerns those individuals, such as Helen's interdisciplinary scientists, who serve as bridges between different constitutive groups. These may be the sites of innovation but also friction and vulnerability (Burt 2004; Powell, Koput, and Smith-Doerr 1996). We also know that flat organizations may embrace radical participation (Chen 2009; Polletta 2002), but they may also reinforce power distinctions and dynamics under the rubric of novelty (Jemielniak 2014; Reagle 2010; Turco 2016). Finally, recent studies of flattened companies demonstrate that even if communication cuts across the organization, hierarchical authority still persists at least for certain kinds of decision-making (Freeland and Sivan 2018; Turco 2016; Zuckerman 2010). As I will describe, both Helen's matrix and Paris's collective as organizational orders integrate hierarchy into their lateral lines of authority, making them mixed-mode "heterarchies" with attendant concerns for the clash of evaluative logics (Stark 2009).

I would be remiss in this introduction if I did not mention an additional, obvious difference between the teams: the Paris robots operate on the ground, while the Helen orbiter loops through a planetary system in

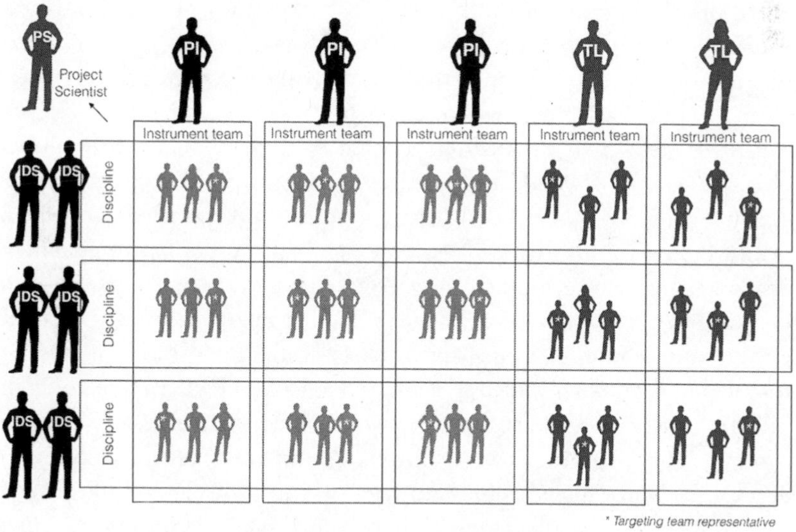

* Targeting team representative

2. Diagram of Helen's organizational form

space. In the course of my research, team members often pointed to this as the reason they have such different organizational forms—because their technical apparatus and positions in the solar system demand it. Certainly, this produces different experiences of working on the missions as well as different local rhythms of time and scale, as I will discuss. A sense of alignment with Earth days and years saturates Paris, where commands are produced and executed with a daily turnaround, while Helenites plan in multiweek segments and think in decadal scales for their collaboration—across "births, deaths, marriages," as one scientist I interviewed put it. However, the robot's type and its environment do not require a particular organizational arrangement. The next vehicle to fly to Mars after Paris deployed a Helen-like organizational chart despite its daily planning schedule, such that during its development, scientists called it "Helen on wheels." Meanwhile, flyby and orbital spacecraft exploring Pluto, Mercury, and the Martian atmosphere have flown with a flat hierarchy under a single PI, like Paris.[5] With such variety, we cannot truly say that organizational and technical forms are mutually predetermined. Members' assertions that their spacecraft simply demands a certain type of interactions from them are the scientific organization's *explanandum*, not its *explanans*.

Nor do operational considerations fully determine interactional norms. Some scientists argue that a ground-based mission has time to delay enough to build consensus while a probe whipping around a planet at millions of miles per second does not. This is true, but it ignores how time is experienced on both teams. On Paris, there was constant fear that their mission could end at any time due to hostile conditions, such as weather events on Mars or political shifts on Earth. Team members hesitated to keep their robots in one spot and only did so when it was impossible to move due to low power or when the aging onboard instruments required multiple days of operation to produce a signal. In such cases, the team faced considerable anxiety over the robots' lack of mobility and agonized over where to move their robots as soon as they were able in order to avoid sudden death—an impending moment they referred to as death by "sniper." Even a decade into the mission, team members exuded a nervous energy that their efforts could end at any time, for instance, initiating a "blood pact" that they would move their vehicle as soon as possible following a series of high priority observations. Meanwhile, when I joined the Helen mission, the team was busy planning a tour that would lead to their craft's intentional—and spectacular—demise eight years later. Scientists displayed little anxiety that it might not last that long, although they were of course concerned about preserving resources such as fuel or protecting the spacecraft's aging parts insofar as this would ensure continuous data collection until its death. The long *durée* of this mission therefore produced a sense of much time in which to accomplish mission goals. The fact that there were so many goals to accomplish within this timeframe, and the necessity to prioritize among them, was the predominant cause for anxiety.

The particulars of space travel, planetary science, or the exploration community do not demand the organizational orders that I describe here. But the sociotechnical organizations that result have a tremendous impact on science, exploration, and individuals. It is for this reason that I will tell not just one but both of these stories here. Each story could easily be a book on its own,[6] yet each team helps us better understand the other. It is through comparative work of this kind that moments that seem particular to one organization are shown to be shared across others with similar forms, while moments that appear universal or natural to one set of team members can be understood as very local indeed. More important, seeing both sides of the story from different sides of the solar system helps us glimpse the organizational orders at work in the production of scientific knowledge. And it is only through telling both sides of the story that young scientists who have had exposure to only one type of collaboration can situate their own experience and grasp the many and varied ways there are of working together. This goes for young executives, nonprofit planners, and engineers in other disciplines as well. For these reasons, I will move back and forth across the solar system in my ethnographic narrative and invoke the reader's patience in telling the critical details.

On Methods and Theorizing

Those details were painstakingly collected over a decade of immersive ethnographic fieldwork with the planetary science community, conducted most intensively between 2006 and 2013, with group work between 2010 and 2013 and intermittent follow-up work through to 2017. I offer additional description of my research practices and how they inform my narrative in the methodological postscript, but certain elements deserve attention up front.

Ethnography is a practiced technique by means of which a researcher embeds themselves in a community over an extended period of time. Through observation and participation, extensive interviewing, writing field notes and memos, the ethnographer develops a more complete picture of what community membership is like in this particular setting. Ethnographic case studies are a primary source of theorizing and material for science studies for which laboratory studies are foundational for understanding the actual, everyday practices of science. This requires a keen attention to details, to be sure, but unlike reportage, my ethnographic perspective was never aimed at getting a scoop on the sites that I study. I am instead interested in tracing the underlying principles of organizational activity, visible through a team's interactions, rituals, and local cultural norms. Like other ethnographers of science, I record the "social facts" (Durkheim, Lukes, and Halls 2013; Latour and Woolgar 1979; Lynch 1985a) that community members share or on which they diverge and how they point to those social facts as motivation for their actions and interactions. This requires capturing and unpacking members' methods for making sense of their world (Durkheim, Lukes, and Halls 2013; Garfinkel 1967; Latour and Woolgar 1979; Lynch 1985a), and putting such practices into organizational context. For this reason, I will deploy many of the tools of ethnomethodology in my analysis: a sociological subfield well suited for analyzing the local, situated production of social order.

Even as my attention will be focused on interactions among team members, many layers of theory will be interwoven throughout my account. We will see the importance of a team's division of labor for how participants feel about each other and their robots (Durkheim 1893), and the continued presence of charisma and hierarchical authority on "flat" teams (Weber 1968). At the same time, as we focus on how decision-making is accomplished, we will witness how power is effected and distributed (Foucault 1977), how social order is maintained while natural order is investigated (Garfinkel, Lynch, and Livingston 1981; Lynch 1985a), and how groups constitute, approve, or deny knowledge claims (Shapin 1994). My analysis weaves these concepts together into the text as the underlying guides for understanding human interactions with machines, and with each other, on both teams. The act of ethnographic writing is one of theorizing as well.

Although one of my participants frequently joked that I was going to write a "kiss and tell" book about his mission, I had a different aim in collecting the details of their interactions and individuals' perspectives on their mission's work. As I explain to planetary scientists, sociology is like space physics. You place an instrument in an environment flooded with all kinds of particles—some charged, some ionized, others free-floating. Using that instrument, you observe where and how those particles move and what happens to them as they move, such as when and where they pick up a charge, whether they suddenly start to spin, whether there are any discernible patterns. Your goal is not to categorize, adjudicate, or pathologize these particles. Instead, observing their unique properties in conjunction with theories of electromagnetism or particle behavior helps you discern the system's underlying *field lines*: the invisible currents that dictate particle direction, spin, charge, and polarity.

Extending this metaphor, organizational ethnography is a question of detecting *field lines* in social worlds as well. This requires describing how, where, and why people—like particles—behave or spin out, move and flow, join and disband; it also requires attention to the specific logics of their local accounts and sense-making, maintaining the particularities of their situated experience and comparing these to other theories and accounts of organizational interaction. Like particles, each individual has their own unique story, but they are also guided by, affected by, and even changed by these underlying field lines in ways that social scientists analyze and understand. Further, field lines are dynamic. At once underlying, directive, and reflexive, as they shape the environment in which particles can move they are also shaped by those particles' movements in turn. In a social science that already has several definitions for "fields," the metaphor of field lines hones the ethnographer's sensitivities to the underlying, active, and interactive dynamics that corral and make sense of actors' situated experiences and accounts. Their detection allows us to move from a thickly described "how" to inferring "why" (Katz 2002).

Collecting these experiences required considerable work. As I will describe in the book, these are multisited teams (Marcus 1995), distributed across centers that may jockey for positioning, and all of which must balance institutional interests with mission work. Thus no single-sited view, no matter how detailed, will enable an organizational ethnographer to fully understand these missions. To best understand the missions from as many perspectives as possible, then, I followed mission scientists to their home institutions, their conferences, and their specialist workshops. I visited nine affiliated Paris sites and relevant Mars conferences and meetings; I attended the Titan jumpstart, Paris science team meetings, and nine PSGs; and I followed up with site visits to as many of the Helen scientists' home institutions as I could, sometimes conducting interviews in the local language. At these meetings, interactions were intense, beginning typically early in the morning with breakfast meetings and talks and lasting until dinner and drinks late into the evening. As a result, being present and available to participate meant that it was often only possible to record impressions in field notes after the week was over. I coded these retrospective notes, photographs, and interview transcripts as part of the full dataset.

I also attempted to visit and interact with as many different organizational parts of the two teams as possible. This meant conducting interviews with people in all positions, from summer research interns to PIs and administrators. It also meant visiting different subteams. On Paris, my interaction with engineers was limited due to export control regulations, but the few opportunities I did glean to talk to engineers about their work assisted me in understanding their perspective as opposed to—or as it was brought into confluence with—the science team. On Helen, this meant visiting as many as seventeen distinct groups within the organization, each with their own local interactional norms and methods of decision-making. I spent time at seven of the twelve instrument teams' meetings or institutions (and sent a team of ethnographers to visit five others, of whom I will say more in the appendix); attended all the working group and targeting teams' meetings, including additional working groups not described in the book; attended specialist conferences; spoke with engineers in the spacecraft office; and sat in a cubicle among members of the project office. And singing in Helen's tongue-in-cheek team choir gave an unprecedented members'-eye view, featuring participants' reflections on their own organization, its concerns, and its triumphs.

Across these sites, I collected more than a hundred interviews from Paris and Helen members. For the sake of clarity, not all interviews are cited in the book. As is common in ethnographic writing, I have relied on certain central characters and moments whose voices and experiences most accurately speak for the organization's workings as I encountered it. However, coding across interviews gave me a sense of these overarching, shared themes as well as organizational distinctions, allowing me to best select voices and moments that spoke to more generalizable concerns. I also interviewed planetary scientists on other teams to situate my own perspective, learning how these two organizations were unique or characteristic of NASA's planetary missions. Some interviews were perfunctory, especially with NASA higher-ups, but many were detailed and long, stretching two hours or more. I found that scientists were eager to talk about the pressures, passions, and people involved in their work. Clocking over a decade of work in the field, I also established long-lasting friendships with certain scientists and instrument operators, returning repeatedly to discuss their work or their perspective on current events, using these reencounters with the field as a check on my analysis.

Noting which stories, strategies, and styles were shared across sites as opposed to which were unique to particular teams or misunderstood by others gave me tremendous insight into the field lines that undergirded both mission organizations. However, this required addressing two complications once I began to *write* about these teams. The first was due to my use of grounded theory, an approach that seeks to develop theory from actors' own accounts of their work. This is considered both empirically rigorous and participatory, as it allows participants to recognize themselves in analytic writing, making them more comfortable with or engaged in the analysis of their daily lives.[7] But as I will describe, Paris's form of talk is characterized by earnestness and politesse, Helen's by irony and disagreement. In an early talk I gave to the Helen team, my use of their empirically observed language made it seem as if I was preferentially treating one mission over the other instead of being a trustworthy reporter. Because these missions are in competition with each other for resources and accolades, this was a mistake.

I therefore made conscientious decisions about my analytical writing that bear discussion here. The first is that *I only use actors' categories when the interpretation is positive across sites*. Hence my use of "consensus," "integration," and "fairness"—all of which are members' terms, emic accounts of their activity—all of which I elevate as analytic terms. I use scare quotes whenever possible to delineate my actors' terminology (including "personality"). When such comparisons could be said to prefer one team over the other, I opt for analytical terms like *collective*, *autonomy*, and *independence*, indicated in italics when introduced. This also led to my development of the framework of *organized science*. While not a turn of phrase that my actors use, it allows the reader to continually return to the conceptual—not preferential—basis for the comparison between sites.

A second complication arose from traveling from subteam to subteam, gathering stories across the missions. On Paris, the homogenizing collective is strong, such that individuals across sites expressed similar concerns, needs, and satisfactions regardless of their location. In an organization as heterogeneous as Helen, I discovered a different problem. As the only individual who was not in an executive role who actively traveled from group to group, my perspective was organizationally placeless. It did not come from this or that instrument team or working group nor did it echo the concerns of the project office, and it certainly did not espouse the virtuosic language of public engagement. Following an initial presentation of my results to the team, some scientists could not understand why I paid any attention to an ostracized colleague's point of view, while others contested the stories I accumulated on one side of the mission simply because they did not align with their own experience elsewhere. To be fair, many members of the team did believe that I captured their work practices and forms of talk accurately: some even accused me of not going far enough. Such reactions are also data about how the organization works. Like people living in different states in the United States of America, Helenites are frequently unaware of how very different it is elsewhere on their mission, as I will describe. But the problem for my writing became clear in a meeting with Helen's project manager, who intoned severely that he wanted to ensure that I portrayed the mission in a way that "all members would agree was accurate."

Unfortunately, there is no singular truth about complex organizations with which "all members would agree." The one point of view that might fit such a bill—a view from the center or from the top—is not one that I am comfortable parroting, as this reifies the collaboration's local power relations instead of disentangling them, contextualizing them, and seeking to understand them in action. Instead, there are many experiences whose similarities and differences speak to the inner workings of the organization: its field lines. On Paris, the similarities are part of the puzzle; on Helen, it is the differences. These voices must be gathered however possible and allowed to express themselves as a function of their organizational positioning and biographical experience. I therefore adhere to Helen's local virtue of polyvocality: giving voice rights to the many unique perspectives that this mission fosters, gathered from extensive, exhaustive work among the team.

This approach has an empirical payoff. Science studies scholars Donna Haraway and Helen Longino have argued that there is no such thing as a singular, impartial, "objective" view of nature. Rather, objectivity in the sciences comes from combining as many different, "partial" viewpoints on an object as one can muster (Haraway 1988; Longino 1990). This must be the case for the social sciences as well. My travels to many different parts of Helen, as well as the comparison of ethnographic data gathered from across and within the site over time, enabled me to accumulate the variety of "partial perspectives" necessary to grasp the workings of the team that I describe here. I do not, therefore, claim an "objective" point of view in the sense that I have located a disembodied and perfect perspective on these teams: far from it. Even the spacecraft teams I study do not claim that their probe's perspective is perfectly objective, affected as it is by cosmic rays, shuddering, or data dropouts in transmission. I do, however, claim a position of *situated objectivity* because of the dedicated accumulation of so many local and different experiences that I gathered within these teams across time, space, and organizational strata.

Readers will also note that I give pseudonyms to the people and missions in this text. I did not come to this decision lightly: after all, half of my bookshelf displays studies of famous scientific sites like the Salk Institute, the Alvin submarine, and the Stanford Linear Accelerator, while the other half features studies of the pseudonymized companies Tech, TechCo, Brit-Arm, and Playco. These conventions speak to the different aims of science studies and organizational sociology as fields that "study up" (Nader 1972). Laboratory ethnographies aim to articulate how the universalizing location and practices of knowledge production are specific to time and place—or as one analyst puts it, how scientists produce "the creation of universality by the circulation of particulars" (O'Connell 1993). Naming study sites and scientists can ground otherwise grandiose claims to universal truth—a core premise for many in science studies. Meanwhile, corporate ethnographies conceal the site of their study, anonymizing the firm as best they can. This enables generalizability to other sites, useful in the business press to be sure. But because analyzing how organizational interactions matter requires discussing conflict as well as specific managerial roles, revealing company names can be cause for shareholder concern and impact on careers. Pseudonyms therefore provide some protection from crisis while reminding us that many other companies *could*—and in fact *do*—display these same organizational dynamics.

While I study a scientific site of work, I share the above concerns with organizational scholarship. Science is a reputation-based economy, and NASA scientists cleave to a narrative of pure science or bold frontiersmen with "the right stuff" in their necessary appeals to the public for support (Hersch 2012; Wolfe 1980). My argument requires extensive discussion of individuals and their positions on the teams, with details about how decisions are actually made. Living with these teams on intimate terms for so long revealed the private interactions behind these public personas. There is little room for personal anonymity once the names of individuals and their venture are public—there is less room yet for a paradisiacal vision of interaction. This would, in any case, be dismissed outright by anyone familiar with organizational life in any field.

Disguising mission identities admittedly produces tensions in the narrative. Unlike banks or toy companies, spacecraft truly are unique: there are no other missions like them. Sociologists of science take seriously the entanglement of scientific work with the natural world, such that concealing too many aspects of the planets and the spacecraft could undermine my argument that organizational arrangements enact knowledge outcomes. I therefore walk this line with tremendous care. I anonymize the individuals, institutions, groups, and meetings under discussion as much as possible when doing so does not compromise my core argument, to buffer against any reputational concerns. I also give pseudonyms to the missions, choosing the names Paris and Helen to reflect NASA's predilection toward mythologically inspired name conventions while maintaining a sense of two overlapping yet separate worlds. While anonymization appears to be a

Sisyphean task, there is no perfect answer to this challenge. My choices are guided by concern for my participants' careers and for the trust they placed in me—a trust I do not seek to violate.

The Plan of the Book

In the first half of the book, I begin with the background and context necessary for what follows: the history and the organizational theory that establishes where these mission collaborations came from. In the following three chapters, I dive into the organizational orders of both teams. I describe how Helen's matrix organization embraces a concept of "integration," such that individuals involved in spacecraft planning must balance many instrument teams' concerns while maintaining subgroup autonomy. The goal is to come up with a plan that not only achieves good science but that all consider to be "fair." Paris, meanwhile, is a charismatic collective, with more in common with religious groups, communes, or tech start-ups. Planning on Paris involves synthesizing different scientists' ideas and needs into a singular vision for their robots' actions, combining instrumental specialties in investigations to produce more than the sum of their parts. The team focuses on achieving unilateral "consensus" on a daily plan that keeps participating members "happy." I elucidate these organizational environments by describing the Helen (chapters 2 and 3) and then the Paris (chapter 4) methods of making decisions about where the robots should go and what they should do. My primary focus in these chapters are the ritual planning meetings and team meetings where scientists surface and resolve conflict over which observations their spacecraft will take. These negotiations produce different observations and commands each time yet are still subject to a specific social order and group understandings that help individuals resolve conflict and shape scientific outcomes—fairness and autonomy, or happiness and consensus. This is the first principle in the framework of *organized science*: that science is conducted by organizations, each with their own practices, local norms, and authority structures. I end part 1 with a chapter on the external tensions that beset planetary science collaborations to show how each team's local orders address common challenges facing all spacecraft teams.

In the second half of the book, I turn to how those organizational orders shape the outcomes of the scientific collaboration—the second principle in the framework of *organized science*. I will repeatedly show how elements that appear to be essential properties of objects—planets, spacecraft, data, and even people—are actually the outcomes of these different organizational orders. Chapter 6 examines the scientific results of each mission to demonstrate parallels and isomorphism between each organization's social form and its scientific production. In chapter 7, I describe how the spacecraft itself is constructed along the lines of the organization, such that it does not challenge the organization's norms but rather reflects and projects each collaboration's social structure back to itself. Chapter 8 follows this discussion by showing how the data acquired by the spacecraft participate in the local social form as well. In chapter 9, I turn to how mission organizations inflect the scientific personas and reputations of individuals participating on these teams.

A short final chapter focuses on the third principle of *organized science*—the concept of the iterative loop—to show how mission products cycle back into the organization as its inputs as well, creating ever tighter coupling between process and product, social order and natural order. As a mechanism, the iterative loop offers new ways of thinking about organizational change and scientific questioning. It also demonstrates that organizational processes and their outcomes are not separate and distinct but interwoven and enacted at the same time, together.

In the conclusion, I elaborate the importance of the concept of *organized science* for science studies and for organizational sociology. This means returning to questions about disciplines, epistemic cultures, and the social shaping of scientific objects and subjects. But it also means thinking forward to the design and development of future teams in science and elsewhere. Even as *organized science* demonstrates how organizational elements shape and constitute scientific facts, it holds lessons for scientific and technical organizations in industry, in academia, and in the nonprofit sector as well.

PART ONE

Orders

ONE

The Context

"Spacecraft don't have social lives."

"Excuse me?" I am giving a presentation at the Helen Project Science Meeting about my research on their team. More than a hundred scientists and engineers affiliated with the mission are assembled in a large auditorium at Spacelabs, seated in rows of folding chairs surrounded by models of famous spacecraft. Principal investigators and leaders of Helen's twelve instrument teams sit behind a U-shaped semicircle of folding tables at the front of the room. The imposing scale models of spacecraft behind them reminds me that many of the attendees worked on the very missions whose images of the planets graced the covers of *National Geographic*, which I read avidly as a child.

My title slide reads "The Social Life of Spacecraft," and it has not yet left the screen when Roger, a principal investigator of one of Helen's instruments in his seventies and a veteran of the missions represented around the room, shoots his hand up with a statement. "Spacecraft don't have social lives," Roger insists. "They're not people."

A few seats over, another instrument team leader, Isabelle, quips that most of the people who *work* on spacecraft teams don't have much of a social life either. Laughter ripples through the audience. It is only my first slide. I wonder if I stand a chance of getting through the rest of my talk.

Addressing "the social life" of objects is a common sociological trope. It demonstrates that something we consider to be an object in its own right is actually imbued with rich meaning and a very human history and is therefore a part of a social world in which we live. Spacecraft are no different. Although NASA is credited for flying them, the probes themselves are typically constructed at a contracting facility—an agency center, a university, or a for-profit company—by a team of hundreds, sometimes thousands, of engineers. This construction is informed by conversation and negotiation between those who build the craft and those who will eventually use it to collect precious data about the planets under study. Spacecraft teams are therefore populated with hundreds of scientists and engineers, academics and managers, secretaries and postdocs.

Mission scientists, for their part, are employed at various universities and research centers and join a spacecraft team to build and operate one or more instruments on board the spacecraft. Selected through a fiercely competitive process of review panels under NASA Headquarters oversight, the teams of scientists that win an instrument contact will build or procure their own hardware and construct flight-ready devices at their home or partner institutions. These devices are eventually mounted onto the spacecraft and launched into flight. Once the probe is off the ground, the science teams must establish operations routines and software to command their instrument, establish the data pipelines for processing and circulating the eventual mission returns, and stand ready to represent their instrument and its needs in mission-wide operational meetings. Meanwhile, an engineering team at the primary center employs the operations engineers responsible for commanding and caring for the craft as a whole.

The spacecraft and its software and hardware therefore bear the indelible stamp of the human teams that built them and the way they locally chose to integrate these elements across instrument teams and institutions. This is difficult to change. While many organizations can upgrade their technologies every few years, the spacecraft is too far away to modify drastically after launch. Promoting, refreshing, or reorganizing scientific personnel is unthinkable: unlike contemporary technology companies where "the re-org" reigns supreme, the contracts issued by NASA at the mission's outset are institutionally enshrined, transforming initial appointments into lifetime sinecures. Further, the path to launch never runs smoothly, so the political problems that each team faces along the way—as well as their proposed solutions to those problems—can exert powerful social effects on the spacecraft's design and development (Pinch and Bijker 1987). Like stratigraphic layers that record past events in rock formations built up over time, the organizational concerns that suffuse these teams from the outset endure throughout the ensuing decades of their collaboration.

In the rest of this chapter, I describe this background and history for the two missions that I will describe in the book. The origin stories of both teams demonstrate the parallel between the organization of the spacecraft and that of its interinstitutional team, producing each mission's unique characteristics as closed, tightly coupled sociotechnical systems. In their background and history, we witness organizational choices that were animated by popular management theories in technology and engineering companies at the end of the twentieth century. We also witness each team's response to the austere budgetary climate at NASA in the 1990s. As each mission comprises its own social environment in which individuals interact and conduct their daily work, the ways in which the spacecraft and its team are interwoven from the outset illuminates each mission as a distinct organizational field site: the first component of *organized science*.

Missions as Organizational Field Sites

NASA is a complex organization. Although the federal space agency presents a unified front to the public, it is in reality a sprawling multisited institution that fosters competition between centers, with employees and contractors dividing their attention between several scientific and technical projects. With its many different centers located around the country, each with their own organizational culture, goals, and priorities, historian Howard McCurdy calls the agency "a confederation of cultures" that "does [not] fit the popular image of a government agency staffed by bureaucrats with a lifetime sinecure" (McCurdy 1994, 52). Studying the scientific side of NASA introduces more heterogeneity besides, as the scientists involved in space missions are employed by universities, government research centers, or private industrial laboratories supported by agency grants.

NASA centers also jostle in competition to be the "lead center" for a particular project, with decentralization throughout the 1980s creating barriers toward multicenter collaboration (McCurdy 1994, 131). Each group also develops its own local logics of decision-making and management (Vaughan 1996, 64). Further, while Spacelabs was the "lead center" for both missions I studied, mission boundaries extended far beyond the lab. The growth of certain universities from centers of excellence for astronomy (documented in McCray 2006) led to key centers for planetary science in states like California, Colorado, Arizona, Rhode Island, and New York (Westwick 2007). The rise in contract work through in the late twentieth century also created many overlapping connections between NASA centers, universities, and private aerospace companies such as Lockheed Martin (McCurdy 1994, 131; Vaughan 1996, 211); the period of my immersion witnessed the rise of new players like SpaceX and Blue Origin. And because NASA is divided near the top of its administrative hierarchy between human and robotic exploration, this makes for a cultural gap—and competition for resources—between those who work with astronauts and the space station on the one hand

and those who work with distant probes on the other. It also made for a knowledge gap: the scientists and engineers I worked with were more likely to know about the latest space telescope than the most recent spacewalk.

Each of these institutions is a distinct entity. Spacelabs has its own matrix organization subject to total quality management and dominated by an engineering culture (Westwick 2007); other centers are known for being more hierarchical or more science oriented. At the European Space Agency, a partner in the Helen mission, national differences also played a role in spacecraft construction, operation, and findings. My fieldwork took me to the offices of the Italian Space Agency on a suburban Roman street, the colossal communications arrays outside of Madrid, a neoclassical observatory on a Paris hilltop, and a stately office park a short drive from a small Dutch town flanked by dozens of national flags, among others. The European project of collaboration within Helen brings these many distinct groups—and sites—into collaborative agreement (Zabusky 1995).

Alongside considerable differences between centers, political elements shape the planetary science community's trajectory through different periods. This should be unsurprising for NASA, although, following the Cold War, presidential proclamations and fiscal shortfalls were more likely than transnational politics to impact space science projects. For instance, when I began my ethnography in 2006, the NASA centers that supported human exploration in Florida and Texas were well-funded by a return-to-the-moon project under George W. Bush administration. Planetary scientists I interviewed described this as an obvious prioritization, as these centers were in Republican states with Bush family governors. Later, the Obama administration closed this project down and incentivized the private sector to take over the development of crewed capsules. Although this followed up on a promise from a prior administration, it also diverted money toward robotics and Earth sciences programs, with key NASA centers located in Democratic strongholds Maryland and California. Funds shifted accordingly. Later, in the wake of the financial crash of 2008–9, I watched as Helenites were asked to report to NASA as to how they would continue their scientific investigations with only half of their expected funding. Extending missions alongside concurrent new missions meant fewer resources too. When I left the Paris mission, the team had ceded its operations rooms to a new mission and was working on a skeleton budget: many scientists admitted giving their time to the project for free. Helen, meanwhile, lost an entire floor of personnel and changed the duration and complexity of spacecraft observational segments to accommodate fewer support staff.

Programmatic pronouncements from NASA Headquarters also have an organizational effect on the community at large. Planetary science as I encountered it in the early twenty-first century was already shaped by the founding of the Mars Program, an innovation in the 1990s that ensured funding for missions to Mars every two years (Conway 2015). Thus supported, the number of Mars scientists grew to a disproportionate number relative to the number of planets available for study. Mars scientists attended their own specialist conferences, had their own advisory panel to report to NASA on prospective missions, and were inundated with data from a series of orbiters, landers, and launches. Although senior scientists on early Mars landers also served on missions to other planets in the solar system, during the time of my study, I knew of only a few scientists whose expertise and mission experience cut across the divide between Mars and the outer planets, located beyond the asteroid belt. The majority of these served on both Paris and Helen concurrently.

As in other "big sciences," questions of patronage and division of funds among these institutions are essential for understanding project trajectories (on astronomy, see Biagioli 1993; McCray 2006). For instance, during my fieldwork, outer planets specialists and many Helenites decided to found their own "program" so that they too could benefit from what they called "the Mars treatment": a guarantee of money and missions over many years to support their science and their students. However, a series of missions in the late twentieth century had already created distinct communities of expertise around Jupiter, Saturn, and their moons. Thus when NASA selected Helen's successor mission, the two competitive teams in the running (called "the outer planets face-off" by a scientist I interviewed) were a mission to Saturn's moon Titan, planned largely by Helenites who had first met on an earlier mission, and a mission to Jupiter's moon Europa, planned largely by alumni of a different prior mission.

Another programmatic change impacted Helen in particular. Following a fiasco in the Mars Program in the mid-2000s, an administrator at NASA initiated a new requirement: all newly proposing PIs must already have PI experience. This catch-22 created new incentives for existing missions. As Spacelabs was grooming a Europa scientist, Francis, to lead its prospective mission, they were concerned that this new proclamation would undermine their chances at the competition. So they put Francis into the lead scientist's role on Helen, replacing Helen's project scientist of almost two decades, Malcolm. As Francis was already locked in the "outer planets face-off" with scientists on Helen who were proposing a competing mission to Titan, this made for some tensions that affected operations and science that I will discuss. Some of the changes that Francis brought to Helen are also illuminating for demonstrating features of *organized science* in action, issues that I will return to as a question of organizational change in the final chapter.

However, the most significant programmatic effort that saturated the context in which both Paris and Helen developed, occurred in the mid-1990s. At the time, America was in the grips of an economic recession, and investment in the space shuttle program throughout the 1980s (known to planetary scientists as "the lost decade" due to the dearth of missions) led to a sense of budgetary bloat at NASA. The agency's then-administrator instituted a new policy for cost saving when building and flying spacecraft: "faster, better, cheaper." This pronouncement created new classes of funding for missions, each with different concomitant social structures.

Helen was built under NASA's "flagship" class, the largest and most expensive type of missions. Flagships typically support up to a dozen instrument PIs and hundreds of scientists spreading the impact and opportunity of scientific work widely. Eschewing these high price tags, NASA in the mid-1990s introduced a new class of missions called Discovery. These would fly for under a billion dollars with a single PI and a small, integrated suite of scientific instruments. Paris was indelibly stamped by Discovery's changes. The mission was proposed repeatedly in the late 1990s, with the PI assembling a team of experts to build each instrument. When these proposals were unsuccessful, the PI made a bid for one of the first Discovery program opportunities, designing the mission to feature a single PI and an interoperable payload to fly on the cheap. It was rejected, but when he turned the proposal around for an opportunity to fly under the Mars Program, the structure remained. Paris bore Discovery's imprint in terms of fiscal footprint and organizational form, even though it flew under a different program. Meanwhile, Helen's flagship status made it large and costly compared to the vision for "faster, better, cheaper." The NASA administrator was reportedly openly hostile to the mission, referring to it as "*Battlestar Galactica*" to emphasize its enormity. Despite the signed international agreements that took place before his tenure, he demanded that Helen scale back or face the chopping block. A difficult period ensued as scientists, engineers, diplomats, and managers alike fought to save their mission. In the end, Helen flew successfully, but at a cost I will describe below.

While NASA focused on the cost-saving measures of the Discovery program, the result was a natural experiment in the organizational sociology of sociotechnical systems. Suddenly, the same community of scientists was expected to operate their collaborations from within differently organized units, alongside spacecraft that conformed to different cost caps and operability requirements. In terms of organization, the council-of-elders approach in the flagship teams stood in marked contrast to the single leader in the PI-led missions, as did the considerable size difference between flagship and Discovery teams. The stark divisions of labor associated with the former class also contrasted with the organic, interoperable instrument suites of the latter. Other shifts were afoot in terms of a sea change in managerial philosophies during the late 1990s, which would affect both missions as well.

Flattening Hierarchies

As NASA moved to smaller, cost-cutting missions, organizations in the technology sector were also making changes: toward lateral and bottom-up decision-making structures, embracing flat hierarchies, matrix management, and corporate culture. Such organizations placed people with different expertise into direct contact in lower levels of the organization's hierarchy to make decisions about their project, instead of sending decision-making authority up a hierarchical chain. Management science at the time abounded with examples of how such project-based teams could produce innovation across a variety of sectors. Responding to the overwhelming success of the software and biotech sectors, these organizational forms spread beyond the start-up world into other domains as well.

Flat organizations in their various guises offer tremendous benefits to multidisciplinary teams in terms of creativity and innovation, but there are also attendant concerns. Project teams require individuals to participate in two communities at once (Vedres and Stark 2010), and when different groups have different ways of assessing value, confusion between competing logics of evaluation can create friction around creative work or exacerbate clashes in "institutional logics" (Girard and Stark 2003; Thornton, Ocasio, and Lounsbury 2012). Other examples place individual experts in the precarious position of brokerage between groups, where they are unlikely to be understood by one or both sides. Because authority structures and attendant expectations for communication and management are unclear, studies of these contemporary, heterogeneous organizations—which Charles Heckscher and colleagues call "postbureaucratic" or "collaborative communities"—describe shared culture as an essential glue that brings people together people across sites, expertise, and local concerns (Heckscher and Adler 2006; Heckscher and Donnellon 1994). Here, culture can be defined as communication practices, interactional norms, common goals, and sense-making techniques specific to a local group. If people have divergent expertise and knowledge that keeps them apart, a common culture, it is assumed, can provide a shared medium for information exchange across these divides. In scientific communities, processes of enculturation also transmit essential, tacit knowledge and skills, as well as machine knowhow, which allow for the production and interpretation of experiments (Collins 1985).

As I have already described, "culture" is not just about communication. In flat companies, "culture" replaces the hierarchical control of bureaucratic management with a form of internalized discipline. As individuals take on the company's goals and strive to achieve them as part of their personal self-actualization, this produces corporate control without a hierarchical chain of command or bureaucratic demands for efficiency (Kunda 2006; Turco 2016; Van Maanen and Barley 1985b). This is especially the case in start-ups that feature a charismatic leader like Steve Jobs, Craig Venter, or Elon Musk, whose vision inspires their employees' near-total commitment. Studies of communication technologies in these contexts also describe corporate control through bottom-up collaborative connections (Heckscher 1994; Turco 2016; Volkoff, Strong, and Elmes 2007). Insert software into the mix and this does not so much disrupt but standardize local ways of interacting among different parts of the corporation (Leonardi 2012a; Leonardi, Nardi, and Kallinikos 2012; Volkoff, Strong, and Elmes 2007). When hierarchy is unclear, technologies can even become a resource for individuals negotiating their power and position in the organization (Barley 1996; Kling 1991). Corporate ethnographer Gideon Kunda, in his study of a matrix organization such as the one adopted by the Helen mission, explains that "information organizing is formally prescribed, and 'culture' replaces 'structure' as an organizing principle to explain reality and guide action" (Kunda 2006, 30). Because shared interactional norms typically replace the formal reporting patterns and decision-making strata of classic bureaucracies in flat organizations, this produces control and commitment through other means.

In the face of NASA's heterogeneous organizational landscape, then, individual missions such as Paris and Helen adopt flat hierarchies with strong local cultures and interactional norms that serve to bring participants together across sites in their collaborative effort. If the relationships between the scientists and the missions I describe and NASA's institutions and programmatic decrees are diverse and set against shifting political conditions, then we can see each mission's organizational order as a way of bridging these divides and producing stability. Indeed, missions can be surprisingly effective at creating a form of group solidarity that transcends institutional distinctions to produce an exceptional level of commitment. European

Space Agency anthropologist Stacia Zabusky even describes "the mission" as something approaching a "sacred journey," complete with elements of transcendence and enchantment. This bore out in my own observations of planetary scientists. Their clothing, badge lanyards, even their bodies are decorated with mission logos, jewelry, or art featuring a favorite planet or probe. I once witnessed a conversation of female engineers in their fifties who were planning to get their first tattoos together: a technical schematic of their spacecraft. A mission's organizational order can exert especially powerful identity and commitment effects on its participants.

Such mission cultures and interactional norms are therefore addressed at not only bringing diverse groups together but bringing them together in order to *make decisions*. During mission operations, the primary decisions that spacecraft teams must make are questions of which observations to take and when to take them. Spacecraft have limited memory, hours of operation, power to support observations, and opportunities to call home to download their data and pick up the next series of commands. On past flagship missions with hierarchical roles, principal investigators and chief engineers negotiated observational priorities; scientists passed information and requests up the chain; and postdocs waited patiently for the data to come down to fuel their scientific papers. On later, lateral teams, however, decisions about what the spacecraft should do and when are frequently made amid the ranks.

Group decision-making is a difficult task. Attempting to produce consensus can take a long time (Polletta 2002); should a group coalesce too tightly, there is danger of groupthink and openness to risk (Neff 2012). Decisions can also exacerbate factions and cultural distinctions between different wings of an organization (Bezrukova, Jehn, and Zanutto 2009; Hart and Vugt 2006; Rico, Molleman, and łdots 2007). Such facts of organizational life appear to be the "primordial conditions" of work for those in the trenches, but they are also byproducts of the way that work is organized and accomplished, as Michael Burawoy argues—that is, local divisions of labor and responsibility over production also establish each group's methods of managing "conflict" versus "consent" in the first place (Burawoy 1979). And the choice of how to divide labor within lateral groups can create silos and other divisive side effects. Groups that specialize too much with limited visibility across practices produce not solidarity but a lack of connection to the societal group, with a likelihood of dispersion and difference (Acker 1990; Durkheim 1893; Heckscher 1994; Paradis, Elise and Albert 2013; Weber 1968).

Importantly, conflict between internal groups is not always negative. Diane Vaughan describes how one matrix team she studied at NASA consistently produced different interpretations among its different branches, leading to frequent clashes among the technical staff. These were locally described as essential to "making good engineering decisions" (Vaughan 1999, 921). This is because, for group members, the affective textures of decision-making are not as important as conformity with the agreed-upon process. Charles Heckscher, a sociologist of postbureaucratic organizations, argues that decentralized groups describe their decisions as legitimate when they adhere to an "interactive process" that team members locally agree achieves the goal of "informed consensus" through "institutionalized dialogue" (Heckscher 1994). As we shall see, such dialogue may be confrontational or calm, and such consensus may be produced in a variety of ways, depending on the local organizational order. Hence each mission's decision-making practices and rituals are of high-value, as they are the place where legitimacy is produced and organizational order is maintained.

Tracing not only how decisions are made, but who makes them and when, reveals that hierarchies and collaborative goals are often unclear, even among the flattest of teams. This is because such organizations typically invoke more than one form of authoritative decision-making. Recent studies have revealed that even "flat" companies typically have an embedded hierarchical authority structure, limiting employees' ability to either impact or comment upon corporate direction (Freeland and Sivan 2018; Turco 2016; Zuckerman 2010). This is evident on "flat" NASA missions too. For instance, most of the scientists on Paris and Helen make decisions in a lateral structure, but the engineers have a more vertical line of authority. Sociologist David Stark calls such organizations "heterarchies" as they possess more than one accountability structure in operation at the same time (2009). When lines of accountability or authority shift, it is important to trace the local narratives that smooth the transition from one form to another. This makes sense of the local organizational order without assuming that all deviations from participatory decision-making are illegitimate, or that the organization's narratives are false.

How members of an organization interact with each other and work together as part of their group's environment can also affect the outcomes of their work. This premise is even well known in computer science as "Conway's Law," which asserts a direct relationship between an organization's communication structure and its software architecture. Sociologists have shown this in action. Studying Hungarian factories moving from socialism to capitalism, Stark describes a case where two competing logics of production and interaction occurred on the same shop floor with the same workers at different times of day. As workers switched between a logic of bureaucratic control to one of worker-ownership over their production (a formalized version of the black market that allowed workers to do their jobs under a different economic frame) this produced different, even competing logics of value and incentives for worker output (Stark 2009). Studying the same transition, Burawoy grounds different ideologies in the factory in different divisions of labor, local interactional norms, and ways of managing conflict (Burawoy 1985, 1979). Ching Kwan Lee also documents a parallel shift in worker and managerial relationships, wages, and outcomes in China's transition toward capitalism (Lee 1999).

This is no different at NASA. Scientists in each mission assert their own epistemic values as a question of the virtues of their particular chosen methods of interacting and decision-making: what they simply call "doing good science." A mission's decision-making about observational choices and priorities dictate which data gets collected and which questions are abandoned or go unexpressed. Group interactional practices for decision-making enact organizational orders at the local level that, such as the ones at the Hungarian or Chinese factories, enable and constrain different outcomes for the processes of scientific work. Even how scientists and engineers divide the work between them and enforce distinctions between these two wings of their collaboration entangles the epistemic and the organizational. The two are inseparable in practice.[1] As mission teams like Paris and Helen assemble under their "flat" rubrics, then, the local organizational orders they develop as methods of resolving conflict and making decisions do double duty: bridging far-flung places, people, and instruments and shaping local responses to their scientific goals.

The Queen of Flagships and a Low-Cost Mission to Mars

Both missions have local origin stories shared among their organization's members (Linde 2009). The contours of Helen's narrative were repeated to me by all members of the team, whether present in the early days of the mission or not. Following the last probe's visit to the Saturn system in the early 1980s, they described, many questions remained unanswered—especially about what lay below the layer of methane haze on Titan, the planet's largest moon. Three friends on this mission from the USA, China, and France began to discuss the possibility of returning, this time with a probe that could penetrate the clouds and land on Titan. The group met throughout the decade to scope their plan, inviting colleagues they had met on the Saturn encounter to formally establish scientific priorities for the mission. They proposed to address multiple targets—Saturn and Titan chief among them but also the planet's rings, the other moons in the system, and the magnetic field environment around the planet. Thus the mission would be multidisciplinary and multitargeted in its scope, aiming to engage hundreds of scientists in its task. The scene was set for a flagship mission to return to the outer solar system.

The mission would also involve an international component. In 1988, the group negotiated a bilateral agreement that including a memorandum of understanding between NASA, the European Space Agency (ESA), and the Italian Space Agency. The Americans would build the Helen spacecraft, ESA would build the accompanying probe to land on Titan, and the Italians would build the communications hardware. Each agency also cross-appointed members of each team from the different member states so that NASA-funded Americans worked on the ESA probe, while ESA-funded scientists joined teams on Helen. Such formal agreements paved the way for meaningful, thorough international partnership in both the scientific and technical aspects of the mission. This founding group formed a tight-knit community of friends who had worked hard to secure their mission together and was exceptionally proud of their successful efforts. Mementos from several decades of work together decorated their office walls, and they spoke consistently of their considerable pride in the international spirit of the mission as its lasting legacy.[2]

Staffing the science team on Helen occurred in response to what was then a relatively standard call for participation on what NASA calls its strategic missions (missions where the target is already defined and scientists compete to join the team), albeit with an international streak. Once NASA and ESA had committed to going to Saturn together, they released an announcement of opportunity to scientists calling for instrument proposals for equipment to fly on the spacecraft. By 1989, the mission payload was officially announced, including twelve instruments on the Helen spacecraft and instruments for the probe selected by ESA. A considerable number of the selected scientists and engineers on the new mission had led instrument teams or trained as graduate students on the prior decade's outer planets flagship mission. The majority of the instruments on board were led by PIs who assembled a scientific team to assist them with instrumental interpretation, augmented with scientists from across the Atlantic. NASA also built several instruments in-house, staffing their teams with scientists selected through individual application: these are called facility instruments. The camera was one such instrument; the radar, radio detector, and an infrared spectrometer were the others. To staff these teams, individual scientists wrote a letter to NASA proposing their interest and experience; their letters were adjudicated by a panel to assemble a team roster and select each group's leader. As was standard practice at this time, scientists belonged to and participated on their appropriate instrument team only, using only that instrument's readings and datasets.

Also part of the mission's origin story is a narrative about Helen's organizational choices. Flagship missions in the 1970s had a hierarchical authority structure, where PIs would argue for instrument time and priorities between themselves at a high level; their decisions filtering down to the rank and file for implementation and analysis.[3] Later examples added interdisciplinary scientists to the mix, hoping to ensure cross conversation between instrumental results. Helen went one step further and embraced a matrix organization, echoing that of Spacelabs more broadly, bringing the collaboration's different types of experts together into task-oriented groups. This produces "structural folds" in the organization (Vedres and Stark 2010), crosscutting groups of scientists who share interest in one or more of the various targets in the Saturn system outlined by the mission; these are led by interdisciplinary scientists who are in a brokerage position between two instrument teams (Burt 2004). Combining these many types of scientists with different investigations into a matrix, Helenites believed, offered the broadest coverage of scientific topics as well as the capability to bring different types of expertise to bear on problems.

The working groups assembled under the matrix are also directly responsible for hosting a target team to determine scientific observations that meet their discipline's goals. Here, the team's organizational order is keenly felt. The mission's chief science planner recalled that this innovation took place following the Jupiter flyby en route to Saturn, where a test-run of processes concluded with a feeling that the allocation of observations was not fair overall. They decided to break the spacecraft's path around the planet into segments during which each working group would have the authority to plan observations. As one science planner put it, "We didn't want a PI [making the decision of what to observe]; we wanted somebody, obviously that was an expert in the discipline, and other team members would respect that or respect their scientific authority in that discipline." At the meeting where this was announced, this planner recalled "crying in meetings . . . and yelling and screaming," and considerable pushback from PIs over their loss of authority. The result was an organization that made decisions about observation allocation within the matrix, at the intersection of discipline and instrument, instead of moving observational decisions up the chain to the PIs and team leaders as on earlier flagships.[4] The PIs and team leaders still formed an executive committee with oversight over mission-wide decisions and topics but did not busy themselves with the daily operational work or planning. This matrixed context is essential for understanding how *organized science* on Helen takes shape.

This sensibility toward distinct ownership over parts of the mission was also visible in the technical development of the spacecraft. In response to the budget climate, the lab proposed that Helen use a new concept for a modular design. A single "bus" (the main part of the spacecraft that carries the instruments) could be built to print for a variety of different missions, saving money by not requiring a bespoke design. System pieces could be chosen, designed, and upgraded without affecting the overall unit, fitted to each mission's purpose. Thus the modular integration that the spacecraft offered was its mark of innovation. The first official demonstration would be to develop Helen alongside another mission concurrently, using the same bus, with follow-up missions expected. Following this modular orientation, the instruments on the spacecraft were designed to operate independently, added to the bus, and integrated into the craft individually. The fields and particles instruments sensing the magnetosphere and the plasma environs would gather data from their onboard locations as the craft flew through space. Meanwhile, the cameras and spectrometers would be mounted on a scan platform that would rotate into place to line up their observations, much like a series of sensors placed atop a distant lazy Susan. Finally, the instrument teams were given responsibility for "remote operations" whereby they could command their instruments from the PIs home institutions, communicating those plans directly to the lab to produce a single timeline to upload to the spacecraft. Helenites call this bringing together of disparate pieces into a sequential order, "integration."[5] Because integration must take place across several organizational fault lines, the process requires considerable attention, which I will describe in the next two chapters.

Meanwhile, emerging from the 1970s, those scientists who studied Mars were not so encouraged as their colleagues who studied the outer planets. Their own recent flagship had incurred billions of dollars of cost overruns and had not detected signs of life on the planet, leading many to abandon the idea of a return (Ezell and Ezell 2009). In the early 1980s, there were intermittent studies for a rover and a sample return follow-up mission, all to no avail (Messeri and Vertesi 2015). This changed in 1996, when analysis of the meteorite ALH000841 presented controversial evidence for past life on Mars. President Clinton authorized NASA to initiate the Mars Program: a mandate that came with an independent line of funding specifically earmarked for Mars under the imperative to "follow the water" and find signs of early habitable conditions on its surface. This enabled Mars science and missions to proceed during an era of frugal cost-saving measures. Yet the combination of continuous funding for the Mars Program and an imperative for low-cost missions resulted in a series of high-profile losses for NASA, including a mission famously lost over the confusion between English and metric units. It was against this background of high stakes and low budgets that the Paris mission was developed.[6]

Designed almost a decade after Helen and benefitting from its push against vertical authority, the Paris spacecraft was designed under a different culture of technical management, expressing an interpretation of "integration" in line with the organizational forms popularized in software companies in the 1990s. The robotic platform was conceived as a holistic unit, with a focus on systems engineering: a style of engineering that focuses not on individual subsystems but on interactions between the pieces as they fit together. Components were designed in an integrated, not modular, fashion. While few scientists recalled this element of the spacecraft's history, they readily explained its extension to the payload of scientific instruments. While the instruments typically take observations one at a time and are constrained by data transfer rates and memory storage, they are designed as a suite with complementary skills. The cameras observe in wavelengths into the near infrared, a spectrometer picks up in the thermal infrared region, two additional spectrometers observe different aspects of elemental and mineralogical chemistry, a tool grinds the surfaces of rocks, and a hand lens takes close-up pictures. Data from these instruments are built to be combinable, and images from the multispectral cameras and engineering cameras frequently form the base map for overlaying data plots, color codes, or other descriptions of the Martian terrain. This enables the ready-made combination of data in order to develop multi-instrumental scientific and technical insights as a first-order proposition.

The science team follows the same principles of interdependence, as described in Parisians' accounts of their local organizational order. Instrument teams are similarly responsible for building, maintaining, and sequencing commands for each instrument, but their membership is porous: mission scientists are encouraged to use any instrument they choose and are repeatedly exhorted to use many instruments together to solve scientific problems. The original science team was proposed as a unit. The PI invited people he worked with well, as well as several of his competitors on prior proposals to work under his leadership. Two later calls for participating scientists opened science team membership up to a wider group with different scientific backgrounds, bringing in men and women with expertise in atmospheres, soil, and geochemistry. Paris also included two small international partnerships:

a German spectrometer and a group of Danish researchers who observed magnets on the robot for dust.

It was under the PI's direction that these individual scientists were encouraged to work closely together, using all onboard instruments, with no internal distinctions between members and no restrictions on data sharing or the production of scientific results. I observed this to actually be the case despite many reasons not to work together, such as competition or international restrictions on participation or stratification between the mission's original scientists and the later additions. Indeed, during my observational period on Paris, all of these members equally participated in building multi-instrumental investigations. There was no functional, hierarchical, or other cultural distinction between early and late joiners, senior or junior scientists, or instrument specialists: member scientists proudly pointed to this element of their order and upheld it under challenge. The team also displayed a strong role for shared rituals and practices that maintain a single organizational culture mission-wide. In all this, the PI's authority is not exerted hierarchically but supports the collective organization through charismatic leadership (Polletta 2002; Rothschild-Whitt 1979; Weber 1968). In this way, the organization of both the human and the spacecraft components of the mission is oriented toward operation as a single, unified group.

Thus both Paris and Helen attempted to move away from a bureaucratic system, whereby decisions are made by a chief officer at the top of a hierarchy, toward lateral forms of accountability and operation that bring individuals together to make decisions across and within the organization (Heckscher and Adler 2006). Helen's matrix solution sees individuals organized into a variety of working groups and instrument teams that are meant to provide visibility up the chain and across the mission. Paris's solution is an emphatically flattened team, led by a charismatic, single PI. Like many lateral organizations with open communicative cultures, hierarchical forms of control persist in both environments (Stark 2009; Turco 2016; Zuckerman 2010): Helen features PIs, project managers, deputies, and interdisciplinary scientists, while Paris has its PI and instrument leads; I will describe the switch between these elements below. But both missions expressly describe their organizational orders as a question of integrating knowledge across instruments and scientific investigations.

Isomorphism and Torque

I have described how each team has different hardware and organizational configurations, speaking to the political and economic climate as well as the managerial philosophies of the time. We might then say that the design of the two spacecraft reflects two very different "styles" of integration in a sociotechnical system (Hughes 1999): one decentralized and modular, the other centralized and interdependent. At least, that is how they were designed.

No one knows what would have happened had Helen had been allowed to fly as it was originally proposed. But in the wake of budget cuts in the 1990s, its twin mission was canceled, along with any prospective use of the modular spacecraft: the mission's innovative technical contribution. More difficult still, its rotating scan platform and articulated antenna were cut in order to cut costs, fixing the optical, radar, and radio instruments directly onto the spacecraft bus. This had an outsized downstream impact on operations.

Notably, *every* member of the Helen team that I spoke to mentioned the loss of the scan platform almost twenty years prior as a defining moment in their mission's history. Many scientists blamed this loss for fewer science opportunities, for difficult negotiations over observational time, and for mission strain between personnel. One scientist recalled for me how he once told the engineer who made the final decision that he had crippled the mission. The engineer replied, "Why didn't the scientists say anything at the time?" The scientist's response was, "Because you put it to us like, 'You can have no scan platform or no mission: take your pick.' "[7] The engineers I interviewed who were present at the review meeting when the scan platform was cut in 1992 articulated their role as one of "saving the mission." As they put it, the issue on the table was, "either fly without a scan platform or don't fly at all." Helenites are happy to have their mission, and some expressed that conflict was likely inevitable on their team. But most people I spoke to asserted that there would have been more and better science, lower cost, and less tension in the negotiations over spacecraft resources: "If we [had] the scan platform, most of the conflicts could have disappeared."[8]

What was so terrible about losing the scan platform? From a technical perspective, everything worked perfectly: Helen flew to the planet, the instruments worked, and the probe made a perfect landing on the moon Titan. From a sociotechnical perspective, though, the loss of this piece of hardware threw the isomorphism of the sociotechnical system out of alignment. From then on, what one instrument team decided to do had a considerable constraining effect on what another instrument team could do. The science teams that were funded to operate independently, the tools for remote operations of independent instruments, and the spacecraft bus with a modular design, all optimized for independence, were forced to operate under the assumption of interdependence instead. This was especially the case with the optical and remote sensing instruments, which now looked out over different parts of the planetary system with different fields of view. This also had an effect on the spacecraft, as it required that the full body of the vehicle—the size of a school bus—had to twist and turn in space to allow each instrument to observe its target. Every time the spacecraft turned to send its data back to Earth, this affected the instruments too.

One could perhaps see this as a case of transforming the modular system into a matrixed one to mirror the organization. That is, the loss of the scan platform turned the spacecraft into an entity that reflected both the independence of instruments on the one hand and the interactional requirements of the matrixed working groups on the other. Regardless, this heterarchical arrangement of the spacecraft's own components put tremendous pressure and even strain on all components of the system. We might describe the situation as one of sociotechnical torque, a phrase with a hybrid origin. The technical term is derived from engineering and refers to how one element of a system can put undue pressure on another, straining or even breaking the system as a whole. Yet it has also been imported into the social sciences to explain the "tremendous strain on individual biographies" that results when individuals and social categories do not align.[9]

Helen's case combines these definitions. Certainly, the cultures of negotiation and matrixed arrangement of team members on Helen produce different scientific biographies, examples of which I will discuss later in this book. But it also exerts effects on the spacecraft's biography. As my fieldwork began, the spacecraft faced an impending shortage of fuel if it continued to use thrusters to change its orientation; later, the "wheels" on the spacecraft that enable it to twist around in space became degraded due to the unexpectedly high number of requests to turn between observations. In response to this technical crisis, the team developed software, observational requirements, work timelines, and spacecraft sequences that attempted to protect the wheels from further degradation. This in turn impacted decision-making. Thus the impact of sociotechnical torque—the heterogeneous and shifting alignments of human, software, and hardware organizational components are experienced sociotechnically too. The trauma of this incident was not limited to the moment the scan platform was cut but reverberated through decades of spacecraft use.[10]

The "Passion" of Mission Work

Such challenges do not dampen individuals' enthusiasm for their mission work. I thought often of Max Weber's classic essay "Science as a Vocation" (Weber 2004) while studying planetary scientists in action. Like a calling, their scientific work expands to fill all aspects of their lives, capturing leisure time and relationships in their thrall as local "total institutions" (Goffman 1961a). Their colleagues are their closest friends. They go to conferences and concerts together. Some are married to other planetary scientists or astronomers. They are always busy planning some new mission, preparing a proposal, heading to the Antarctic or a desert to collect samples, or staying at work late to conduct an experiment in their lab overnight. A scientist once showed me the photos of an archeological site that he took at a Helen meeting in Greece, where he used the same digital techniques to remove the tourists from the foreground of the shot as he did to process images of the planets he studied. It was often hard to discern a difference between professional and personal lives.

Planetary scientists also travel incessantly. I often joked that they acquired the same number of air miles traveling to meetings on Earth that it would take to get to the planets they studied. They vacationed near their meetings and conferences, sometimes with their families in tow. One scientist in a senior position even planned his honeymoon to coincide with a planetary science gathering. There was little time for sight-seeing. At meetings in Annapolis, Barcelona, Cagliari, Denver, Houston, or Oxford, our day began at 8 or 8:30 in the morning and ended at 5 or 6 in the afternoon. We then wandered off at the end of a full day of discussion about planetary science first to dinner and then drinks—where more discussions of science and mission politics ensued until late in the evening. Hobbies were interspersed with work too, such as interest in rocketry, photography, or tinkering. Some especially energetic scientists ran, rowed, or cycled in races past middle age and well into their seventies. Roger took long lunch breaks from Helen meetings for marathon training, and I could only get enough time to interview the Paris PI or his deputy when I joined them on a bike ride or during a run.

Shared stories also swirled about the early days of their missions. There was the exciting story of Helen's discovery of plumes of liquid water emanating from the moon Enceladus, a legendary tale I heard several times. The deputy PI of the magnetometer, one of the smallest instruments on board, flew all the way from the UK to Spacelabs to argue that they should turn the ship around and go back to Enceladus to investigate these strange readings. The scientists who told me this story frequently admitted to me that most of the instruments had detected something strange at Enceladus (including the camera team lead, who also claimed priority in the discovery), but they had not told other instrument teams about their unusual readings or taken the step of going to management to request a change in flight plan: for reasons I will describe later in the book. On Paris, those who were on the prime mission before the participating scientists joined in recalled working together at the lab on the Martian clock (Mirmalek 2020), with unusually dry humor born from collective exhaustion. One group of scientists recalled looking for a beer in town when their shift ended early on a Sunday morning. When they asked for a round of lagers at a local restaurant, they were surprised when a group of elderly ladies dressed for church at a table next to them followed their example. Such oft-told stories blended the thrills of exploration and discovery with a sense of the necessary total commitment, the continuous melding of work and life on project time (Shih 2004), and the all-encompassing nature of mission work.

It would therefore be disingenuous to describe this organizational ethnography as a description of yet another job, one that people leave at the end of the workday in order to begin their second shift at home or engage in their hobbies. In my travels in planetary science, mission members' lives and life courses are unavoidably altered by their participation. Even at the most mundane moment of planning or during heated negotiations when stakes are high, commitment is total, enthusiasm is high, and energy is excited. Being *on a mission* is a question of moving back and forth between an everyday attenuation to spacecraft power, bytes, and resources on the one hand and the sweeping sense that you are a part of something greater than yourself and greater than any one person on the other. This is the *why* of participation in planetary science missions. We cannot forget this human, elevating, even transcendent aspect of the experience of mission work, even as we turn to the details of daily work, addressing the important question of *how* organized science on these missions is done.

TWO

The Integrators

"Welcome to Helen! We put the fun in dysfunctional." Betty stood in the doorway of the office of the engineer I was interviewing, interrupting our conversation. One of the mission's valuable science planners, a job that bridged science and engineering roles, Betty's energy filled the room as she extended her hand with a wide smile and a twinkle in her eye. I noticed the spacecraft logo embroidered on the breast of her thick-weave mission T-shirt and her long skirt brushing the tops of her sneakers, what she would later describe to me as her "uniform." Were it not my first day on the job, I might have taken her comment with the familiar smile of a group member, one who knew how to marry a sense of deep commitment and pride in their mission with a spirited sense of fun and self-deprecation. Over the course of the next several years, I would witness this culture up close and come to internalize its rhythms as a part of the Helen family, such that I ceased to notice it at all. On that first day, though, I was baffled.

The Helen offices are located in the windowless building at Spacelabs, upstairs from the mission control board. Colorful murals line the outer walls, describing instruments, orbits, and rings in Warholesque hues and swooping brush strokes (figure 3). The floor is laid out in something of divided concentric circles with offices in the outer ring, then gray cubicles, and finally a couple of conference rooms in the middle of the floor. In these rooms, gray carpet pile climbs the walls and large screens sit at one end. Moving between the science side and the engineering side, including software engineering, requires swiping a badge to go through heavy doors. The spacecraft office, manned by the chief engineer and her crew, is downstairs, where the ceilings are lower and the same gray cubicles are laid out in a maze on the floor. Another group sits upstairs, responsible for sending commands to the spacecraft.

3. Murals on the wall of Helen's project office. Author's photo.

Helenites who work in the project office experience the mission here, especially gathered around the tables in these conference rooms. Some are scientists employed at Spacelabs who walk across the lab to join meetings in person; others are science planners who facilitate the meetings at which spacecraft observations are planned and scientific priorities negotiated, each of whom has a cubicle or sometimes a wall office with a door on the floor and decorations of mission posters, NASA award certificates, and public relations giveaways.

Teammates who aren't employed by Spacelabs experience the mission differently. When they come to the lab, they attend meetings held at the large public auditorium, with additional group sessions in smaller meeting rooms near the cafeteria. Otherwise, the vast majority of Helenites are located in offices and laboratories across America and around the world. They are in university offices with linoleum floors and large computer screens and in cold rooms, clean rooms, and laboratories accessible with swipe keys and pin pad locks. They are in a private lab in a Boulder, Colorado, office park—bright and airy with blonde wood, large windows, and computer terminals ringing the room. They work down narrow, whitewashed hallways at another NASA center in researcher offices where they dial into a workroom computer that no one sits at. Or they work in a suite of offices in Rome overlooking a seventeenth-century university courtyard, newly renovated with tall windows and smart modern fixtures or in a research station on the outskirts of that ancient city featuring the 1970s' brutalist architecture, perching heavy concrete blocks atop metal cylinders, winding along corridors to find flight instruments carefully stored on dusty shelves behind operational instruments that simulate activity millions of miles away. There are Helenites in sparsely decorated offices in London colleges, behind tinted windows in a storied university building in upstate New York, and at nerve centers for instrument work at universities in Iowa and Arizona. Some of the work of Helen even feels placeless. When I asked a Helenite at the University of Michigan if I could visit his lab, he shrugged and said that there wasn't much to see there but his computer: he regularly tunneled into remote high-performance computing servers elsewhere to run his models. No wonder that Helenites call themselves a "virtual organization."[1]

Although in its strict sense, the virtual organization refers to the distribution of instrument command centers across the United States, many layers of virtual copresence knit together the Helen community. Attending conference calls was part of the fabric of daily life on the team. While I sat in on many of the meetings I will describe in conference rooms at Spacelabs, the majority of the participants phoned in from their offices elsewhere. The teleconference conversation was piped in through speakers mounted in the corners, the voices of distant colleagues floating through the room. The rooms themselves had long tables in the center with a heavy wooden top, gray wheelie chairs circling the tables, and padded chairs lining the walls for overflow seating. One room even featured a photographic "who's who" laid out in an organization chart on the wall (figure 4).

The gray-carpeted walls, the grandiose auditorium, and the invisible teleconference lines all serve as the setting for the interactions among Helenites from their many roles and positions. This heterogeneity is a reminder that the team must bring together many different and sometimes competing elements into a single plan for execution on board the spacecraft. In doing so, they attempt to retain as many different voices and experiences as possible. Helenites call this process *integration*. Stacia Zabusky explains that at the European Space Agency, "the disorder of diversity is in fact valued by participants . . . who refuse to be implicated in hegemonic assertions of consensus" (Zabusky 1995, 11). In the same way, hearing, retaining, and

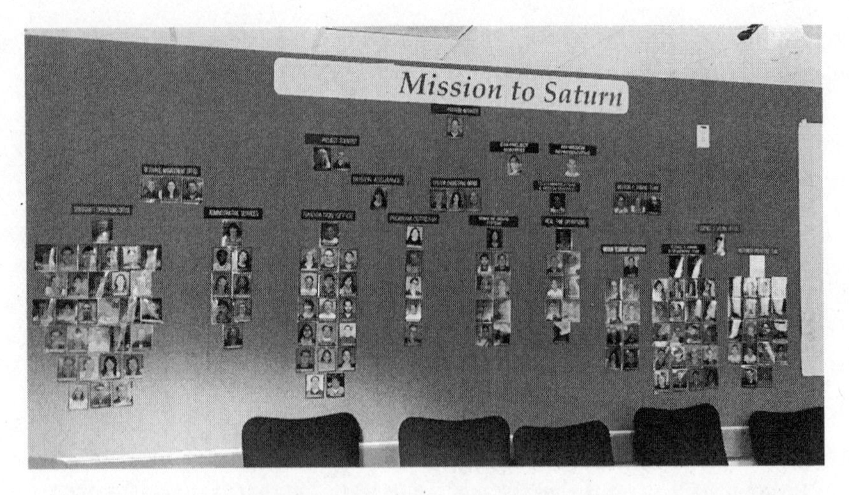

4. The organizational chart on the wall of Helen's conference room. Author's photo.

working with a plurality of needs and requests is the goal of *integration* on Helen.

To make decisions, Helenites bring members together across their heterogeneous, multi-instrumental, and multidisciplinary team in two chief types of ritualized meetings. The first I will describe are target team meetings, held on a bimonthly schedule, at which instrument team representatives assemble a scientific plan for the spacecraft to execute in the coming weeks and months. Meanwhile, scientific priorities are evaluated in the overarching discipline working groups then raised to the level of the entire team for a kind of decision-making integration at the mission-wide Project Science Group meeting. I will use these decision-making examples to describe the Helen team's cultural focus on fairness and autonomy as they approach the challenges of integration. Before I begin, however, it is necessary to review the scientific questions at stake in the planning meetings in order to appreciate the complexities of integration in the matrix.

A Division of (Scientific) Labor

As François—a tall, well-dressed, senior ESA manager—once described it in an interview, "the international collaboration and multi-target investigation is really the achievement of Helen." Practically speaking, this means that conducting an ethnography of Helen is much like trying to study something as diverse and distributed as the United States of America. The twelve instrument teams and crosscutting working groups and the project office's scientists and engineers each have distinct local "idiocultures"—local practices, knowledge, and interactional norms (cf. Fine 2012)—in addition to the shared concerns, rituals, and practices of the mission as a whole. In my travels across the team, I met many individuals whom I observed to be key participants in one part of the mission yet who were unknown to individuals elsewhere on the team. This is a function of Helen's division of labor, where responsibility for science questions and planning are distributed among both scientific instrument teams as well as targeted parts of the Saturn system, with limited visibility or reciprocity between groups (Durkheim 1893).

Each working group on Helen is concerned with its own piece of the scientific puzzle. The Atmospheres group is interested in Saturn's atmospheric composition and dynamics. Using mathematical modeling and physical equations, they are trying to determine the gas giant's rotational rate, which is still unknown, as well as if there are different rotational rates for different layers of cloud on the planet. They are also fascinated by a strange hexagonal cloud structure discovered at the northern pole. Meanwhile, the Rings group is concerned with the composition and dynamics of Saturn's rings. They use harmonic ratios and observations of tiny moons in the ring plane to estimate rotational parameters and use imaging, infrared, radar, and ultraviolet scans to make claims as to the composition of each ring. Small moons are also the concern of the Satellites group, which is concerned with all but one of the moons in the planetary system. Of these moons, some of the largest have earned relatively close flybys for higher resolution imaging and mapping to understand their surface composition and texture. Scientists in this group also detect the moons' internal composition using radar and their external environment using physical instruments. One of them looks exclusively at the tiniest moons, which are rarely seen up close and require "pirouettes" in space in order to photograph them.

One moon gets its own group: Titan. The largest moon in the Saturn system, nearly the size of Mars, Titan has a thick atmosphere of hydrocarbon gases like methane, ethane, and benzene. Given that carbon is central to life on Earth, this moon is therefore of considerable interest. Titan also has its own cycle of clouds, lakes, and rain like Earth's own water cycle. But its soupy atmosphere makes it nearly impossible to photograph the moon's surface. This group therefore relies heavily on radar and multispectral imaging to peer through the clouds and understand the terrain and topography, as well as the atmospheric probe, built by ESA, which Helen carried to Titan and deposited on the moon's surface early on in its mission. Titan is of such interest to so many Helenites that their working group meetings are usually held when no other groups are meeting to facilitate wide attendance. The group is also quite cohesive, putting on regular conferences and workshops about Titan that invite all Helenites and others outside the mission studying Titan to meet together. The probe team also meets every few years to share new results.

There were several questions of special interest related to Titan while I was observing Helen. One concerned the moon's seasons and variation in the outlines of its methane lakes: this relied heavily on the radar instrument but also involved images of clouds and spectral imaging of the surface. Another question involved its internal composition and density. This required many, many passes from the radio instrument, and when I visited the radio science group in Rome, I met the group of graduate students hard at work on interpreting these flybys. Finally, the radar team was busy investigating a possible discovery of ice volcanoes on the moon, called cryovolcanoes, and scientists were wondering whether the moon had the internal characteristics to produce these surface features or if they could simply have been produced through erosion due to methane rain. As both the radar and radio instruments require many others to be turned off during their flybys, their observational requests were often hotly contested—although several Helenites pointed out to me that since the radar PI was also the head of Spacelabs that team usually got what they needed.

Finally, the physics group—abbreviated PAM for "particles and magnetospheres"—was a group with strong ties. The PAM instruments are called "in situ" instruments because they detect the particles, plasma, dust, and magnetic environment around the spacecraft as it flies through space. They are usually turned on during flybys as they do not impact each other's ability to take instrument readings. Through these many, many observations, PAM scientists piece together a comprehensive picture of Saturn's plasma field and its magnetic environment. PAM scientists do not work alone: as they put it, knowing how plasma works requires piecing together measurements from each constitutive instrument. You can't know what the particles are doing if you don't know what the currents of the underlying field lines are and vice versa, but you'll know the former from one of the particle instruments and the latter from the magnetometer. PAM scientists therefore have their own data sharing agreement, placing slightly downgraded versions of their data into a shared repository that all could access, in a "postcommunitarian" vein typical to physics collaborations (Knorr-Cetina 1999; Shrum, Genuth, and Chompalov 2007). At PAM meetings, scientific presentations wove together many instruments in the physics suite to address the problems of the planetary environment. Some of these scientists also participate in astrophysics communities with an interest in, for instance, the cosmic microwave background or interstellar dust and brought their observations to bear on those issues as well.

Of course, not everything in the Saturn system falls neatly into disciplinary or instrumental categories. The moon Enceladus, itself under gravitational strain from Saturn and the moon Dione, spews water into space, bombarding other moons with particles and forming a large and distant ring around the planet. This in turn affects the plasma and magnetic fields in the system, even producing a planetary aurora. By the time of my work on Helen, a few meetings had formed that brought together members of these different groups around a topic such as plumes and magnetospheres (PAM and Satellites) or Saturn's aurora (PAM and Atmospheres). I will describe these interactions in more detail later in the book. For the most part, however, at discipline working group meetings, these different instrumental scientists would give five or ten-minute talks updating the others in the room on their progress on the questions described above. With the few exceptions of the PAM meetings and presentations by interdisciplinary scientists—ten scientists permitted by contract to work with two instrument teams' datasets—these presentations typically focused on a single instrument's results.

Helen's instrumentation brings specific strengths to bear on these questions. The camera takes photographs, measuring light intensity or capturing geological terrains, helping identify spokes in the rotating rings, craters on the icy moons, and clouds on Titan. One infrared spectrometer breaks down light past the visible spectrum, giving clues to composition of Titan's surface features or Saturn's clouds, while another infrared detector records thermal information, whether the hot spots on Enceladus from which the plumes erupt or the ways that the ring particles retain heat, giving clues to their material composition. Scientists working with the ultraviolet spectrometer love stellar occultations: moments when their instrument can glimpse a star through the atmosphere of Saturn on the edge of the planet, the rings, or Enceladus's plume, using the difference in spectra to determine their targets' composition and features. Together, these instruments form the remote sensing palette, mounted on one side of the craft.

Two instruments use the antenna dish on the spacecraft. The radar instrument generates black and white images that describe Titan's topography as if through a glass darkly, while the radio uses radio waves between the spacecraft and Earth to determine if any of the moons possess a form of slight wobble that would indicate they have an ocean or molten core beneath their surface crust. Both of these instruments can also be used to determine differences and gravitational effects in the ring plane as well. The long boom on the spacecraft detects the planet's magnetic field. Distributed on the body of the spacecraft are the "in situ" instruments. Among them, a dust detector scoops up interstellar dust, a charged particle device looks for positive or negative electrons, another instrument scoops up energetic ions, and yet another collects radio waves as indicators of the plasma environment. Because these instruments do not produce visual data (in local parlance, they produce "squiggly lines" instead of "pretty pictures"), they experiment with other ways of communicating their findings: the radio wave instrument releases sound files to "hear" the sounds of Saturn, its lightning, and audible shocks across field current lines.

Each instrument's capabilities can typically be applied to all five discipline working groups, so they all have a representative who participates in planning observations in targeting meetings. During my immersion, each instrument team itself met once or twice a year, largely to host science talks of five to ten minutes to report on individual scientific tasks using that instrument. At these meetings, I noted that each instrument team was its own social unit with its own local concerns and ways of working. The thermal spectrometer team, for instance, was actively debating which of four competing calibration algorithms to use on its data. The infrared instrument, on the other hand, had only a few scientists actively releasing calibration algorithms, with new versions regularly updated online.

Some groups actively police who can work with their instrument and their data. For instance, the camera team leader insisted that keeping careful track of who was working on what problem was necessary to protect the team's scientists, especially juniors, from being scooped. On the other hand, the radar team created a category of member called "team affiliate" to invite others—typically those well known in the Titan community, in which radar plays a hefty role—to join their meetings and work with their data. As such, radar meetings became something of a hub for Titan science, its membership roster overlapping significantly with the working group and external meetings. Some teams require all coinvestigators to be named as coauthors on papers, others require the PI's name to be listed, others still require their instrument team credited as an author, while others have no requirements about authorship at all. These variations took place across PI-led teams and facility instruments with no ready pattern I could discern. I will discuss these issues in more detail later in the book.

Finally, each group had different international collaborators. The dust instrument was provided by Germany and enrolled many German scientists, and due to a long-standing collaboration, the ultraviolet spectrometer also featured many Germans. Most of the Italians worked on the radio, radar, and infrared instruments, while French collaborators were common on the thermal instrument. British, Greek, and Hungarian scientists cemented ties among the physics suite, following the nationalities of certain PIs and interdisciplinary scientists. These national participations were not exclusionary, and many were facilitated under the same international memorandum between NASA and ESA. They usually followed linkages established because an instrument was provided by a particular European nation or because of prior connections between American and European colleagues.

This overview of Helen science gives a sense of the extraordinary heterogeneity of the mission. Each group is itself distinct from the others with its own norms for interaction. But each group is also internally heterogeneous, as, consistent with the matrix organization, the discipline groups incorporate many different instruments, and the instrument teams incorporate different disciplines. Recall that local processes and transactions that take place in different parts of a matrix organization's structure can create limited visibility between parts of the organization as well as uncertainty and difficulty in communicating across divides (Vaughan 1999, 916). The challenge of *integration*—producing a timeline for spacecraft activity—takes place amid this extraordinary heterogeneity, situated perspective, and uncertainty. Setting up a plan for scientific observations that includes many different and distinct needs and bringing together this many different types of scientists requires considerable work. This is the organizational work of mission planning that takes place in and across several parts of the mission team.

Integrating within Subgroups: The Target Teams

Scientists belonging to the discipline working groups—Atmospheres, Rings, Satellites, Titan, and PAM (Physics)—also staff each group's targeting team.[2] These groups' job is to take a segment of spacecraft activity—while it is flying past the rings or Titan for instance—and establish which observations the spacecraft will take. Targeting team meetings are the site for the fine-grained negotiation of observations and allocation of resources among members of that group, laying out the plan for the science achieved during that segment. Will it point a spectrometer at the rings, capture in situ information about magnetic fields or charged particles, or snap photographs while whipping around a moon in an orbital maneuver? To decide, each instrument team sends a designated representative to each of the targeting teams. Usually this is a different scientist for each meeting, although in one or two cases the same scientist served as the representative on all five of the teleconference calls.

Targeting team meetings usually take place on the phone as well as in the gray-carpeted conference rooms. People file into their regular seats around the table, with one Helenite describing meetings as "like church" to express their ritual character. Meetings ranged from an hour to three hours in duration and typically took place once every two weeks. In addition to the instrument representatives and a senior scientist from the hosting discipline working group, meetings are facilitated or supported by a science planner, a role that bridges engineering and scientific concerns.[3] The science planners themselves form a team on Helen, even if they are usually seen in action deployed to these different meeting groups. They even play softball together in Spacelabs' amateur league, with the project secretary regularly celebrating their team's achievements in status updates on Facebook.

There is no way to describe a typical targeting team meeting because each team has their own way of working together toward integration. This is especially apparent when a plan is oversubscribed—a moment when more people want to observe than there is time or space to observe in. This constraint gives rise to most of the conflict at meetings. For instance, the Rings group typically asks its scientists to simply input their observational requests into the science planning software for discussion at their targeting meeting later on. Plans at this initial meeting are often oversubscribed, with scientists from even the same instrument team inputting conflicting requests for the same period. The Rings targeting meetings, therefore, aim to figure out which of these observations should go ahead based on spacecraft timelines and a general sense of whose turn it is to offer up an observation. This question dominates a typical Rings targeting team meeting that I observe from Noel's office, the science planner who supports this group—and who has nicknamed me "ringlet" due to my regular appearance as a hanger-on at his meetings. Noel listens intently to the conversation on the teleconference line, his lanky frame leaned back in his chair, his long legs crossed at the ankles, and his arms crossed behind his head. I imagine the other scientists and engineers on the phone line at the same time, whether at their desks, in their cars, or in their labs thousands of miles away.

The problem today is that the group needs to cut 150 megabytes from their observational plan to fit into the spacecraft's memory allocation, but "a few weeks ago [they] went through this and cut out all the fat already," so there is not much left to trim. Noel focuses on the end of the segment to see what they can do, saying, "I dunno, let's see. Plasma, I mean, you guys cut already. You're still collecting a little under nine megabits." There is also a thermal calibration maneuver, but Noel says he's "reluctant to cut [it] because we're going into a Titan flyby," and calibration acquired during this segment is necessary to interpret those results.

At this point, Perry, the thermal representative on the line, volunteers: "I'll cut a hundred megabytes." To which Noel asks, "Are you sure?" "Yeah, I can cut the hundred." Because thermal is a secondary observer on this flyby and has requested observations of two targets, there is room to make cuts. But Noel still needs another fifty and suggests that the ultraviolet instrument "wasn't part of our last discussion, so they haven't cut anything." Perhaps it is their turn now. "It looks doable," says Noel. He rattles off the number of megabytes they have planned for observing the plumes and different moonlets, saying, "It looks to me that there's fifty megabits there that we could cut." An ultraviolet representative defends one of these observations: "The plume is at very high phase, so we'd like to leave it alone." Noel continues to go down the list, inquiring about a thermal observation of propellers in the rings and then about a camera observation that has already been trimmed, but these do not offer much opportunity for cuts. Finally, Noel declares, "I'm happy to take the one hundred from Perry and . . . if ultraviolet wants to cut another twenty, I'll take that right now. Then we'll have enough cut here."[4] I note how he uses "we," "you," and "they" to refer to different teams and how Perry metonymously stands for the thermal instrument, parallel to Noel's call to "ultraviolet."

In addition to managing time and bytes, observations must also take into account the spacecraft's pointing and its use of either propellant ("delta-V") or rotation using the craft's reaction wheels. These are all limited resources that must be treated with care. For instance, during one of their meetings the Satellites targeting team was concerned with an observation that had been planned in advance of a new project-wide requirement. A group of scientists and science planners sat around the table in the windowless room, laptops open and various colorful Excel spreadsheets on their screens. Almost all of them were women, their dress tailored but informal, their hair worn long around their shoulders. They had initially promised the flyby to the radio science instrument, but the project required them to fly by the moon using thrusters, which would upset the instrument's delicate readings. Rebecca, a veteran mission scientist and the group's science leader, called flying on thrusters "the way you'd do it if you were an engineer"—meaning that the science opportunity would be severely limited. Gwen, another scientist with many missions under her belt and a long history of Helen science planning, added, "It's the way you'd do it if you don't push back and you don't think there are any consequences." "But we're not gonna get any of our science,"

Rebecca worried. Both Gwen and Rebecca normally exuded an aura of confidence, even brusqueness, a direct style of interaction associated with their seniority and experience. Now the group—including Karen, a sharp early-career scientist who was both Gwen's protégée and Rebecca's deputy, and Rose, a science planner with a dual science and engineering PhD—began to discuss alternatives but found them to be limited:

GWEN: What if we have to give up that flyby and plan something else? That's going to ripple all the way through to [the extended mission]. And every observation we make, radio is gonna wanna get a replacement.

KAREN: Well they're not going to get one because they're never going to get another south polar flyby [i.e., this observational opportunity is unique].

REBECCA: We have to talk to [a flight engineer].

ROSE: It could be that they looked at it and said well, [it's trivial to fix].

REBECCA: We really need this; it's going to really screw up our science [if we don't get it].[5]

The list of fixes included tweaking the tour ("Let's not even ask; it would make people hate us"), allocating the flyby to a different instrument (to the detriment of radio, who needed another flyby for their gravity measurement), talking to the scientists involved, talking to an engineer to see if they could fix it, and appealing to the project on the basis that their observation required precious fuel. But as Gwen put it, "I don't wanna make it so complicated that [the engineer] has to spend a week designing it because we don't have [those] kind[s] of resources." Resolving this problem required invoking the pointing and capabilities of the spacecraft, but these were appeals to organizational roles too.

The Atmospheres targeting team managed this kind of situation differently. They met weekly and their meetings were highly ritualized. George, Emily, and Max were among the supporting science planners, usually wearing their thick-weave mission T-shirts and jeans, and kicked each meeting off by distributing a handout with many pages of color-coded tables indicating requested observations and sites of conflict. George and Max were also members of mission instrument teams. The Atmospheres targeting team often attempted to assign priorities to each of the scientific observations offered by the instrument teams to ensure that only the most important or best observations made it into the spacecraft timeline. But because each team has their own requests that are considered "top priority," and those came up against other teams' priority observations, it was still difficult to decide who should secure the observational time. The targeting team confronted this when someone asked the camera representative, a soft-spoken veteran scientist named Rod, if they could cut all images of the moon Iapetus:

ROD: That's kind of a hard question; it's kind of the standard question we ask of everybody, but . . . How do you answer that? Because it's up against something that's higher priority—that someone decided was higher priority. And the originator of the request? That observation is the highest priority for that guy.

GEORGE: That's true, but if he says I'd like to get five of these Iapetus [images] back in, then we could look at who that would impact and see if they would be willing to trade.

EMILY: And who that would impact would be thermal with their [observations].

GEORGE: We could then go back to thermal and say, "Would you mind giving up more time?"

THERMAL REPRESENTATIVE: We like the timeline the way it is.

GEORGE: I'm sure you do.[6]

Note that the hope for "trades" between the thermal instrument and the camera rely on assessing priorities that may or may not be commensurate between groups.

PAM instruments are typically always on, so conversations at the PAM observation planning meetings revolved around whether the magnetometer needed a calibration roll or if an instrument needed the others turned off briefly for an observation or calibration. These were mainly questions of coordination: as all instruments shared goals and data, competition was not visible outright. The Titan group took a different approach. In their case, some instruments had to be turned off entirely in order to get precious radar or radio data to resolve their questions. In response to reportedly heated debates in early planning sessions, Titan's targeting team chose to meet once at the outset of each mission period for three days to divide up the entire tour among the instrument teams. This way, once weekly planning came around, the question of who would observe was already resolved, and the meeting focused on fine-tuning.

I attended this meeting, called "the jumpstart," in April 2009. It was a three-day event hosted by an East Coast laboratory in the United States. The group planned to allocate each flyby of Titan in the tour from 2010 until 2017. A spreadsheet was projected on the screen at the front of the room, with a hard copy circulated in the room. Each row indicated a prospective flyby the moon over the next ten years, assigned a number, and coordinated with the orbital segment, year, and other relevant details to fix it in time and in space.[7] Columns indicated start time, end time, which instrument would observe as a prime observer, operational modes if necessary, and comments. Each pass also already had a tentative assignment of an observational request from one or more instruments, highlighted in yellow if more than one instrument wanted the observation at the time.

Betty chaired the meeting, exuding her typical boundless enthusiasm for Helen—especially Titan, her charge. She opened with, "What we thought we would do is sort of go around the room and annotate the spreadsheet to give an idea of what observations have been allocated." She then asked, "Is there any discussion we want to have before we just jump into the spreadsheet?" There was some departure from format by jumping straight to a conflict that was evident on the list, over how to allocate a flyby wanted by radar, to observe a northern lake, and radio, to get more information about gravity. Both were high-value questions, so the group decided to call in Cliff, one of the interdisciplinary scientists, to adjudicate. Over the speakerphone in the room, Cliff decided for radar. The radio representative protested, but the decision was accepted by the rest of the group. Upon resolving this initial dispute, the group went through line by line over three days, filling in the spreadsheet with details about each instrument's requests, ascertaining whether the flyby could support a secondary observation from another instrument (called a "rider"), and indicating essential altitude tweak requests to pass along to the navigational engineers. I noted how looking over the allocations at a first glance helped them ascertain whether each group got what they needed, determine de facto priorities, and also assess the fairness of the allocations over all the instruments. With this primary allocation accomplished, at the biweekly Titan targeting telecons, the group referred back to this spreadsheet to determine which instrument would be observing in that segment and then spent the rest of the hour or so figuring out the details instead of negotiating or prioritizing.

Integrating across the Mission Team: The PSG

Targeting teams involve small groups in focused discussion, but integration takes place at scale across all of these disparate teams as well. This typically takes place at the Project Science Group meeting, or PSG, a meeting format that crystallized early in the project and remained relatively stable for close to twenty years.[8] The entire roster of mission scientists is invited. Every year the team meets twice at Spacelabs and once in a European city for a week. The meetings have assumed a ritual character with a similar structure each time. Mondays and Fridays are reserved for instrument teams or emergent groups to host meetings on their own agenda, while Tuesday through Thursday are reserved for group-wide work on a common problem. The first plenary session on Tuesday establishes the goals for the meeting and includes updates from the project manager, Everett; the NASA Headquarters representative, Kenneth; and from members of the project office like Francis, deputy project scientist Victoria, and others giving an update on science planning, spacecraft health, media relations, and data archiving.[9]

PSG meetings offer an opportunity for the project office to seek input from the assembled scientists on a decision of import to the entire mission. Throughout Wednesday, participants meet in their working groups where, alongside a few science talks, scientists tackle the charge from the project office. That evening, there is a social event that brings all participants together, although in practice, people mostly socialize in their existing groups. When the meeting is at Spacelabs, the project secretary always chooses an upbeat restaurant in the town nearby with good opportunities for mixing and mingling. When in Europe, the social event involves regional cuisine and talent, local wine (or beer), and tours of nearby monuments. By Thursday morning, the group reconvenes in plenary, and the leaders of each discipline group report back about their specific task. Each group also nominates a few scientists for showcase talks to update the group on a few projects happening across the team. The meeting ends at 3:00 p.m., at which point the project's executive council (the PIs and interdisciplinary scientists, team leaders, and senior project office personnel) meet to discuss the outcomes of the working group presentations and make a project-wide decision to address the problem. Some team members loiter outside waiting for their friends or for news of the decisions, but most disperse. Outside discussions of happenings at this private meeting are fiercely taboo; a summary of the executive committee meeting circulates by email to the team weeks later.

The ritual of the PSG meeting is so consistent that after visiting a few of these meetings, one gets a sense of timelessness on the mission. The meetings exude calm. Even when budgets are being slashed, or an instrument has suffered a fault, project manager Everett remains cool-headed, calm and affirming, saying things like "Things are going pretty great" or, in times of trouble, "I have no doubt that things are going to be A-OK." Kenneth, Helen's representative at NASA Headquarters, similarly lets the scientists in the room know that their mission is highly valued by the agency and that he is doing what he can to ensure it is well represented in funding conversations, even if he cannot give details due to embargoes. Even the group problem, established in advance by the project office, is part of the ritual, its management requiring discussion among discipline working groups and reporting in color-coded spreadsheets.[10] The problem is always keenly felt and requires everyone's attention and input and therefore serves as a way of orienting disparate groups on the mission around a unified question.

A perspicuous example of this is tour selection, a problem that occupied several PSG meetings throughout 2008 and 2009 as Helen moved into its final mission phase. The task was to select the path—or "tour"—that the spacecraft would travel for the next seven years, from 2010 to 2017. I observed the final selection meeting in January 2009. By this time, each working group had produced lists of requested observations and assigned each request with a priority order of 1, 2, or 3. That spreadsheet was handed off to the team's navigational engineers, who used it to generate seven possible tours. The task of this PSG meeting was to go back over each of these seven tours in the working group sessions and get each group to rate tours to indicate their preference, reporting back to the team on Thursday morning. The product that would be reported was a master spreadsheet listing each tour option in the columns and scientific priorities along the rows. Each team had to fill in the corresponding box at the nexus of tour and priority with a single color—red, yellow, or green—to indicate their rating. This would eventually become "one combined spreadsheet" to guide the mission executive's final decision, which is made among the PIs and team leaders with the project office leadership alone.[11]

On Wednesday afternoon, I joined the Titan working group in a gray lecture hall at Spacelabs, much expanded from its targeting team to include official representatives from each instrument and interdisciplinary scientists sitting around a horseshoe of tables up front, with rows of chairs behind for others who came in and out of the meeting. The Titan group's science planner Betty kicked off the discussion by summarizing the purpose of the process and its history:

> The Titan targeting group approach was to take the prioritized spreadsheet that had been worked on very hard [by the whole group] . . . [and to say] OK, who [i.e., which instrument] contributes to this objective? They [i.e., the instruments] got their own row in this massive spreadsheet I've been talking about. . . . What is their [i.e., each instrument's] piece of the pie, so to speak, in terms of how they're going to achieve this objective? Over to the right, there are twenty rows that are, "What are the [instrument] priorities?" . . . Now we're at the point where each instrument is going to assess how well this part of the tour [*gestures to columns*] achieves this objective [*gestures to row*].[12]

In other words, unlike the targeting team meetings, where tour segment requests were broken down among instrument teams, today's purpose was to work together across these assembled teams to assess Titan priorities and later to integrate that assessment with those from other working groups. The meeting proceeded by looking at the tours and priorities and assigning colors, going through the rows one by one with some variety of "For tour one, priority one . . . looking at that for this group, is [it] green, or yellow, or red?" Betty called upon different instrument representatives to voice their opinions and inputted the information into the projected spreadsheet. When a radar representative indicated that version six was "the worst of five," she turned the corresponding box red. But it soon became clear that red, yellow, and green did not adequately express the range of opinions in the room. Scientists hedged their accounts by suggesting that a particular observation should be ranked below a green-ranked one but not as far down as the yellow range. Betty started to suggest intermediary colors, such as "lemon-lime" and "orange." Soon the spreadsheet lit up in a rainbow of colors. At one point, a scientist declared, "I want to change the color of that box. It was an orange, but it should be a pink." The three colors—red, yellow, and green—requested by the project simply did not cover the range of opinions and subtle variation between instrument appraisals that needed to be captured from the different teams in the room (figure 5).

Even as they filled in individual cells, the group had their eye on the bigger picture. Betty addressed the radar team who assigned one tour an orange despite it working best for those on the radio team interested in Titan's gravity. She explained, "We're going to have three reds out of four tours, and we have to understand the significance of those [local rankings] in terms of what the overall rankings are." This was on the one hand a question of ensuring that radio science could get their priority observations. But Betty herself was a radio team affiliate trained in gravity science. Lest this seem to be a conflict of interest, she explained that she wanted to "go through all the reds and oranges" on the spreadsheet, regardless of instrument or observation at stake, to "understand the significance of the science that is lost if those [tours] are not selected."

By Wednesday afternoon, word started to filter out across the mission that tour number seven was preferred by most of the groups. Rather than hang up their boots and go home, each group then began to discuss how to "protect our observations" following this eventual selection. Yet the conflict between radar and gravity persisted in the Titan group, where the gravity scientists preferred version seven, but several radar scientists were concerned they would not be able to image the moon's northern seas. Ultimately, the

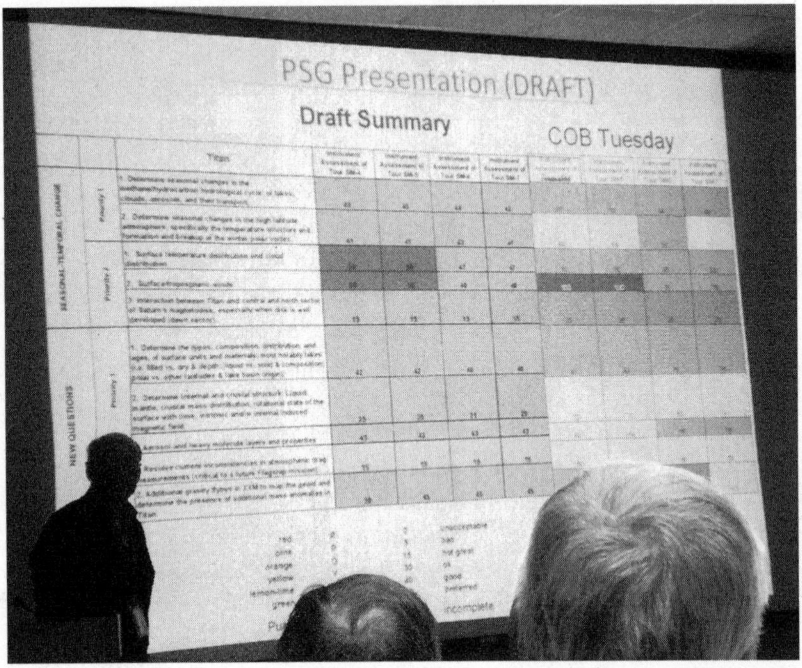

5. The Titan group colors their tour planning spreadsheet. Author's photo.

Titan group opted to argue *against* version seven when they returned to plenary. As scientists gathered in the auditorium and each group presented its support of version seven, the mood of the project office personnel began to lighten in expectation of this emerging agreement. So when the Titan group took to the floor to argue for version *six*, scientists across the team spoke up to question them. They also inquired as to the meaning of the rainbow spreadsheet. When asked if radio preferred version six or seven a gravity scientist answered, "We [radio] have a red for six," indicating conflict within the Titan group.

She was quickly quieted by her colleague, who spoke to her in a hushed voice. The observations for Titan science had not yet been divvied up, and since radio relied upon many other instruments turning off entirely while it took its gravity measurements, its team members could not afford to lose their colleagues' goodwill if they hoped to argue successfully for future observational opportunities. Everyone knew that the radar team leader was also Spacelabs' director, so it would be impolitic to argue against them. Plus, many of those arguing for tour six were advocates of the proposed mission to

Titan, competing against the new project scientist Francis's own proposal for a follow-up mission to Europa. There were many fault lines underfoot.

Managing the eventual decision was a question of articulating what was off-limits as much as what was conversationally fair game in order to determine what would be an appropriate trade. Francis tackled the consideration about the new mission (the "outer planets face-off") immediately with the following: "Clearly, we have to do a good job on Titan. We know that Titan is a very high priority; we know that there's a flagship mission that has to be considered." When Isabelle, the camera team leader, asked, "If the Europa mission is [chosen] over the Titan mission, would we still be worried about this?" he replied, "This is still the only mission that will look at that lake before [the next] flagship goes, whenever it goes." He sought agreement by wondering if version seven would somehow be acceptable to radar with a small change. Betty responded that the lake that could not be viewed from version seven was a priority objective that the tour designers did not include in their calculations. But it emerged that the other tours were also suboptimal. The Titan team left the floor to some confusion. A tour designer commented that "things don't seem like quite as foregone a conclusion" as they had before, while scientists murmured about agreeing upon a tour that was perhaps at least "equally bad for everyone."

After all groups had finished presenting, Francis invited the acting leader of the radar team, Terrance, and an interdisciplinary scientist in charge of the Titan group, Cliff, to "take any more time to make your case." At first Terrance restricted his comments to Helen priorities, explaining that not being able to see this lake would restrict the team from being able to satisfy a mission-wide priority for the solstice mission of change detection throughout seasons. But he also suggested that future missions that might want to land in the lake would need the radar's essential map. Other instrument leaders spoke up—Isabelle from the cameras, Leon from infrared—to suggest that they could assist with the requisite mapping on tour seven. The Titan radar group countered that the resolution, the light, or the number of available observations wouldn't be good enough to see the lake or to map potential landing sites.[13] Leon threw up his hands in mock jest. "You might as well throw infrared off the spacecraft!" He chuckled. "You guys have figured it all out."

Unable to achieve integration among the assembled group, Francis opted to invoke mission hierarchy. He called Arthur, a distinguished interdisciplinary scientist who bridged both infrared and radar and was a member of the proposed mission to Titan. Arthur had already left for his home university to teach a class but phoned back into the meeting to adjudicate between the options on the table, his voice booming over the speakers at the assembled scientists in the hall. While the observation was important, Arthur admitted, the rest of the team's science in his opinion outweighed this part of the Titan case—that is, going for version six would have an outsized and therefore unfair impact on the other teams. In recompense for radar giving up this important observation, he affirmed that infrared could step in. The group acquiesced and the team coalesced around version seven. In the next chapter, I will describe how such moments of a higher-up weighing in do not compromise matrix decision-making but are rather endorsed by the team as part of their "interactive process" (Heckscher 1994). In the meanwhile, we will turn to two cultural elements that play a key role in the process of integration at all levels on Helen: autonomy and fair trades.

The Tools of Integration: Autonomy and Fair Trades

In group decision-making in PSG meetings and in target team meetings, several observations emerge. One is the fierce *autonomy* of the heterogeneous subteams on Helen. Each not only works very differently but also defends its right to do so. Target team members have a sense that each group does it differently, but only a few people who observe multiple groups at once grapple with the considerable differences in process amid the ranks on the mission. A science planner I spoke to discovered this when newly assigned to support two planning groups. One group managed an oversubscribed plan by asking everyone to take an equal percentage cut to their data, but the other rotated responsibility by volunteering to take the entire data cut themselves, like buying rounds in a British bar (as in the Rings example above):

> When I tried [one targeting team's] "percentage cuts across the board" [technique] in [the other team], people complained that some instruments did not put reasonable first estimates in our database initially, so asking for a percentage cut from each team unfairly penalized teams that did not start out with inflated data volume. In the [first targeting team], I once tried the volunteered cuts method [the other group's method], and . . . [the chair] noticed and said, "Did we change our procedures?! Is there a reason that we're not doing the percentage cuts that we all agreed to?" . . . I didn't appreciate that the conventions in each group were so ingrained.[14]

This emphasis on autonomy over target team process extends to changes that emerge throughout the project at the PSG. For example, early in my observations of Helen, the mission was considering what to do about allowing a second instrument to "ride along" and observe at the same time as the instrument to which the observational period was initially allocated. Categories of "rider" status were proposed to help each team standardize their treatment of these observations across the board in a way that provided some project-wide consistently. But as the proposal was brought back to each of the discipline groups at the PSG meeting that year, each one turned it down. Some didn't want to encourage more riders by making the process too easy, others didn't see the categorization fitting their particular needs but as being better suited for a different target team, some wondered where the money for planning riders would come from, and yet another group decided that people should just use the planning software better. The mission-wide conversation instead turned to how rider management required better communication between the technicians implementing instrumental observations and others on the team. Groups could therefore continue their own local ways of allocating primary and secondary observations.

Even this process of introducing new procedures was ritualized. During another team-wide discussion about spacecraft maneuvers that could cut down on the strain on the wheels (called a "y-bias"), each planning group was required to implement y-biases in their plans—but each one did so differently. This created complications for software engineers attempting to update the science planning software to include new elements like y-biases and riders. Because no standard could be implemented, no single indicator or set of fields could be implemented in the software. Asked to address a question of standardization, each team came around to a decision that allowed everyone to do it their own way.

An additional element of autonomy pervades PSG and targeting team meetings: instrument team autonomy. Throughout the planning process, the emphasis is on handoffs between one instrument's observations and another's, finding time for each in the spacecraft's precious time and pointing attitudes. The spacecraft typically moves from one type of observation to another—dust measurements, propeller observations, stellar occultations—and from one observational campaign to another. A thermal representative could volunteer her own instrument's observations for cutting but could not suggest that the ultraviolet representative make cuts or changes in a target team meeting or decide whether a tour achieved the plasma instrument team's objectives.

This element of autonomy was affirmed through linguistic associations that linked individual people with the instrument they represented. "We" and "you" were used colloquially between instrument representatives to designate their distinct affiliations, goals, and independence, as in the thermal rep in the Atmospheres meeting asserting, "We like the timeline the way it is," or as in Karen calling radio "they." In the snapshot of the Rings meeting above, Perry is thermal, and in the PSG discussion, radio is the "we" that has a red for tour six. I will never forget a moment during the Titan jumpstart when Betty asked Edgar, one of the group's most well-known scientists, how long he needed to warm up. It was not a particularly chilly day, but the instrument Edgar was there to represent needed to warm up before it collected data, and this had to be integrated into the observational timeline too. This kind of pronoun use, a form of *instrumental metonymy*, is so common it is invisible to team members, serving to demarcate lines of accountability and autonomy among the many different groups and representatives working together in the room or on the phone.

Helenites worked to integrate these autonomous systems not just through respect of individual particulars but also through a local concept of "fair trades." This concept originated early in the Helen mission, during its development phase. In conversation with an economist at the local technical university, an expert in game theory, the project office set up a resource trading board among its PIs: an online site where they could exchange money, data rates, or mass among their instruments (Wessen and Porter 1998). For instance, a PI with dollars to spare who needed more mass for their instrument could instantiate a trade with another PI who was strapped for cash but had extra mass to give. At the time, PIs made enthusiastic use of this system to shape and scope their instruments during the development phase, and they appreciated that the rules for engagement were well laid out. The board was established to set up a neutral marketplace, but true to the performativity of markets (MacKenzie 2006), it naturalized a zero-sum economic system with cultural effects within the community. Even decades later, this sensibility toward capturing finite resources, strategizing, and ascertaining "fair" exchanges within a zero-sum environment continued to suffuse Helenites' negotiations for resources and time on their operating spacecraft. As I will describe, it also articulated a system that members felt they could potentially game to win.[15]

Fair trades are mobilized as the conventions that each targeting team uses to decide on the appropriate allocation of resources. Note how in the excerpt from the Atmospheres target team, George hopes for a "trade" between the camera and thermal instrument. This quality of "fairness" aligns with team autonomy, as it is associated with making "everyone . . . feel like they had their fair chance to influence the decision":

> Decisions have to be made, but in a group this big, what you want to make sure is that you fairly get an opportunity for each side to air what they think is the compelling argument and then make the decision.[16]

Although fairness is an overarching principle, it is adjudicated locally using different methods. Recall that taking a "percentage cut across the board" was considered fair in one target team but unfair in another team unless they agreed to other changes to their process. This further cements the autonomy of different instruments as trades are made between them for observational time and spacecraft resources. In the trade at stake in the Atmospheres meeting, the thermal representative's comments made it clear that the trade would be contested as unfair if the group were to declare the camera's observations to be a higher priority than their own. In the tour selection, Arthur ruled that a choice based on one group's priorities would cause undue strain. At a project-wide discussion, the PIs wondered if a 10 percent cut to observational time across the board would result in "some disciplines [getting] four percent and other [getting] forty."[17] Trades between scientists and engineers were articulated along the lines of what propulsion expert Eugene phrased as "We'll give you this flyby, but it's going to cost you half a kilogram of propellant" that would affect science later on; in other cases, scientists joked about "baking cookies" for the engineers in exchange for using so much fuel in an observation.[18] Throughout these trades, a turn of phrase I heard repeatedly on the mission explained how scientists and planners alike resolved these sorts of conflicts: "We can't make everyone happy, but we can make everyone *equally* unhappy" (emphasis mine).[19] I heard this statement so many times that I lost track of its attribution: it was simply a shared form of talk on the team.

Fair trades may involve immediate trade-offs or baked goods, but they also invoke social capital for the management of future trades—that is, individuals gave up observations with the expectation that this will elicit a certain amount of "goodwill" that can be used the next time around when a different, more high-priority observation is at stake. Hence the radio instrument team member's concern that his colleagues' outburst could later affect their group's goodwill in the distribution of resources. This also occurred in the Rings target meeting, where the thermal representative offered a large cut and was thanked for it, while Noel suggested it was high time for the ultraviolet team to volunteer, and in the Satellites meeting, where the group was concerned that canceling a radio flyby would mean having to slot them in at a later date, in exchange for another flyby that was already allocated. As the soft-spoken scientist who led the Rings targeting team, Walter, put it in an interview,

> It's fairly rare that somebody goes in [to a targeting meeting] and says well, "I really wanted to do this, but I can see your observation is much more important than mine. So I gracefully decline." Occasionally, it happens. But not too

> often. It's more often: "I give up on this. I have something else similar I'd like to do in the next [round]." "Stay out of the road and let me have the time to do this newest observation, [and in return, I'll] go away and let you do your Tethys observation." . . . If you hit them over the head with something like this, and they have only a few moments to react, they're going to defend their own turf. If you give them a few days to think about it . . . they'll usually be more reasonable and likely to say, well, "Yeah, this . . . would be the fourth time I've done this observation. And I'm going to do it again on the next orbit so I can see that something [else is] more important." So most of the time, we get a consensus out of sort of arguing this.[20]

The "consensus" that Walter describes arises from locally approved solutions to the problems of negotiation based on such trades. The group does not reach consensus on the question of whether images or thermal scans of Iapetus are more important. Rather, the group comes to agreement that their local allocation of spacecraft resources has occurred "fairly"—that is, accountable to the groups' autonomous ways of making decisions and to each instrument team's autonomy over their own observations. Local solutions to problems of fairness, as Walter points out, can also involve putting one's own social capital on the line in exchange for an observation later down the road, with the expectation that this will be fairly repaid in a future trade. This notion of "fair trades" as a method of managing integration between autonomous instrument teams is central to all cultures of planning on Helen. These related cultural practices play an important role in understanding how *organized science* gets done on the mission team, upholding group autonomy through the harmonizing process of integration.

Conclusion

Whether in small group meetings or across the entire science team, the work of "integration" on Helen is one of respecting different teams' needs and requirements and finding space and time for each of these constitutive groups to conduct their respective observations. To achieve this, a shared emphasis on autonomy and fairness animates the work of integrating among a heterogeneous team. Local negotiation norms and talk that associates instruments and individuals are resources that Helenites deploy to contain conflict around shared resources while ensuring that many different groups' concerns are respected and integrated. In cases where conflict emerges, the group must seek out fair trades either among individuals who are willing to trade or through an appeal to hierarchy, as in invoking an interdisciplinary scientist. In this, Helen exhibits similarities with many complex firms whose subunits display competing priorities and overarching goals.

This snapshot of Helen's observation allocation processes illuminates several building blocks for the concept of *organized science*. Science planning and division of resources on Helen is a question of allowing many scientific flowers to bloom independently. The task of adjudicating priorities among such different groups of scientists with such different kinds of scientific questions, instruments, and observational targets is a complex one, with no obvious answer. Small wonder, then, that individuals develop local cultural practices and rationales for decision-making to resolve these conflicts and that they staunchly defend their methods as the way of ensuring the fair distribution of resources across heterogeneous—even incomparable—instrumentation and objectives. These mechanisms are enshrined in interactional norms such as instrumental metonymy and in ensuring that continued observational allocations on Helen follow local rules of fair play.

Still, the political considerations of the tour selection, the strategizing among working groups for observational allocations, or the invocation of hierarchy to resolve conflict require further shared resources. Alongside these guiding principles are other cultural codes and forms of talk that bring additional complexity to the twin issues of fairness and autonomy. The next chapter explores these cultural norms in more detail as they surface in other decision-making moments, addressing the challenges of conflict resolution.

THREE

The Resolutions

To ensure a "fair" outcome, an essential element of decision-making on Helen is recognition of both *who* is making the decision and *where* within the organization that decision will take place. Both of these situations may be unclear due to the mix of hierarchies and accountabilities common to matrix teams. The difficulties of working within a matrix organization in which workers report to more than one manager are well known (Davis and Lawrence 1977, 1979; Galbraith 2009; Miles and Snow 1992), but there can also be confusions as to the locus of decision-making authority when there are so many possible groups and leaders to appeal to.

The scientific side of Helen's organization features a matrix of PIs and team leaders atop the scientific team hierarchy and interdisciplinary scientists atop the working group hierarchy: these are assembled as something of an executive board model. The project scientist (Malcolm, or later, Francis) is not technically the leader of this group but instead fills a service role to assist the PIs and interdisciplinary scientists in coming to intergroup decisions and communicating with the engineers, reporting to the PIs like a CEO reports to a board of directors. The team leaders of the four facility instruments also have responsibility for their teams but limited authority over their members. They commonly described to me the challenges of trying to manage a group of scientists each with their own independent investigations as akin to "herding cats." Among the engineers, there is a clear hierarchy of roles with a project manager at the top. Money and resources flow from the project manager—who is somewhat tautologically described as "good with budgets." In line with Shrum and colleagues' (2007) category of leaderless scientific collaborations and Turco's (2016) description of start-up company TechCo, it was easy to be confused as to who could legitimately make a "fair" decision and when.

Accusations of unfairness in the organization revealed that when even when *where to decide* was clear, *who would decide* could be fraught. After all, Helenites are members of both an instrument team and a working group, and project science members may also be instrument team members. So in whose interest is an individual team member speaking when they enter into a negotiation? This was apparent in the prior chapter, in the tour selection example where Betty had to avoid accusations of unfairness due to her affiliation with radio by applying the same principles to all members of the group. It was also apparent when George, a thermal team affiliate and leading the Atmospheres planning discussion, asked Rod to broker a trade between the thermal instrument's and the camera's observations. In such cases where allegiances are unclear, how could any negotiation produce fair allocations?

Helenites refer to this situation as "wearing multiple hats": a phrase common in matrix organizations (Galbraith 2009) that recalls the shifting concepts of value associated with heterarchy or the "multiple ordering frameworks" that can exist in postbureaucratic collaborations (Knorr-Cetina 1999; Stark 2009). Following a Titan targeting meeting, Betty used this turn of phrase to issue a concern about my own ethnographic progress:

> It must be hard for you because we wear so many different hats, and how do you know which we're wearing? Say you were to follow Ali all day. Well, there's times when he's wearing his radio hat, like in Titan or Rings, then there's times in the radio meetings where he's speaking on behalf of Titan . . . and it'll just switch like that. And I know because we've worked together for twenty years . . . but how do you know?

Wearing multiple hats publicly and neutrally articulates any potential conflicts of interest, as well as the understood need for all interlocutors to be familiar with the many potential layers of social ties in an interaction. One's long-standing personal knowledge of team members and the deep well of contextual knowledge are necessary to circumscribe any given interaction and to "know" when someone is wearing their instrument "hat" or their working group "hat." This challenges any newcomer to understand what is going on in any given meeting and outsiders to accept local negotiations as fully "fair."

It also produces some very strange interactions. For instance, Rod is one of three mission scientists who is a coinvestigator on two independently selected instrument teams—the ultraviolet instrument and the camera team—but is not an interdisciplinary scientist. During a targeting meeting when the science planner, George, called upon Rod to describe an upcoming ultraviolet observation in which the camera was planning to ride along, the humorous exchange went like this:

GEORGE: OK, so Rod, for your [ultraviolet observation], are you doing any others?
ROD: Yes, the camera is a rider.
REORGE: So—you'll coordinate with yourself?
ROD: Yeah, I've already had a detailed conversation with myself about that.
(*Laughter*)
BETTY: Did you agree?
ROD: No. It wasn't easy. I finally acquiesced to the other side.[1]

Two types of conflict resolution on Helen manage the multiple-hat situation of the matrix organization's heterarchy by invoking different *where*s and *who*s. One is a form of trades and interactions *between* working groups or target teams, implemented by the project to avoid conflicts between groups in advance. The *where* of these decision-making rituals is within the matrix, among its varied branches, but *who* is involved includes people with different cultures of allocating fair observations—who, additionally, may be wearing multiple hats. The other moment of conflict resolution involves moments when decisions must move up the chain to some form of leadership, enacting the heterarchy by engaging individuals who are believed to be above the fray. In both cases, *where* the decision takes place within the organization as well as *who* makes it and which hats they are wearing are essential considerations in the production of "fair trades" and autonomy. As in the case of integration, these ritual resolutions also feature accompanying interactional norms and cultural forms of talk that naturalize these techniques of decision-making on the team.

Negotiating within the Matrix

What happens when autonomous groups with their own notions of fairness must make decisions among themselves? This occurred when the project came up with a new process for managing conflicts between target teams by using what they called PIEs: preintegrated events. First introduced when I joined the mission in 2009, preintegration was meant to resolve a recurring problem. Although each of Helen's orbital segments was confined to a specific working group and its associated planning team, there were opportunities to do scientific work that crossed over between segments. In such cases, members of one target team would attempt observational opportunities or trades with *another* target team altogether. For instance, a Satellites scientist interested in looking at the small moon Tethys might find a tantalizing opportunity to take pictures of it right in the middle of a Rings segment. In this case, scientists from the Satellites targeting team might approach the Rings targeting team to request time during the Rings segment to conduct an observation—one that might very well be of little interest to the Rings group. This was experienced as a continuing issue that required resolution on a case-by-case basis, but these piecemeal resolutions could not be said to be fairly allocated overall. Given that each group had its own working definition of fair trades and observational allocation, coming up with a global solution or injunction could be frustrated from the start.

The solution that emerged from the science planning office was to establish a list of highest priority requests from each discipline that would be "preintegrated"—that is, held in place in the spacecraft timeline in advance. This would allow a targeting team to claim an observational period over another target team's segment before the latter group even began to integrate their own observations. If there were many requests in a single period, the project could use this information to allocate that whole segment to that group. The project therefore requested a list of these high priority science observations immediately after the tour was selected. At this point, only the spacecraft's route through the Saturn system had been decided, not how the project would divide up the timeline among the constitutive groups. Each targeting team was granted the same number of PIE hours to request. In theory, the working groups that hosted the target teams had only to look at their top priority science and decide on up to four hundred hours of observations over the coming seven-year period that were essential to their science. No PIE trades were allowed.

From the administrative point of view in Helen's project office, the PIE situation enabled the ready dissolution of conflict from the get-go by eliminating intergroup requests and making for a transparent process of dividing up the tour. From the scientists' point of view, the PIEs presented a complex set of opportunities and constraints for negotiation. If a discipline group laid out several consecutive PIEs and the project turned that period into a segment to assign to that working group's target team, then these PIEs would be poorly spent, as it would reduce the overall hours of observational requests that the group could make of *other* groups. To maximize their observational opportunities, then, each group had to *best guess in advance* which sections would be allocated to whom based on a shared priorities list and balance their own PIEs carefully so as not to squander any opportunities to observe during other groups' segments. These implications became clear as each group began meeting not only to plan their immediate segments of observational activity but also to look over the coming tour outline to determine possible PIEs.

These trade-offs were neatly articulated in a Rings working group meeting where the group attempted to strategize an approach to getting a seventy-two-hour long observation for one of their group members, a dust scientist, into the spacecraft timeline. They were unsure if they should suggest this as a PIE; after all, if Rings got the whole segment, that would waste seventy-two hours of PIE requests to the project *and* open them up for PIE requests from other groups during that segment:

IAN (INTERDISCIPLINARY SCIENTIST): If Rings gets that whole segment . . . that's different than these PIEs. And the PIE has a four-hundred-hour maximum [for] each working group [for] the entire mission. So if this one dust instrument observation [Rings PIE request] amounts to seventy-two hours, it's just a nonstarter.

WALTER: Actually the whole thing here is to request this as either a Rings or a PAM segment and not as a PIE. Because if the . . . instrument does not need thirteen hours . . . [and] they need continuous or periodic coverage throughout this orbit, they could presumably fit their observation around any presumed Satellites [potential PIE] requests.

IAN: I imagine that will happen. But remember, if it's not a Rings segment, if it's a PAM segment . . . they're not guaranteed at getting anything at all. I think we should suggest a small number of PIE hours that [accomplishes the observation] in an appropriate way. . . . Keep in mind we only have four hundred hours of which we've already used something like thirty. . . . You get an idea of how precious these [PIE] hours are.

SCIENCE PLANNER: How to get enough PIE hours in there to capture it [as a segment] for Rings? . . . You have to be first and say Rings wants it.[2]

The process required intergroup strategizing as the Rings group had to consider what the Satellites group might do in that period or potentially ally with the PAM group to see if they could request the segment and hold the time aside for the dust observation.

Meanwhile, the Atmospheres targeting team was also wondering how best to proceed. The strategic contingency of PIE planning, segment requests, and other groups' actions meant that whether something should be promoted to a PIE was not necessarily clear until the other pieces on the chessboard were moved. For instance, as the plasma instrument's representative to the majority of the targeting teams, Daniel suggested that the onus would be on the Atmospheres and PAM targeting team to "work around what other people want"—to the consternation of his colleagues on the Atmospheres targeting team who had hopes for their own observations. Recognizing that Daniel was also a representative on PAM's targeting team, a colleague worried that leaking knowledge about the home team's plans would ruin the local advantage. Still the group tried hard to guess what other groups might do, drawing on their representatives' multiple memberships in other groups such as Rings or Satellites or on hearsay about other teams. One suggested that Rings was about to ask for an observation, and "as long as Rings only asks for one, we can hold 'em to it." Another brought up the conversation in a rival targeting team: "In Satellites, there has been a lot of discussion. There are a lot of PIEs but there are some places where a lot of them are concentrated, and those are the ones where we want real segments." When a scientist proposed an important observation and the meeting chair asked, "Do you have a specific desire?" The scientist countered, "You mean, a *scientific* desire?" The chair shrugged: "It could be scientific, could be strategic." "That's the criteria I've been looking for," was the response. He confessed that it technically didn't matter when they requested the observation, but "strategically, I guess [this observation] gives us more control." Keeping track of PIE requests and strategy at the same time took a toll on the science planners involved as the group agreed it was "really important that the negotiators have this information in their heads when they go to negotiate." George, knowing that he would be doing the negotiating with other groups, bemoaned the fact that there wasn't a larger column for "notes" in his planning spreadsheet.[3]

In this exchange, wearing multiple hats proves beneficial to groups, as people relied on their matrixed networks to bring them information about what other target teams or working groups were doing in order to best strategize their own next steps. However, in such cases, fairness may be in question due to the multiple allegiances in the room. Establishing a competitive frame with other teams in order to produce local group cohesion was a common way to mitigate these effects (consistent with social identity theory, see Lamont and Molnár 2002; Tajfel 1982). I frequently heard some variant of "Our group is the best group" or "The other groups don't work nearly as well together as we do." One targeting team seemed convinced that everyone else was always "picking on our [group]," and Betty closed the Titan jumpstart meeting with "This is how Titan works. We are in *agreement.*

We *agree*. We are a *united front*. It is now us against the project if they decide to change any of this. We are now *Titan*, Titan science" (her emphasis). This sort of insider-outsider boundary work is effective in producing a sense of affiliation in organizations with distinct subcultures, as ethnographer John Weeks found when the bank workers he studied claimed morale to be higher in their local branch than among the organization as a whole (Weeks 2004, 124). During or leading up to moments of conflict resolution across Helen's matrix, these assertions combatted the multiple-hat problem by asserting everyone's primary allegiance to their colleagues in the room, thus supporting the notion that the eventual outcome would be fair.

Note here the emergence of another form of talk on Helen: battle and political metaphors. Groups spoke frequently of strategizing their position and needing to know what other working groups and targeting teams were doing in order to better assert their position. They reached for political metaphors to describe ongoing negotiations among actors with distinct goals and needs. As one scientist put it, "[Helen is] like Congress . . . sometimes you wanna align yourself with this guy to get this observation, sometimes you need to align yourself with those other people." Another explained, "It's all about which tribe you're in. . . . Helen is like the United States, where everyone has their own state and their own tribe and they do things differently."

Alongside politics talk were constant jokes about going to battle. This was apparent in the targeting meetings, where team members assemble a "straw man plan" and then "batter it down." Observations are "threatened" and must be "protected," and individuals assert that they are "prepared to go to battle for that observation" or "fall on [their] swords" in order to keep it in the plan or that they risked a "kick in the teeth" from their colleagues. Mission members even nicknamed the end of the mission the "death spiral," the "climb to death," and the "gallows staircase." Although I never heard anyone raise their voice in a meeting or in a PSG, I was frequently advised that "this group will end up yelling at each other by the end of the day" or that "there's going to be a lot of sociology at this meeting." At the outset of the Titan jumpstart, a well-mannered and collegial three-day conversation, people joked about "being beaten into submission," and when Daniel offered support for someone else's observation, Betty admonished him, "Don't let everybody know who your allies are!" His colleague Grant requested a "kiss and cry room" to make up after arguments, and in the middle of another debate over observations, someone asked if the results would be published as "a magnetometry paper or a sociology paper." This form of joking was natural, easy banter, a language for circumscribing and rendering with humor the conflicts that necessarily arise during negotiation. It also pointed to the inherent challenges of negotiating amid so many autonomous groups.

Strategizing also caused a problem for the PIE process, as individuals complained that they could not strategize to both claim segments and argue for PIEs at the same time. The science planners therefore requested a preliminary submission at the end of August 2009 to allocate segments and introduced a second round of planning where groups revise their lists a little based on direct conversations with other groups to gain preapproval for their PIE requests in another group's segments. There was precious little time for this intergroup negotiation, but science planners took advantage of the opportunity to make trades between and on behalf of their targeting teams. For instance, shortly before the final list of requests was due, George and Max sat down with Karen to request Atmospheres observations during the Satellites period. Over two hours, they went down George's list one item at a time, the three of them poring over the segment notes and their notes from their targeting team's existing observational requests:

GEORGE: OK. [Year] 2011, [day] 274.

KAREN: OK. (*Consults chart*). PAM flyby of [a target; *meaning that the targeting team had allocated the observational period to the physicists among them*].

GEORGE: PAM flyby you said? And this one [PIE] in particular would occur at 16:00 hours.

KAREN: Day 274, hour 16:00? Yeah, I don't know, it's really not so good. (*George crosses it off on his sheet*)

GEORGE: Next one. [Year] 2011, [day] 292, 12:00 hours.

KAREN: Ah, same thing as the last one, where Jordan has these three eclipses that he really wants [to observe], and they go boom-boom-boom [e.g., one after the other].

GEORGE: And he wants all three of them?

KAREN: Yes, he does.

GEORGE: (Sighs) I should talk to him about that. . . . Out of eighteen possibilities, we haven't even found one yet!

Toward the end of the meeting, Walter, a Rings representative, happened to walk by. He stopped in his tracks in mock shock, crying, "Karen, what are you doing with these guys? They're Atmospheres!" Without skipping a beat, Max replied wryly,

> Walter, it's 1940 and this (*gestures to himself and George, the Atmospheres members*) is Germany and that's (*points to Karen*) Russia, and you're England. And we've just made a pact to divide up Europe and keep you out!

Walter laughed and reminded Max that by the end of the war, the alliances changed and England ended up on top. The joke drew on the battle and political metaphors common to the team to draw attention to the contentious practices of observation planning, the suspicions over the PIE process, and the behind-the-scenes trades, but it also demonstrated the difficulties associated with diplomatic negotiation. After all, in two hours, they had located only one or two possibilities for the Atmospheres group to observe during Satellites time. As the meeting ended and Karen walked away, having successfully protected her group's observational priorities from another group's request, George sighed. "I've never heard a woman say no so often," he said.[4]

When the final PIE allocation meeting kicked off a few months later, the science planners gathered around the wooden table. Group members had brought in pies for the meeting and everyone was impressed when Max confessed he had baked his himself. Betty bragged at how smoothly the Titan group came to consensus: "That's how Titan is, pretty noncontroversial." This both played to Betty's penchant for trumpeting Titan's successes as well as offering a joke: everyone knew that before the jumpstart was introduced, Titan was considered an especially volatile group.[5] Rose from Satellites shook her fists in mock jest: "Damn you with your single target!" Their targeted moons were many, sometimes tiny, moving fast, and often in conflict with other observations. Betty agreed, "It makes it easier." She then regaled the group with a story of the time when deputy project scientist Victoria had visited a Titan targeting meeting hoping for a chance to fit in a Rings observation during a Titan segment. The "leading opponent" as Betty recalled it was the PI of Victoria's own instrument team. "He just shot her down: What do you want *that* for, Victoria?" The group laughed at this example in which decision-making appealed to people wearing so many multiple hats, with conflicting hierarchical roles besides.[6]

As the team members went down their lists, they found that most conflict had indeed already been resolved outside of the meeting, displaced or agreed upon in invisible moments and interactions such as the one between Karen, George, and Max. It also became evident that there would be a relatively equal number of observational hours allocated as PIEs in the end, and no single group would end up having more PIE hours than anyone else.

The mood lightened and participants helped themselves to more pie. Only the Satellites targeting team felt the outcome was unfair—despite the fact that the PIE process was in part developed to assist them in the first place. During tour planning, Satellites had rated several of their small moon observations as low priority because past experience led them to believe that they would not be able to get them during others' segments. But when the project introduced PIEs and demanded those observations be drawn from teams' top priorities, the rules changed under their feet. As Satellites targeting team leader Rebecca quipped, "If we knew that it was going to play out this way, we would have put them on [the priority list] differently."[7]

Invoking the Hierarchy

The PIE discussion came to a resolution within the matrix, but if this were not the case it would have been elevated to the project level. Conflicts could occur within the target teams or at the PSG, when the project scientist or targeting team leader could call upon an interdisciplinary scientist to adjudicate. A science planner once explained this typical form of decision-making to me as "the strategy of authority: let's go to the old wise gray hair and get his or her opinion." Rarely did such decisions go up to the mission's chief scientist, the project scientist. On the one hand, this is because the role plays a more representative function than a directive one. On the other, it may be the result of imprinting (Stinchcomb 1965) from its initial project scientist, Malcolm, who insisted upon resolutions within the matrix and upheld the role's inability to make or enforce decisions. This was a structural feature of the role: the project scientist is more of an advocacy role, arguing on behalf of science requirements to the engineering team, and has no technical authority over the assembly of PI's and team leads, each of whom have their own contracts with NASA. Given that he occupied a position of limited authority but one with considerable responsibility, most of the scientists I interviewed described Malcolm as "completely hands off" or someone who "hated making decisions." As one put it,

> There were a lot of times when people hoped that [Malcolm] would have gotten to stand up and make a decision and sort out some issue that was pitting one group against another group. Malcolm's style is to just stand back and say, "Well, you guys argue . . . tell me if you can't make a decision." But everybody knew that Malcolm wouldn't make the decision anyway. . . . Once or twice, he made a decision. But he made it clear that he didn't want to be called upon to adjudicate disputes between people.

Yet another scientist who arrived late to the mission explained feeling like the "[PSG] meeting was disorganized. I was confused as to who was in charge. I didn't understand why Malcolm wasn't up there showing leadership." For his part, Malcolm explained to me that he was inspired in his reticence to make decisions by a conversation with the same economist who set up the resource trading board. As he understood it, given the complexity and the pace of decision-making necessary on Helen, consistently good decisions would be impossible to achieve, so the only logical conclusion was that decision-making should be avoided entirely. As Malcolm put it,

> The rule was the PSG operates by consensus. Form a consensus, and then we'd do that. If you cannot form a consensus, I [as project scientist] would make the decision of what to do. So [the economist's] idea was when you get presented with a decision, go through and look at the possible outcomes and pick the worst one. Then when the next decision comes along, pick the worst outcome. And, he says, you probably won't get a third one.

Malcolm recalled deploying this strategy when the Rings and the Atmospheres groups came to loggerheads over a set of proposed conflicting observations involving the radar instrument. He asked each side to write up a single-page "white paper" describing the value of their science but could not see an advantage of one set of observations over another ("They looked equal to me, absolutely equal"). But he noticed that among the radar observations was an interesting opportunity to collect physics data. He sided with the radar team not because of the science prioritization among Rings or Atmospheres, but because of this PAM opportunity. The scientists were incredulous. When another conflict erupted between Atmospheres and Satellites targeting teams and a senior scientist came to Malcolm to request resolution, Malcolm had only to hint that "there might be some valuable PAM data that can be obtained" for the group to opt to sort it out on their own.

Malcolm described this process as "coming to consensus," a form of the word in harmony with Helen's notion of integration through negotiated agreement and fair trade, which produces the "consensual legitimacy" necessary for decision-making in a postbureaucracy (Heckscher 1994). In the following chapter, I will describe another form of consensus that produces strong social solidarity. This form of consensus at Helen was one of deciding to decide within the matrix, not by ascending a hierarchical ladder. But because individuals had to work it out on their own and place their own social capital on the line to do so, this contributed to a social environment that many described as fractious or, in some cases, "unfair." Thus in the absence of a single scientific arbiter, individuals could and did appeal to Helen's heterarchical environment—one in which there was more than one line of accountability. They did this in the targeting teams by asking the interdisciplinary scientists to weigh in, but they could also appeal to the one role that truly was at the top of a hierarchical line of authority: the project manager.

This occurred regularly in conflicts that emerged between project science and engineering. Despite addressing quite different concerns each time and often involving different constitutive groups, these adjudication meetings displayed a common, ritualized format with respect to how conflict was addressed and resolved. As one example, in mid-2009, the Satellites group planned several high-priority observations of Enceladus, the moon sending jets of liquid water out into space, to better specify where they came from and what they were made of. This included a close pass by the moon's south pole where the team would use the radio instrument to detect Enceladus's gravity in order to determine whether the moon had a subsurface ocean that was fueling the plume. This would be flanked by an observation using the thermal instrument and then the cameras to determine if there were any surface features or heat associated with the plume. To do these measurements, however, the spacecraft would have to fly very close to the plume itself. Because using the spacecraft's hydrazine thrusters would upset the radio's delicate gravitational sensitivity, Helen would need to rely on its degrading reaction wheels to point the craft and rotate between instruments. There was as yet no reliable estimate for the plume's density to determine if this would cause any injury to the craft or to the wheels; the best guess would only be available following a mass spectrometer flyby mere weeks before.

The issue came to a head in two meetings held on the same day in July.[8] First the Satellites team met for more than three hours in the morning in the windowless meeting suite to discuss how to convince the project to support their observation despite the risk. They resolved to plan *two* possible timelines—one using the wheels as preferred by the scientists, the other farther away and safer in case the plume posed a danger—and to make the decision which to deploy when the plume data came in, as late as possible. At two o'clock, mission higher-ups filed into the same meeting room in the middle of the Helen operations center at Spacelabs. Heading into the meeting, several people in the hallway intimated to me that this would be "it." More than one of them predicted, "It's going to be science versus engineering!"[9]

In terms of the layout of the room, it was indeed "science versus engineering." Malcolm's successor, Francis, sat on one side of the table near the head, his long salt-and-pepper hair pulled back into a ponytail and a distressed leather briefcase at his feet. By his side was his deputy, Victoria, sitting tall in her usual tailored suit, her blonde hair loose around her shoulders. The science planners and scientists on the Satellites target team sat farther down the row, and the interdisciplinary scientists chairing the overarching working group phoned in from their home institutions on the teleconference line. The chief engineer, Connie, sat on the other side of the table with her deputies and other staff engineers in mission T-shirts and jeans. As project manager, Everett sat impassively at the head of the table, the de facto arbiter of the decision. In case it was unclear, the photographs of these individuals in the room were positioned in a hierarchical pyramid on the organization chart pinned to the room's carpeted wall in the background, with Everett's photo at the top (figure 4).[10]

Everett's opening remarks made clear that as far as he was concerned, this was an engineering and project management decision that came down to "thruster performance and workload"—elements that fell under his purview. He was perturbed that Francis had invited several scientists and requested to make a presentation on the importance of the science as planned. Francis nonetheless launched into his PowerPoint slides with the following: "What I'd like to do is to convey to you that the gravity of Enceladus over the pole is one of the fundamental elements of discovery at Saturn." He then invited other important and high-status scientists on the mission, including the interdisciplinary scientists, to speak up as to the value of the observation.

Everett responded, "This is always 'the greatest,' 'the best,' 'the only.' . . . That's usually what is stated by Science." And Connie stated outright, "I don't need to be convinced of how good the science is." Still the engineers listened politely. Connie asked if the experiment could be done with thrusters, which could point the spacecraft equally well and protect the fragile wheels. Antonio, the scientist heading up the radio experiment, explained that it couldn't: "Every time you fire a thruster . . . that destroys the experiment." The results also had to be comparable to prior readings, which were taken without thrusters. The Satellites targeting team then proposed their solution to carry forward two timelines until the plume results were in. Rather than seeing this as a solution, the engineers on the team reeled at the double workload. Everett broke through the chatter:

> The thing that I'd wanted to get out of this meeting was what the scope of the effort is, in the science planning and for the [spacecraft engineering] team. . . . I'm concerned about the strain we're already putting on the troops. . . . That's what I was looking to get out of this.

Thus he turned the conversation to engineering, staffing, and planning concerns—the project manager's domain—and away from scientific issues involving project science. He turned to Connie to ask about her workforce constraints. Connie explained that she had just lost three of her staff members—one passed away, one transferred to another project, and another took maternity leave—and wouldn't have enough manpower to make changes at the last minute. The Satellites targeting team scientists offered to "spread out" the science planning for their two timelines as far in advance as possible, to give Connie's team enough time to get the work done. Grant, the mass spectrometer representative on the phone line, explained that he could get results down from the last flyby together as quickly as possible and would work with Connie's team to focus their efforts: a collaboration she welcomed.

Everett sat back in his chair during much of the back and forth between Connie and the scientists about the specifics of the maneuver. Then he said, "Well, this is all very interesting. I'm not sure where we're going to go with this." He turned to Connie:

CONNIE: If you ask me to do it, we'll do it. That's what always happens. I told you last night; I told you last week. . . . It's a whole lot of extra work on a team that's already missing two and a half people—

EVERETT: —I know that you told me that.

CONNIE: You're going to do it. You're either going to do it or you're not—

EVERETT: —I need some information from you before I can—

CONNIE: It's going to be a lot of work, most of it is going to have to be redone. . . . It's really the piece going into [two flybys from now] that's more the concern. . . . There's a plasma instrument survey . . . [and] then there's camera plume observations. I can guarantee you that those are going to be complicated.

EVERETT: If the science teams agree to [cut those observations], does that solve your problem?

CONNIE: Of course, that simplifies it.

EVERETT: Is that a price you'd be willing to pay if we're going to do this?

At this point, the meeting pivoted. Everett and Connie's discussion made clear that they have already talked about the issue such that the conclusion may even be foregone ("That's what always happens")—that he was seeking her input ("some information from you") but that the decision was in his court ("You're either going to do it or you're not"). The exchange enacted

Everett's bureaucratic authority as the decision-making figure at the top of a hierarchy. Then he turned to what Connie would be willing to trade. She admits she would be willing to trade a cut in the science observation requests on either side of the troublesome observation, which would reduce the labor on her crew's side in exchange for carrying forward the two timelines and implementing the flyby as planned if the prior observations showed it was safe.

As soon as this opening was revealed, the scientists in the room leapt at it. A young science planner checked the schedule and noted which instruments would be affected. From the phone line, Arthur indicated that there would be "a little politicking to do" to assuage those PIs, so Francis offered to phone the PIs directly and negotiate the trade. Everett intoned, "I think we're clear. . . . [There's] consensus on the science community that there's going to be a period where we just sit Earth-pointed and don't put anything else [other observations] in." In exchange for this important scientific observation, the spacecraft would remain silent and not perform other observations for a period to satisfy the engineers. Victoria, the deputy project scientist, approved: "That's a fair trade." James, the Satellites working group interdisciplinary scientist and a veteran of mission teams since the 1970s, agreed: "Alright, it seems you have the right decision here so I'm gonna chime off," and he hung up.

But Connie was still unhappy. It would be a lot of work for her team. As people gathered their things to leave, Arthur spoke up on the phone: "Connie we could all hear your pain over the phone, even the speakerphone. You're obviously so critical to this mission and your team is so critical to this mission that we don't make this request lightly. . . . It's clear to us that you're under a lot of stress. . . . Is there any way we could get your guy [that you lost] back?" Connie shot back that there was no way to train anyone up in time. The maternity leave replacement was good but wouldn't be ready in time either (prompting someone to chuckle that a line about "no procreation'" should be "added to the program policy"). A scientist in the room offered that Antonio was planning to visit for two months later that summer: this would be a "silver lining" because he could work directly with Connie's team to help with the observation. Then Antonio sweetened the deal: "Connie, you have to give some lectures to my students. . . . I will invite you to Rome and pay your trip." Connie brightened. "I think that's a fair trade," she concluded. The meeting adjourned, with Francis and Victoria discussing on their way out of the room how best to bring the situation up with the PIs whose observations would be cut and wondering which "other kind of carrot" they could offer as an incentive for the trade.[11]

Note that "fairness" is upheld here not despite but *because* of Everett's adjudication. The "fair trade" involved planning a difficult flyby in exchange for a mute spacecraft on the next pass and a trip to Italy and was established through adjudication by a member higher on the organizational scale. In Everett's comments throughout, and especially in his exchange with Connie, the authority of his person is associated with his office, what Max Weber identifies as bureaucratic authority (Weber 1968). As the engineering manager, this appeal to bureaucratic authority works by moving upward along one of the organization's hierarchical lines. Helen's local heterarchy, then, includes both the lateral scientific matrix (with its members' multiple hats) and the vertical engineering hierarchy as resources for the resolution of conflict. Even as the potential appeal to bureaucratic authority adds to the number of venues in which problems may be resolved, it also calms any anxiety about the "fairness" of decisions that might otherwise be made by individuals wearing too many "hats" to be impartial about the outcome.

I saw this adjudication ritual repeat several times while embedded on the mission. The next time occurred only a few weeks later, when the results from the mass spectrometer flyby came in. The instrument's scientists and the spacecraft office team came up with two different results for the plume's density, diverging by orders of magnitude and leaving little possible resolution for a course of action for the next flyby. Once again, we met in the windowless meeting room with Everett at the head of the table. Once again, both sides presented their arguments: Grant speaking over the phone line for the scientists, and a navigation expert leaping up to gesture at the projected slides in the room to explain the engineering results. Once again, Everett adjudicated, and once again the engineers agreed that it was OK to proceed and fly through at the lower altitude. Again, this ritual resolved the intergroup conflict and produced a local notion of "fairness."

Ultimately, the ritual character of this conflict resolution is important as it produces something of a tautological sense of fairness. The adjudicator—whether project manager or interdisciplinary scientist—is fair because he[12] decides what is fair. The issue that caused so much concern and worry beforehand is resolved and subsides. When another issue arises and demands attention, the cycle repeats. So effective is this form of conflict resolution on the mission that I found individuals quickly forgot that the conflict even took place or downplayed its importance in subsequent interactions despite the fact that they were noticeably upset about it at the time. Thus intergroup conflicts on Helen that go up the chain do not disrupt the social order but rather participate in it, providing something of a sense of rhythm, a background ebb and flow to the mission.

The Importance of Humor

Conflict resolution can cause friction in any organization. Helenites dissolve these tensions through recourse to their local sense of humor: a wry style of ironic commentary and self-deprecation that pokes fun at the many layers of complexity in the organization even as it celebrates the successes and importance of their work. I have already mentioned the battle and political metaphors that are rife on the team. Far from being a source of concern or animosity, these were part of the informal banter and were often met with uproarious laughter, as in Walter's World War II metaphor for the PIE trades. Internal project acronyms contributed to these ongoing jokes, lightening the pressures of integration work. The spreadsheet used for observation planning is called the SPASS (Science Planning Attitude SpreadSheet) and is pronounced "spazz," a slang word for "freak out"; other project science tools are labeled SPAM and SPLAT. Reporting to the PSG on the success of the PIE process, one science planner decorated his informational slides with photographs of railway disasters, drawing laughter from the assembled scientists (figure 6). During the rider classification discussion, people spoke of "rough riders" and "easy riders" to chuckles among the PSG. Helenites looked forward every year to their pumpkin carving contest, whose winners often commented on the mission's current context. Concerns over budget cuts and reassignment of young science planners to other missions at the lab animated the prize-winning pumpkin the year I was there, which depicted talented young planners like Rose and Noel as rats abandoning Helen's sinking ship.

This sense of humor could even be operationalized on the mission. Rough riders and easy riders made their way into mission software as observational categories. Science planners enthusiastically baked pie to bring to the PIE meetings, where pie-related puns abounded. Inspired by the PIE process, Max and George from the Atmospheres group even began floating the idea of CAKEs, a series of instrumental observations of Saturn's atmosphere that could be taken at different times during the tour. The acronym worked in their favor: purposefully recalling the PIEs, CAKEs leveraged the intergroup negotiation networks established through the PIE process, which may have made other targeting teams more ready to accept them despite not being a project-levied requirement. I will return to the implication of these CAKEs later in the book.

This sense of humor is institutionalized in the team's chorus, the Helen Virtual Choir, the name itself referring to Helen's "virtual organization." Betty enthusiastically invited me to join when she found out I had a background

6. Affective ambiguity in evaluating the PIE process. Author's photo.

in music. The group consists of exceptionally talented lyricists, with a songbook containing rewritten lyrics to hundreds of famous show tunes and popular songs about the daily work of mission planning, compiled over many years. There was "The Sound of Science," sung to Simon and Garfunkel's "The Sound of Silence," with a line about teleconference meetings that always got a laugh:

> And in the midnight meeting called
> Ten thousand people, maybe more
> People talking without thinking
> People typing without muting

There was "Somewhere with Rings" sung to the tune of "Somewhere That's Green" from *Little Shop of Horrors* ("Well Helen's the greatest / But we're working with semi-sadists" or "We'll keep the mission flying / Just as long as Congress lets") and a song listing each of the planet's moons to the tune of "Gary, Indiana" from *The Music Man* that concluded "We'll see 'em all, if we can" (given that the Satellites targeting team frequently requested observations of these moons during other teams' segments, "seeing 'em all" potentially glossed this tension with levity). Songs could also describe mission milestones, such as when we belted out "This is the dawning of the age of the Equinox" to a tune from the 1960s musical *Hair* to commemorate the start of the planet's equinox season and the mission's extension and then again a few years later to welcome the mission's final phase. Other songs described external pressures on the team. "Perseverance" included the lines "Don't let yourself be swayed by negative funds or politicians," while "How much is that spacecraft in the window?" tackled budget cuts and the loss of the scan platform. Holiday parties made for extra fodder. "All I Want's to Pass the Review" was written to a popular Mariah Carey Christmas hit as Helen entered a nail-biting NASA review that placed its funding on the line, and when concerns about the wheel-saving maneuver called a "y-bias" entered into the ritual adjudication meetings described above, the group crooned, Nat King Cole-style, "I'm dreaming of a y-bias."[13]

As sociologist Erving Goffman describes in his essay about black humor among physicians, joking about the difficult relational situations that one finds oneself in can help produce and maintain role distance (Goffman 1967). This is an essential task on Helen given that instrumental metonymy makes it difficult to fully step away from the roles that must repeatedly assert and resolve conflict. Because this metonymy could too easily accrue negative personality traits to individuals, such jokes also allow time and space for repair so that people can walk away from meetings without hard feelings. During a lunchtime conversation at a PSG meeting, Grant explained, "When I leave the room, those arguments stay in the room."

> These are the people you can have a huge argument with and then when you leave the room at the end of the day you go out to dinner together. . . . I have some very good friends on the mission who would kick me in the teeth if I wanted to do an observation that would [conflict with theirs].[14]

Walter also described this as both the result of—and the practice that enabled—working together for so long. As he put it, "We do all know one another pretty well. We know one another's weaknesses and strengths and things like that. . . . We're not random people thrown together in a room for the first time."

Essential to the humor on Helen, then, is a form of *affective ambiguity*: an ability to hold both the tensions and the successes of mission work in balance. One is supposed to celebrate mission milestones while singing about precarious funding, to speak affectionately about a spreadsheet called SPASS, to have friends who could kick you in the teeth, and to be jocular about world wars. In this way, Helenites' humor eases the frictions involved in matrix work such as wearing multiple hats and working continually with conflict. The combination of positive and negative affect is not limited to the jokes on the team but is also visible in many forms of talk common to the mission. This was especially the case in frequent talk about the mission being a "dysfunctional family," a common turn of phrase that invoked all the mixed emotions that a large family gathering might imply. As Gwen put it,

> So there's definitely a family feeling on Helen. I mean we've been together nineteen years. You know very well what your cousin's personality is like. You love him anyway. You know? Or not (laughs). Maybe that's too strong of a word, but you know what to expect from your colleagues. And there's no getting rid of them just like there's no getting rid of that crazy cousin. They're going to be with you no matter what. They're not going to get fired; they're not going to go away. So you learn to live with each other in the same way that a family learns to cope with each other.[15]

While "dysfunctionality" could also be wielded internally within the team by those dissatisfied with the process, this form of talk largely provided a shared vocabulary for articulating and accepting and managing and mitigating the unavoidable conflicts that arise between autonomous teams. Talk about battles and strategy, screaming and politicking, even family reunions is therefore reminiscent of ethnographer John Weeks's study of BritArm bank, wherein employees spoke disparagingly of the bank not as an evaluative decree on the bank's poor performance but as part of a ritual that produced membership and solidarity among employees (Weeks 2004). Such *conflict-talk*, if it takes shape in locally appropriate ways, can bring people together in the complexities of their work.

At the same time, jokes only work if a team member is in the know. Being too direct or too literal on Helen limits one's effectiveness in the context of a negotiation or a meeting. These forms of talk can also only be wielded with authority by members of the group. When outsiders inquire about the "politics," "dysfunctionality," or "screaming" that they had heard about through their colleagues, Helenites grow defensive of their mission, resorting to a uniformly positive language to describe the "success," "passion," and "family"-like context of their work. Everett took the time to point this out to me, telling me not to take this conflict as evidence of social unrest or difficulty:

> You undoubtedly are gonna hear a lot of arguments, heated debates about strongly held positions for people's science . . . but I think the kinda debates you're hearing, the kinda competition, the kinda verbal scraps you're gonna hear, are just a consequence of the nature of the spacecraft and the nature of the mission, and it's perfectly normal and to be expected.[16]

François too suggested that "we should really make an effort to try to promote Helen as a very successful, very challenging, but . . . very successful mission although it had some difficulties [i.e., the loss of the scan platform and articulated antenna]. . . . I think it's a challenge but it's working." Reminiscent of Vaughan's (1999) observations of a matrix team at NASA or Stark and Girard's (2003) notion of "creative friction," the conflicts associated with a matrix organization—featuring decision-making between subunits with distinct styles and limited information exchange—can be posed as a positive feature of the organization. Of course, how individuals speak about the mission *within* the mission is very different than the public-facing vocabulary that mission leaders and representatives use to assure funding agencies of mission success. Internally, however, affective ambiguity does important work for the team through asserting the role distance that brackets the complexities of conflict resolution. And like other common forms of talk, it also demonstrates membership on the team.

Decision-Making and Organized Science

Taken together, these two chapters have given an overview of the practices, rituals, and forms of talk that govern decision-making and conflict resolution in the large, complex matrix organization that is the Helen spacecraft team. I described the mission's focus on integration whether through harmonizing across the many different needs of smaller autonomous groups or through formulating a plan through that artfully sequences individual scientific observations through "fair trades." I have also shown the variety of ways of making decisions that uphold these values of autonomy, integration, and fairness. The organization itself must uphold considerable scientific heterogeneity, bringing together multiple goals and types of expertise, with adjudication meetings assembling different subteams from among the scientists and engineers at each turn. The practices of conflict resolution detailed in this chapter illuminate the complexity of collaboration within a matrix that, like many other lateral or flat organizations, also requires periodic appeal to authority. Malcolm's legacy persists in finding local solutions among individual scientists and science planners amid the matrix, which, as in the PIE process, requires strategic thinking on behalf of the groups and awareness of the "multiple hats" of one's conegotiators in order to arrive at a "fair" solution. At the same time, members can move conflict up the chain to an arbiter who does not wear "multiple hats." Both of these approaches make use of shared project resources—shared forms of talk about battle and politics or ritualized meetings—in order to produce a decision. The decision produced through either of these mechanisms is ultimately judged to be "fair" in accordance with the process of fair trades and upholding autonomy.

The group's approach to decision-making and conflict management, then, helps illuminate this scientific collaboration's organizational order. The dual appeal either to hierarchical authority or to local negotiation demonstrates the underlying heterarchy of Helen's organizational environment. Interactional modes such as rituals or a shared sense of humor uphold both the process and outcomes of decision-making, while talk of battle and strategy further reifies the autonomous divisions between groups. Even in intergroup negotiation among individuals wearing multiple hats, a place where we might expect to find boundary objects, trading zones, creoles, or other forms of interdisciplinary cross talk (Galison 1997; Star and Griesemer 1989; de Vaan, Vedres, and Stark 2015), Helenites maintain organizational distinctions between instruments and groups.

The logic of integration and its associated values of autonomy and fair trading therefore go beyond the local task of observation planning: it permeates the organization's decision-making, enabling modes of collaboration across trusted social groups within the matrix. Solutions cannot be asserted unless each instrument scientist has the opportunity to argue for their observations to find a "fair" place in the timeline. Of course, this restricts the ability for groups to collaborate in other ways, for example, through bringing more than one instrument to bear on a problem; at least, Helen reserves this ability for only certain members of its community, the interdisciplinary scientists, who hold their own brokerage positions in the matrix. These elements are central to Helen's production of *organized science* as a question of finding ways for many different groups to work together in a sociotechnical system with limited resources, all the while maintaining their autonomy and independence. As we will see, it also forms a tapestry of scientific observations of the Saturn system that emerge from singular instruments and their respective teams.

speaking with one alone. This does not happen at the expense of individual needs, however. Rather, *organized science* on Paris is a question of assembling a variety of viewpoints into a science plan that goes beyond mission participants' individual parameters, synthesizing them so completely that they cannot be readily disentangled.

In this chapter, I will describe three rituals for producing consensus in which collectivist imperatives inflect the planning process from the ground up: the Tactical Planning meetings, "campaign" planning in the science strategy meetings, and group interactions in the All Hands meetings. Turning to the role of hierarchical authority in this lateral group, I will conclude with a discussion of charisma in the collectivist organization, a style of scientific collaboration that Shrum and colleagues would otherwise describe as thoroughly "participatory" (Shrum, Genuth, and Chompalov 2007). Throughout, my goal is to illuminate how this small, single-PI mission works as a unique, local, social system, as a comparative case to the Helen project.

Comparative case methods are a common technique in the social sciences as they reveal social mechanisms at play across sites. But comparing missions can be a fraught task for participants. Because of the competitive context of work in planetary science between Mars scientists and outer planets specialists, comparisons between missions can be leveraged by scientists against their fellow community members in order to argue for or against funding or expressed support. Animated by this concern, perhaps, several scientists I spoke to reminded me that Helen was simply too big for Paris's processes. As one scientist on both missions put it, the closest thing to Paris on Helen, in his experience, was participating in the Titan working group: a more relevant comparison, others suggested, was between Helen's instrument teams or working groups and the Paris mission, as Paris could fit into Helen several times over. Importantly, the processes and practices I describe on Paris are not unique to that spacecraft, and its participants are not uncommonly friendly or harmonious. Paris's practices are also common among activist groups, communes, open source communities, start-ups, or religious orders: other social units with a heavy emphasis on collectivism, shared goals, and social unification and an appeal to charismatic leadership for direction (Andreas 2007; Junker 2014; Zablocki 1980). Still, the comparison to Helen is essential as it will demonstrate how these different organizational approaches to decision-making and interaction effect different kinds of scientific results. To appreciate this comparison, then, we must leave behind the world of targeting teams, battle metaphors, and integration and adopt a different vocabulary altogether. We must leave behind the world of multiple hats for one of uniting behind a common front, a world of "we," "you," and instrumental metonymy for one of just "we," and a world of affective ambiguity for earnest talk of being "happy."

Collective Planning

Like Helen's targeting teams, the ritual planning meeting on the Paris mission that brings together scientists and engineers to establish a plan for the following day on Mars is the Tactical Planning meeting, or TacPlan. This is a meeting at which the scientists who want observations and the engineers who must code these observations and send them to the robot work directly together to craft a plan. Each meeting has a scientist "chair" in charge of shepherding the immediate plan through to completion, alongside a scientist in charge of reminding the group of their long-term plans. Scientists on the mission volunteer for technical roles typically performed by engineers or instrument technicians, taking turns caring for the instrument, discussing detailed operations plans with the engineering team at Spacelabs, and submitting code to the spacecraft. All of these positions rotate among small groups of chairs, long-term planners, and technical leads, typically over periods of two to six weeks. There is no direct equivalent role to the "science planner" on Helen, although one person, a software secretary, is in charge of inputting observational times and parameters into the team's software to minimize confusion. This individual's screen is projected and visible online for all to see and follow along in real time as the spacecraft plan takes shape throughout the meeting.

Like the targeting team meetings, the TacPlan meeting is highly ritualized and scripted. The entire meeting takes place in under an hour. It opens with a series of reports, beginning with a long-term planning report that reminds the group of their current location and their agreed-upon priorities for exploration in the area. Next, specialist engineers alert members on the line as to the robot's current power and operating parameters. The chair then opens the floor to a period for suggesting observations from anyone on the phone line. This includes science theme leads who can speak up for thematic, crosscutting, or long-term observations. Once all the requests are on the table and entered into the software, the chair begins a process of conversation with each scientist and representative to trim their observations so that they fit into the daily fluctuating limit of time, power, downlink capacity, and memory. In contrast to "integration," Parisians call this process "accommodation."

Let's join a typical TacPlan meeting in 2007 to observe this in action. March 27 was an ordinary day on Mars in which two scientists had observational requests. Ben, a specialist in geology, was interested in understanding more about two rocks nearby using multispectral imaging. He requested a camera image of each rock, taken through all the camera's thirteen filters but only using half the image frame as well as two thermal observations of both rocks and the soil around them using the thermal imager. The camera filters observer through different wavelengths of light, while the thermal spectrometer peers into the infrared. The PI, Jeremy, happened to be serving as meeting chair that day, and in that role, he informed Ben that there was enough memory on board that day to take a full-frame image, not just a half frame—that is, Ben could technically double his observations without compromising onboard memory, a limited resource. Despite this embarrassment of riches—or perhaps because of it—Ben declined the opportunity, saying, "I just hate getting in the mode of 'we have the [memory], and then when we get to [our next high priority target nearby], we don't have enough [memory] to take pictures.'" But Jeremy prodded, "If you think it would help the science, Ben, we should make that a full frame." Ben acquiesced.

The software secretary on duty removed a routine calibration observation from the plan to fit in the two thermal observations alongside the photos, but the timing proved a challenge. Ben was disappointed that this meant the thermal observations might not go forward after all but said, "We fit in what we fit in, so I'm good with it." At this point, the representative from the Atmospheres working group asked if the morning was full of observations yet: the chair said it wasn't, and he should feel free to "go crazy" and make many requests. The Atmospheres representative considered a few options and then requested a short observation to leave room for other requests. "Is that a pass from Atmospheres?" asked Jeremy. "Back to you, Ben." But Ben said, "You know what, I think I'll take a pass as well with the intent to save bits for when we get to the [next target]."

In this moment, the scientists are engaged in an earnest and genteel exchange, politely declining their own observational time in favor of demonstrating their commitment to help others or preserve resources for future opportunities. Such an interactional style recalls historian of science Steve Shapin's description of "gentlemanly" interactions among noblemen in the early Royal Society or organizational ethnographer Calvin Morrill's description of "honor" among conflicting corporate executives (Morrill 1995; Shapin 1994). This satisfied the social order of the team but did not suggest an outcome that would benefit everyone. As meeting chair, Jeremy therefore asked the software secretary to move the observations around so that the atmospheric work happened in the afternoon and the camera and thermal work happened in the morning. As he put it, "We gotta make sure that Ben gets his thermal observations exactly where he wants them." This meant that the Atmospheres representative would have to give up an observation, but he replied, "It's a pity to lose that, but I think I'd rather get the afternoon one." In the end, everyone's observations were accommodated on the spacecraft with no bytes or minutes to spare. Jeremy let out a low whistle. "Very good," he said.

This example is a case where two individuals made requests that needed to be massaged into place. But it is also frequently the case that the plan is oversubscribed, with too many requests at once. Helenites manage this problem by fairly distributing cuts to observational requests, rotating responsibility for canceling observations, or prioritizing among different options. On Paris, the robot's memory and power capacity for the day forms the starting point for a group-wide conversation about how to craft robotic activity within those parameters to fulfill shared goals. The terminology also draws on economic talk but on accounting instead of game theory, as TacPlan chairs speak of "tallying" bits and "bookkeeping" observations. When onboard memory is tight or power allocations are low, all requests may be subject to trimming by the group—led by the chair—to make sure that they stretch all available memory in the "bit bucket" to "accommodate" as much as possible.

If an observation is cut entirely, that could ruin a graduate student's dissertation; but if it can be trimmed in some way, this allows more time and space for other observations. Trimming therefore enables the team to massage observations to suit shared strategic goals, making trimming a collective concern and practice. To identify what can be trimmed, the TacPlan chair will usually ask exactly what the observation is *for*, in order to best tailor it to a specific need and make room for other requests. For example, faced with a request for a camera observation using both stereo cameras, a TacPlan chair tried to articulate the purpose of the observation:

> CHAIR: So lemme ask about these [camera images] . . . I think what we want them most for is . . . identifying with confidence where there is exposed rock.[1]

Identifying what the image is *for* permits the chair and whoever requested the photograph to trim the photograph to just its predicted context of use in order to conserve bits. In the above case, this identification allowed the group to trim the requested observation to best satisfy the scientific need. To do so, they may opt to use only a few of the camera filters, or trim the image resolution (measured in bits-per-pixel), assuming this does not affect the image's ultimate legibility or utility. This was the case one day when the camera operator and scientist discussed how best to weigh the bit constraints against camera resolution.

CAMERA PLANNER: To get the best focus, going with three [bits per pixel] is the best bet if we can afford it.

SCIENTIST: If you do need to drop back to one [bit per pixel], the 20 millimeter is the best position.[2]

Note the continual use of polite self-abnegation. Here, the camera planner, an engineering position, argues for higher resolution even as the scientist offers a further possible downgrade to the image by shifting to a different position. In still other cases, the chair will assist in trimming an observation back just enough so that it can be used by the requesting scientist as well as others who may need the image, such as to overlay a spectroscopy reading or a microscopic image. And if trimming will not do the trick and it truly comes down to one observation or another, the team apologizes to the member who requested the observation that cannot be accommodated and adds the request to their "liens list." On Helen, the "liens list" was once described to me as a "place where observations go to die," but on Paris, a slide with this list of pending observations is shown at the outset of every TacPlan meeting so that, as James explained it, "people understand we're gonna get to 'em." Liens list observations were often slotted into a day's plan as it developed, filling in time between robot activities or overhead satellite passes. I do not recall seeing a lien that was never fulfilled. Making space for others' requests and ensuring everyone feels cared for is inherent to the practice of "accommodation" via "consensus," in contrast to "fairness" or "integration."

Consensus-Talk

There are several aspects of the TacPlan meeting of note for how they underline the team's sense of collectivism, as representatives from across the entire team—scientists, technicians, and engineers—must craft, approve, code, and uplink a plan together over the course of one or two days. Note above the use of the word "we" throughout. Paris team members do not have instrumental metonymy and do not use "us" and "them" to denote different parts of the team. Instead, they use "we" to refer to their entire collective, including their robots. Note also the multi-instrumental work. The observations that Ben requested that March day were designed along with a thermal scientist at that instrument's home institution. Working together, the two scientists wanted to be sure that the thermal data could be easily overlaid on the visual data to provide a wider range of information by combining their spectral range and to provide visual context for this data. The atmospheric experiments combined photographs, particle spectrometer measurements, and dust measurements from magnets and calibration targets on the robot to estimate daily atmospheric conditions. Further, observations could be, and indeed often were, requested by anyone on the team, from graduate students and postdocs to senior scholars. Ben, his colleague, and the Atmospheres representative were all staff scientists; their observations were regularly accommodated alongside those of senior scholars and tenured faculty. Even engineers were welcome to pipe up, as I witnessed several times: in one case, prompting Jeremy to declare enthusiastically, "I love it when engineers make good science suggestions!" Instrumental affiliation was not a concern. When the scientist who built the particle spectrometer requested his first photograph from the camera, his colleagues cheered, whooped, and offered hearty congratulations to him on the phone line.

Observation planning on Paris, then, is a process that achieves many goals and satisfies many individuals through "accommodation," interdependence, and collectivism. Members collectively approved plans that not only fit observational requests exactly into robotic time, power, and memory limits but also satisfied as many scientists and as many goals as possible *at the same time*—hence the satisfaction over fitting in both Ben's and the Atmospheres representative's observations or finding what was called "two for one" science when one instrument's observation could be used by many different parties for different purposes. Of course, observational "fit" and finding as many opportunities to observe as possible also featured on Helen. But as one member of both mission teams expressed the difference,

> At the end of the [Paris] meeting, you want people to have a sense of ownership of the plan. That's why I kept asking at the [recent] meeting, "Are there any other comments, are there any other comments?" . . . It's the whole empowerment thing, the team needs to feel like they're part of the process, and they're getting their two cents in and we're doing the right thing and we'll get the other stuff that we can't get [today] as part of the future [plans]. . . . That's the most important thing. Because if . . . everyone comes in with their own discipline-oriented or pet peeve kind of things then it's chaos, total chaos.

Note here how the consensus process is associated with language such as ownership, empowerment, and a sense of inclusion in the process. It is also purposefully opposed to a language of ownership, "fairness," or discipline orientation common to "integration"—here described as "total chaos" despite the latter approach's other virtues such as autonomy or independence. This same scientist, a few years later, reflected on the differences between Paris and Helen directly, drawing upon the forms of talk common to each:

> Walking into a situation [on Helen], and being wary of the fact that there were loaded guns all over the place, that people were anxious to fire. And in a Helen meeting, I would never walk in without both guns on and my sword. In a Paris meeting, I would leave them in my hotel. I mean, [on Paris], I'd be totally stunned if there were a confrontation that was not an honest argument, you know, that we could debate.

Guns, swords, and firing are consistent with Helen's battle metaphors but sit awkwardly against the softer forms of talk on Paris, which this scientist describes in gentlemanly fashion as "honest arguments." Like affective ambiguity on Helen, these forms of talk are essential to the production of collective social order: a form of *consensus-talk*.

Consensus-talk deploys high degrees of positive affect and genteel interactional exchanges in an effort to secure participants' collective assent and continual commitment. Incorporating multiple requirements did not require making trades with social capital on the line but appealing to and maintaining a font of shared, collective goodwill. I already mentioned the earnest approach to discussion, featuring politeness, care for others' concerns, and even self-abnegation. Jokes were rare and always lighthearted, Helen's wry commentary and affective ambiguity strikingly absent. A Helen Virtual Choir member who had been assigned to both missions once suggested writing a song about the robot's troublesome and limited memory to the tune of Monty Python's "Spam" ("flash, flash, flash, flash"). The chorister was astonished when the Paris team's response was one of horror and anxiety instead of hearty laughter, concluding that their team had "no sense of humor."

Like on Helen, this Paris team members' form of talk could also be rallied to support insider-outsider boundaries by intimating that sustaining consensus and continual group participation were of utmost priority (Lamont and Molnár 2002; Tajfel 1982). Teammates described the Paris mission to me as uniquely harmonious, describing other teams as characterized by hierarchy (a negative feature to them, given their flat team) or by fragmentation among types of scientists or between scientists and engineers. James described a mission he was on in the 1970s as having an "old-school,

British system of 'don't speak unless spoken to,'" while Jeremy recalled waiting breathlessly outside the room for his graduate advisor to argue for acquiring his thesis data against a mission's other PIs. Adam described a spacecraft team with thirteen instruments and thirteen PIs as being so fraught that "it's a wonder the spacecraft didn't fly apart into thirteen pieces!" Such stories cemented team loyalty and commitment among members as well as adherence to local processes and norms.

Nowhere is the emphasis on positive affect and its link to collectivist commitment more evident than the closing call-and-response pair at the end of each TacPlan meeting. In this ritual, the chair asks every representative on the line if they are "happy" with the plan. The appropriate answer is "I'm happy."[3] Of course, there is no reason to assume that this directly corresponds to happiness. Several team members explained to me that this phrase, "I'm happy," operated as a shorthand for a variety of statements from "I'm satisfied" to "I have all the information I need" to "I feel like I've been listened to." We might also see this as a form of emotional labor, wherein workers display the emotions that are considered acceptable to their trade and the production of their organizational culture (Hochschild 1979; Van Maanen and Kunda 1989). Yet even if the reaffirmation of happiness did not express emotion per se, it did represent the successful achievement of what sociologist Randall Collins calls an interaction ritual chain: shared, mutual focus in a collective assembly of people that produces strong emotions. Collins describes how the "emotional energy" derived from the successful completion of group rituals affirms membership and cements social ties (Collins 2004).

Without an "I'm happy" from all members, the team does not proceed with the plan. On the one or two occasions out of hundreds of TacPlan meetings I observed when a scientist or an engineer expressed concern or dissatisfaction, I saw TacPlan chairs go back to the plan and rework it until their colleague could declare their full assent. The final "I'm happy" ritual was therefore as much a sign of group membership and complicity in the planning process as it is an expression of satisfaction with the plan at hand. It is also a sign that the purpose of the meeting—making a safe, scientifically robust plan of robotic activity—has been fulfilled. For instance, at an in-person team gathering, I witnessed David, an early-career scientist, chair the TacPlan. It was a stressful time as there was a dust storm on the planet and robotic activity was limited due to low levels of solar power. My notes recorded that "the room is kind of tense" and that Jeremy, in charge of the gathering (but not the TacPlan meeting), was "hanging in there" but "looked tired." In response to this tension and concerned about going over the allotted hour for the meeting, David stated that he would skip reading out the plan at the end and asking for members' assent. Jeremy interrupted in a soft voice from his seat among the science team: "David, don't rush through this. Take as much time as you need. [The next meeting] can start when we're done." David nodded. "OK." He went through the final steps, and when an engineer piped up on the line to tell him, "We need to wrap this up, we got [people] in here waiting [for the next meeting to start]," he answered calmly, "I know, I recognize. We'll finish this up right."

Of course, there is no reason to assume that the correct answer to "Are you happy?" is, repeatedly, "I'm happy." The turn of phrase takes on new meaning in contrast with that commonly used on Helen: "We can't make everyone happy, but we can make everyone equally unhappy." I noted this difference in my interview with Betty, a devoted Helenite, when I once tried to explain to her how Paris worked. After I described the Paris closing ritual, she nodded and said, to show she understood what I was saying, "Everyone is equally unhappy; you've done your job well, or equally happy-ish." "No, no, no," I corrected. "They never say equally unhappy." "Oh, really," said Betty skeptically. "Nobody ever says they're equally unhappy," I pressed. "They all say they're happy." Betty continued to try to put this in Helen terms: "So, the one who didn't get what they wanted, they're happy because they've been convinced that the other science is more important?" Not exactly, I explained. She tried again: "Yeah, OK, so I was—my issue was fair. I had my fair chance to argue for it, and I didn't win today, but that's OK."[4] This too was incorrect: there is no "winning" on Paris, and there is no "fairness" either. This fundamental confusion of the Paris process of collective observation crafting and genteel consensus-talk in decision-making, coming from a Helenite embedded in a system of integration involving fair trades, science prioritization, and zero-sum allocations, only highlights the essential culture clash between the two mission teams.

Crafting Campaigns

At the time of my observations, Paris scientists met once or twice a year in person and followed up with a weekly teleconference meeting to share work-in-progress results. Scientists and students from all parts of the mission signed up to give these talks. These were typically about work that was ongoing or in early stages in order to invite scientific questions, ideas, and prospective coauthors from across the team. I never saw completed or publishable studies presented at these meetings, or talks that I had also seen at a scientific conference; raw, fresh, or newly processed data was the order of the day. A chief purpose of these meetings was to use these preliminary analyses to produce ideas for new observations that could solve the problems at hand. For instance, if a rock showed strange features in camera imagery, the scientists might propose following up using the infrared instrument or other spectroscopy equipment. They might also suggest grinding the surface of the rock before imaging it again to see if it was coated with a different material on the outside than the inside. This technique was called "hitting it with all we've got."

To encourage this kind of work, the instruments and their data suites are all designed as interoperable from the outset, with communal data sharing tools and norms built into the spacecraft, data processing software, and patterns of interaction. One scientist called this a "science network" essential for generating better explanations:

> You shouldn't limit yourself to one instrument—that's the most foolish thing you can do. . . . You're starting with one instrument; [then] you use all the data, all the instruments. You use topography, you use local geology, then you use something from the orbiter, then you use something with [a] lab experiment. That's kind of the science network.

Conducted in this way, the science meetings led to problem solving that fused instruments and specialties. In one case I witnessed, two scientists realized at a science meeting that they were independently working on the same dataset; they started emailing and talking on the phone frequently to copublish their results. This extended across the science and engineering divide as well. When their robot got stuck in a dune, scientists with expertise in imagery, spectroscopy, and simulation modeling respectively teamed up with an engineer at Spacelabs with expertise in physical model-building to figure out the characteristics of the soil and solve the problem. Certainly, as one young Paris scientist put it candidly, "your colleagues, they're your competition, they have control over your future, for example, your grants." But while some might see this as evidence of underlying competition, for Paris members, it was evidence that not playing by the rules of shared work could later prove disastrous in peer evaluations.

The multi-instrumental approach is built into observation planning through what Paris team members call "campaigns." A campaign is a series of observations using many different instruments that addresses a single target or scientific question, bringing together many participants around a scientific problem that they believe is important. It therefore does the work of aligning many scientists and instruments in advance to produce a series of collective observations. Thus campaigns are yet another method of building consensus by ensuring that team members feel "empowered" and "part of the process," frequently achieving the goal of "two for one science."

One typical example I followed had its start at the January 2007 science team meeting in an auditorium with high vaulted ceilings at the local technical university. I sat in on a discussion with a group of atmospheric and soil scientists who wanted to analyze a series of dark streaks emerging from the northeast corner of a crater that the robot was examining. The conversation was focused but everyone was at ease, with scientists proposing a variety of observations with a collegial affect to figure out what the dust streaks, visible from orbit, were and how they were being formed. A few weeks later on a science telecon meeting at the end of March, one of the scientists, Hugo, put up a series of slides describing "an outline of the wind streaks campaign," which followed up on the conversation in January. Hugo went to graduate school at the same time as Francis; he also wore his long hair in a ponytail and partner-danced as a hobby.

The campaign proposed to use multispectral photographs to build a comparison map of the surroundings on both sides of the streaks, choose a target based on that map to use the hand lens camera and spectrometers to get a closer look, then drive the length of the streak to find its source. The long-term planning leader chairing the meeting questioned Hugo whether this was of interest to the rest of the team, not just a small group of dust scientists. After all, they had to justify driving away from high-value science at the crater's rim. Hugo made the case that there would be "two good sources of science for the price of one": there would be information about the source of the streak and the ability to drive up to the crater edge, and it would only be a short drive. He also put up a list of hypotheses about the dust streak that the group planned to test.

In response, the other scientists jumped in to recommend more observations, not fewer. Shouldn't they sample outside the streak too for a good comparison? How about hand lens cameras to check for particle size difference? Scientists soon began to offer their opinions on the hypotheses—"It's hypothesis two, I guarantee!" "Trust your gut!"—and Jeremy asked for "all hands on deck among the dirt folk." Although the group had planned their campaign to be appropriately modest to allow for others' needs, their peers countered this polite self-abnegation by making the wind streaks a respected shared concern. Jeremy elicited expertise from chemists, geophysicists, and soil scientists on the team alongside the atmospheric and geological scientists who proposed the observations to begin with, telling them to use as many instruments as possible to "hit it with all we've got." Following this discussion, dust streak observational requests began to show up in the "liens list" slide deck shown at the outset of the TacPlan meeting under the title "Wind Streak Campaign." In the summer of that year, Hugo gave a summary presentation of preliminary results. He introduced his talk with the following: "The main purpose [of this presentation] is to fill up the author list here as we put together a manuscript." This was not a joke: team-wide science presentations were meant to solicit active feedback, new ideas for exploration or analysis, and more expertise to incorporate into a paper.

The wind streak campaign never made headline news, even if it was the subject of several papers and science talks. But this example of a simple set of observations over a few interplanetary months is indicative of a larger pattern of cross-instrumental work. It shows how the group links strategic, long-term planning to the everyday work of planning by identifying observational opportunities in advance, shaping them with the help of the team, deploying them in an instance in the field, and then working with multiple experts on interpretation. These "campaigns"—and there were countless such examples—relied on and drew together scientists across varying types of expertise as well as the whole suite of instruments to address common problems in a way that could not easily be disentangled. Under a multi-instrumental campaign, it makes little sense to cut one instrument's observations in favor of another's, if all such observations are addressing the same issue and satisfying many scientists in doing so. This contributes to the collectivist sense of common problems and common work as well as to the immediate goals of crafting consensus.

Collective Accountability

A powerful organizational narrative that brings together work processes and practices, emotional energy, and social solidarity draws on logics of care. This was often a familial framework, extended to the robots who could be compared to "children" or aging grandparents (Schairer 2006). But it extended to other team members as well. I was surprised at how readily the extension of the word "family" included me in the team and its events. They would invite me to events, group photos, or meetings with a warm smile and a welcome sweep of the hand. If I wondered aloud whether I had the right to be there, they would say, "Why not? You're part of the family."[5] Several years later, I visited the team after a longtime member had contracted a serious illness. Usually, team members maintain the sanctity of role-related boundaries and the expertise associated with them, so I was surprised to observe people offering gentle correctives on the phone line or pinging each other behind the scenes using chat to suggest that others pipe up to help the sick member with their role. Jeremy muted the microphone on his phone and addressed me directly. "This is [the sick team member's] life" and "we are, the Paris team is, [their] family." Forbidding participation in the family through sick leave could make the illness even more unbearable: hence the extensive use of backstage work to help the team member save face (Goffman 1959) while preserving the family sensibility and checking their work for accuracy at the same time.

Talk of care was also assiduously followed with talk of accountability, or as team members put it, a form of collective "responsibility." This was especially clear in the first-ever All Hands meeting called by the mission. When faulty commands were uploaded to the spacecraft in 2007, Glen, the mission's project manager, canceled all meetings and operations for two days and sent his staff on an enforced long-weekend vacation. This was not a disciplinary action: as he put it, something was wrong with the process or else the team would not have slipped up. It was time to stop, take stock, and get everyone to evaluate the process to see what went wrong and how to fix it.

The following week, Glen and Jeremy scheduled an open meeting and invited all members of the team, scientists, and engineers. The video feed showed a full house at the lab and the teleconference line was so full that some callers could not get a line in. Glen began with an opening statement that reiterated the values of the community. This involved a story made popular in total quality managerial literature in the 1990s about the importance of full complicity and participation in the process and shared responsibility:

> In Japan, anyone on the assembly line can stop the process. They have these cords at every station and the entire assembly line can stop instantly. . . . We all share . . . a responsibility to the health and safety of the Paris robots, and we should all be willing and motivated to ask that question, to raise our hand and say I don't know, I don't understand that. . . . You all have the power and the responsibility to hold the process if you have any questions or concerns or just need extra time to work through something.

The Japanese auto industry story circulated in management circles outside of the lab as a familiar part of the managerial framework that they had adopted over a decade before to improve production under hierarchical management (also cited in Turco 2016). But it took on additional valence in the context of the collectivist organization, where it stood not for bringing responsibility for quality control to the bottom of the pyramid but for collective accountability.[6] The manager encouraged everyone to contribute their perspective with the statement "Be candid and frank in your discussion" and ruminated that it may not only be technical procedures that needed some revision but "maybe what we need to do between now and Monday [when we restart the robot] is in our hearts and our minds." Such statements recalled both the team's shared cultural associations and their collective emotional energy.

Jeremy took the floor, speaking over the teleconference line. He described how as he was "losing sleep" over this series of errors, he started wondering what had changed since the outset of the mission from the perspective of several years in. "If anything's changed in the way the operations process runs is that today it seems a little more casual than it used to," he declared. He described an issue of *Aviation Week* that included "a wonderful article about us":

> I remember [the reporter] remarking to me at the time, "I've never seen a better team than this," . . . and he was including in this the flight planners from the [space] shuttle. His article stresses the rigor of our process. It also stressed the intellectual rigor of the questioning that we all go through when we are trying something very challenging. . . . He was here for the entire uplink process that he documented in this article, [and] he said we were like a fighter squadron at the top of our game, and I'd like to get that back.

As such, he advocated "a return to first principles." More rigor in following the process would bring the team back to this level of operations where they would be "like a fighter squadron at the top of [their] game." His elision of ritualized interactions—"the rigor of our process"—with "the intellectual rigor of our questioning" reasserted the collectivist imperative for all members to speak up, participate, make their voices heard, and listen to their colleagues. He also underlined the authority that Glen sought to invest in all team members, extending it from the engineering team to the science team as well:

> Science team folks, when Glen made that analogy to the Japanese production for automobiles, that applies to you guys too. . . . We have [secretaries] on this team who have probably sat through more [engineering meetings] than many of the mission managers [an engineering role]. . . . You have to be focused on your job . . . but if you see something that looks funny, you are empowered just like everyone else to pull the cord, to ask questions. . . . It applies to everybody as part of the process.

The requirement that speaking up and asking questions applies to "everybody as part of the process" reaffirms the combination of strict operational roles as well as social flattening on the team. On the one hand, scientists and engineers are both subject to these rules, even if they are in different roles. On the other, everyone *can and should* question each other regardless of status or role. His example of support staff, including secretaries with no experience in engineering, questioning professional mission managers also emphasized the flat hierarchy.

Over the next two hours, scientists and engineers alike spoke up with their concerns and opinions. Many echoed this perspective of paying attention because "one person may catch something that someone else might have taken for granted," frequently phrasing this as "everybody is required to watch everybody else's back." One engineer blamed the "complacency" on "too much automation" and suggested checking each sequence by hand to be safe. But several engineers also raised concerns about the rising project workload. Calling it "a can-do" team, one voiced the concern that "we've lost the balance between what we can do and what we should do, and I think we need to take a breath and think about how that affects the whole process." Another agreed that it's a "can-do, can-do, can-do" team "but at the expense of mistakes." Jeremy raised the idea that they might include a role call at a quick "tag-up" meeting immediately before the TacPlan meeting "so that everybody on the line has an awareness of whether or not there's someone in that process who's gonna need extra time. . . . [And to] make sure that we have a really clear understanding of who's doing what. . . . Sometimes out here [via remote access], it's really hard to tell." Throughout the conversation, people echoed what others had said before, picking up on vocabulary such as the repetition of "can-do" versus "complacency" or beginning their statements with "I agree with what has already been said."

The conversation lasted more than two hours, with Glen at its close reporting that he had taken several pages of notes and would follow up on the issues raised and work with the team to implement them. He then concluded by restating the team's values of collectivism:

> We are a single team, we are all one team, we are all participants in that team. We share the responsibility for these robots, keeping them healthy and safe. . . . But we are not alone in what we do. If we need help, we should ask for help. If we have a question, we should voice that. . . . I want, whatever problem we have, I want us to fix it. . . . You, all of you, really are the finest team in the solar system, there is no one better than you guys. . . . I'm privileged to be your project manager and there is no other team that I would want, to have to do what you're doing. But even the greatest team might trip and stumble once in a while. . . . We'll pick ourselves up, we'll brush ourselves off.

In these comments, Glen positions his leadership role as one of service to his team, inverting the engineering hierarchy. The team is unified under their shared responsibility for their spacecraft, relying on each other to check their work: collectivism once again. Alongside the notion that "anyone can pull the cord," this local form of talk further emphasizes the flattened hierarchy on a mission where support staff, scientists, and senior engineers can and should question each other with impunity—even with congratulations.

Consensus and Charisma

I have so far described the Paris team as possessing rituals and interactional norms that produce what they call "consensus": a collectivist, interdependent way of conducting spacecraft planning, with strong social solidarity and shared responsibility for outcomes. I have also described Paris as a relatively flat organization, wherein the processes of planning rely on strict roles and expertise but also synthesize between them to such a degree that status or other markers of differentiation do not play a role in the achievement of scientific goals. While one or two campaigns on Helen—such as monitoring clouds on Titan or acquiring CAKES—had a similar effect in terms of allowing a series of observations to occur over time and across organizational boundaries, they were not widespread, first order, instrumentally integrated, or everyday, like multi-instrumental campaigns on Paris. Helenites share responsibilities too, although their primary responsibilities are to their specific instrument, expertise, or other hats they might wear. The notion that an error could be *everybody's* fault (including the project secretary's!) for not having caught it in advance is a peculiarly Paris phenomenon. However, like Helenites who could appeal to bureaucratic authority in moments of uncertainty, in this case, a different hierarchical authority in the Paris organizational landscape plays a key role in the production of this social order: Jeremy, the team's single, charismatic PI.

The PI represents the apex of a hierarchy on the mission according to its organizational chart and flow of financial resources. But Jeremy rarely makes or enforces decisions or adjudicates between conflicting sides. Instead, a different kind of social relation is at play. This is first of all evident in the near-superhuman terms by means of which members of the mission frequently describe their leader. Teammates whisper in awe that he only sleeps four hours a night or sometimes not at all. Even his vacations are the stuff of legend. One year, a story circulated about him facing down a polar bear while on an arctic exploration ship (the bear was on an iceberg some distance away, he explained when I asked). Another year, he and his family hiked the Silk Road trail in China, leading to a colloquialism in which "being in Kashgar" became a team metaphor for being thoroughly electronically unreachable, a Herculean impossibility it seemed only Jeremy could attain. His mentorship extended beyond his students and postdocs. An up-and-coming young engineer asked him to be his best man in his wedding party. And a department manager proudly carried her bike helmet into the office daily, a sign of his support of her taking up cycling to improve her health. She never failed to relay to him her current mileage, to his wholehearted congratulations.

In person, Jeremy's enthusiasm is infectious. Wiry and energetic, even exuberant, he is frequently featured in public events or lectures on television and in film talking passionately about his mission. Of course, this alone is not enough to qualify for charismatic leadership. Founding sociologist Max Weber described charismatic leaders as "the bearers of specific gifts of body and mind that were considered supernatural (in the sense that not everybody could have access to them)." Such a leader is not seen to be professionally compensated for his work but rather "enjoys loyalty and authority by virtue of a mission believed to be embodied in him" (Weber 1968, 1117). Although members of these communities describe their leader as "charismatic" as if this were an innate characteristic, the sociological view on charisma asks us to see this description as a way of assembling and structuring a social group—that is, charismatic leaders are seen as charismatic by virtue of the authority that their acolytes invest in them and the social order that assembles around them. Thus "the attribution of charismatic qualities . . . is a response to great ordering power" (Shils 1965, 204; see also Friedland 1964).

Understanding the Paris PI as a charismatic leader not only makes sense of the dramatic—even overblown—stories about Jeremy. It also demonstrates an additional way in which the Paris team's consensus-based collective operates. Like communes, religious communities, or technology start-ups (here, recall the cult of personality that develops around individuals such as Steve Jobs or Elon Musk in their own charismatic organizations), the Paris team can powerfully transform individuals' commitment to the work from a singular or strategic interest into an enterprise of passion for the organization's mission (Zablocki 1980). Individuals work toward shared goals not through a rational mandate or through evaluative measures that incentivize their work but through an emotional commitment to the leader, their mission, and each other.

Although frequently described with respect to spiritualism, activist groups, or tribal authority structures (Andreas 2007; Friedland 1964; Junker 2014), science studies scholars have argued that charisma may be especially important in modern technoscientific organizations where functional lines of authority are unclear (Thorpe and Shapin 2000) and collective social imaginaries undergird technological interactions (Ames 2019). In the postbureaucratic organization or the multi-institutional context of a space mission, hierarchies sit awkwardly amid crosscutting groups that are meant to share expertise, and the space agency's military heritage and bureaucratic structure must often be reconciled with scientists' rejection of militarism or even simply being told what to do. The prior chapter on Helen describes how this awkwardness can be experienced as a question of where the decision will take place: if irreconcilable within the matrix organization, then it is taken up the hierarchy to an interdisciplinary scientist or to the project manager. The response to this uncertainty on the Paris team is an appeal to charismatic authority, experienced as consistent with the team's collectivism and its commitment to consensus. Indeed, the Paris team—like internet startups—can trumpet its "flat hierarchy" and experience such strong local social solidarity *because* it is both a charismatic organization and a collective.

Charismatic authority does not lack social organization, even if it obscures the role of organizational structure, processes, and interactions through focus on the group leader. Such organizations may enroll a system of "disciples" that a leader puts into place "united by discipleship and loyalty and chosen according to personal charismatic qualification" (Weber 1968, 1119). Jeremy assembled the backbone of his team based on personal connections, mentorship, and competition to simultaneously produce "the best people" (according to him) and "the best science." The roles and rules associated with participation on Paris are not hierarchical positions in which individuals move up a ladder but rather create a form of organic specialization such that acting in the best interest of the broader group is the appropriate and legitimate form of activity on the team (Durkheim 1893). Paris team members regularly ascribe their group's harmonious functioning to Jeremy, who purposefully architected his mission to adhere to participatory principles where all would be involved. Also a member of Helen, he established Paris's organizational order in explicit contrast to his experiences on prior missions.

As Weber describes, charismatic authority can be fragile as the leader must be able to continually demonstrate that they are capable of caring for their flock. Team members described to me how their PI "works magic" on Capitol Hill and "protects us" from NASA politics, exposing a relationship of care alongside a belief in the miraculous ability to deflect the headaches of NASA politics. The Paris PI is no less devoted to his team, calling them "the best group of people you could ever hope to go to Mars with." Like Thorpe and Shapin's description of Robert Oppenheimer, who reportedly knew every person at his Los Alamos National Laboratory regardless of rank, the PI answers their every email, question, and phone call and stays sensitive to the needs of many individuals in the group (even the sociologist's!). When a rival interpretation of mission results was presented at a conference, the PI reportedly rejected it with a vehemence that some found out of character for his otherwise positive and laid-back style but that was entirely in line with his "people over process" care for the flock.

Charismatic Shaping

How does charisma work within a collectivist organization? Just as hierarchy sits alongside group-level decision making on Helen, in my observation, charisma sits alongside collectivism in the production of consensus as a potential appeal for decision-making. Most often this occurs through encouraging collectivist interactions among team members, working in a leadership role to detect outcomes that can most effectively keep everybody "happy." Note how Jeremy did this while acting as chair at the TacPlan meeting above, figuring out the right combination of observations and time to satisfy his team members, or in the wind streak observation, asking for "all hands on deck." In both cases, this leadership role was not overtly directive but demonstrated how group interactions ought to occur. Indeed, his charismatic leadership was rarely experienced through overt decision-making or direction. When he did not have a stake in the argument or when passions about scientific discoveries were running high, Jeremy described his team as "driving around, having a great time," and was emphatically hands off about decisions. Even when he had an idea he would always open the floor to discussion and ask to hear others' opinions. If his opinion was strong, he might say, "Let me offer a counterargument," and make his case among the rest.

Such moments led more than one scientist to explain to me that the team could talk about the options until everyone had their say, but at the end of the day "[they'd] do whatever Jeremy wants to do" or, as another put it, "Jeremy *consensused us*." However, no one I spoke to complained about their leader's ability to influence the outcome of decisions. The majority were decidedly unconcerned as they felt he knew best or chalked it up to being on a single-PI mission. When I pressed one scientist about this situation, he just shrugged. "It's his mission," he explained. "We're all here because of him." He was happy for the opportunity to propose observations that moved forward and unfazed when that didn't happen. Another who found this approach too heavy-handed didn't leave the mission but shifted to work on the robot team without his powerful personality in the room to sway the group. Late in my ethnography, I talked to another Paris scientist who was running a new project team. When his team members asked what his "management philosophy" was and he announced, "We're going to operate by consensus," a teammate responded, "Consensus? I hate consensus! That's just another way of saying we're gonna do what only this one guy wants to do." The team member thought about it and decided his colleague was right: "So I said, OK, I think what I mean by consensus is informed decision-making. And what that means to me is if we have to make a decision, then I ask everyone what they think."[7]

So tightly interlinked were these modes of interaction that slippage points between consensus and charisma on Paris were extremely subtle. For instance, one day while waiting for a TacPlan meeting to begin, I made small talk by asking Jeremy what he thought the robot should do next. There were two opposing views at stake. He professed that he thought one of the two options was better than the other, but he was interested in hearing what other people thought. He was chairing the meeting that day and followed the meeting ritual by opening the decision up to comments and encouraging all sides to speak up. In the end, the team agreed on his preferred direction, and as people signed off from the teleconference line stating, "I'm happy," I suggested to him that he must be especially pleased with the outcome because it was his preferred choice. He looked genuinely confused, as if he had forgotten that he once expressed an opinion. When I reminded him of our earlier conversation, he shook his head vigorously and corrected me firmly. *No*. He was happy with the outcome because *the team* had decided *together* that this was the right thing to do. He had had nothing to do with it.

This may actually have been the case. After all, the mix of team processes and interactions made it nearly impossible to disentangle what role charisma might play in that case or any other. The practice of consensus as it was operationalized on the team involved the meeting chair (only occasionally the PI, as the role rotated among the group) skillfully managing team discussion so that the many sides of any debate would come around to agreeing on a single possible option as best for the robots' health, resources, or ability to solve shared scientific problems. This way everyone felt like they were heard and "part of the process" in coming to the decision, especially when such decisions involved finding a third way, a path that both sides at loggerheads could agree to. The process also served to further reinforce members' associations between consensus, "the best science," and doing "what's best for the robot." Thus if Jeremy expressed an opinion, this opinion, like any other, won out because it was also considered to be best for the robot and best for science. In practice, this cycle reinforced Jeremy's charismatic authority, as it positioned him as an individual with the unique scope of expertise over the robot's scientific and technical capabilities to objectively judge what is both best for the robot and best for science. Like Polletta's description of participatory democracy in civil rights movements, these conversations lead to a positive sense of leadership as a form of guidance or tutelage, not control or moral authority (Polletta 2002). Such moments therefore became more reason to trust in his leadership, not less.

Another instance I observed was subtler still. This occurred at a science team meeting a week after the All Hands meeting described above. Jeremy hosted a follow-up conversation with Glen in the room. He opened the discussion with a broad request for contributions, associating the participation of all team members with the safety of the robot, the "best science," and the happiness of the team:

> What I'm looking for is everyone's suggestion on absolutely anything that we can do to make the operations process work better. Better from the perspective of keeping the robots safer, better from the perspective of getting better science done on Mars, better from the perspective of making it a better experience for the people on the team—less stressful, more happy. So we're looking for all of the suggestions we can possibly find.

He then summarized his takeaway notes from the recent All Hands, including "add[ing] more structure" to the tag-up meeting before the TacPlan meeting and including a "preplan" meeting that had been dropped over time. These were originally his own suggestions, put forward as part of regaining the team's original "rigor" but consistent with team interactional norms they had been picked up, reiterated by subsequent speakers, and incorporated into later contributions. The next scientist to speak up affirmed Jeremy's point about preplan meetings before stating his own point; others also began by agreeing to prior statements before offering their own. As team members picked up on each other's vocabulary and reiterated each other's points, individual suggestions became part of the flow and therefore the natural outcomes of the meeting without being associated with any one person. For his part, Jeremy actively engaged what his team members proposed as they made suggestions from the floor, weaving them together to build a unified conversation that appeared, seamlessly, to produce communal agreement.

A pivotal moment occurred early in the conversation when a mission team member on both Paris and Helen, James, offered, "The biggest problem is facing up to the fact that this is a real can-do group. No one says 'I can't' and that's good, but it does lead to this sort of situation [of people being overloaded]. We don't want to pour water on that [enthusiasm]" by restricting planning days. Another scientist, Adam, also discouraged cutting back on the planning, even though it would mean getting out of work and home to his family on time. As he put it,

> There was a lot of discussion from people at Spacelabs worrying that we're trying to do too much in many cases . . . but I didn't want to go too far in another direction. . . . I think that to water things down, to protect ourselves by planning things in sort of a bland, not ambitious way of doing it will encourage complacency to creep back in faster than ever. This is a team with a lot of pride, and the way that they know that they're really good at what they're doing is every once in a while taking on an ambitious plan where the team itself can understand what the gains are, putting in the extra work. . . . You don't wanna pour water on the pride of the team and the can-do attitude because that's why a lot of people wanna be on Mars.

Note how Adam restates James's turns of phrase such as watering things down or pouring water on enthusiasm and pride, and James himself restates the "can-do" conversation that occurred among the engineers during the All Hands itself. As the discussion turned to address plan complexity, Jeremy returned again to questions of process, suggesting that preplan meetings and TacPlan rigor would help prevent complacency. Alexa, another scientist, affirmed this idea as adopting an attitude of "not working harder but working smarter" instead of adding bureaucracy.

It was only in reviewing the transcript of this meeting many years later that I was able to detect the subtle conversational shaping that took place throughout the discussion. Where the engineers had complained that the "can-do, can-do" attitude was leading them to make mistakes due to overwork, the science meeting discussion associated *losing* this "can-do" attitude with complacency—and therefore making mistakes. The scientists linked such potential complacency to poor team performance, potential hazards to the robots, and poor science. Jeremy's shift from talking about "complexity" to talking about "process" was not a non sequitur but a way of reframing the problem of errors as one that could be resolved through rigor and principle instead of the team taking their foot off the gas. This effectively steered the outcome of the discussion away from one in which doing less science was an option—that is, the assembled group came to associate their team's "can-do" attitude as the *opposite* of—and the *solution* to—the creep of "complacency," a considerable conversational achievement. Of course, this conversational shaping could be seen as a question of the PI manipulating the team to "get his way." But there was no scientific finding or glory at stake. Instead, in the background was a continual, essential consideration: the strength of a positive relationship between scientists and engineers, which for this team at least must be upheld at all costs in order to ensure continued consensus. The effect of this charismatic shaping was a form of closure (Pinch and Bijker 1987) to the problem at hand that evaded group fragmentation. Far from undermining collectivism, then, charisma worked in this instance as unifying, underlying force that energized team members with a shared mandate toward their work.

Occasionally, Jeremy was required to make nonconsensual decisions to steer the group's direction. Over two years, I witnessed only *two* moments where this was overtly the case, occurring over the tension between exploring new terrain on the one hand and better characterizing local findings on the other. At one science team meeting, the group was discussing either moving toward the next target before the winter approached with its loss of solar power or staying in place for a few more observations. But the meeting was interrupted when it turned out that a group of student musicians had booked the room for a concert starting that evening. Faced with a truncated conversation and the inability to hear everyone out or come to consensus through usual means, Jeremy announced that he was going to "call it" and decreed that the robot would spend only three more days in its current location before moving onward. In another case, a protracted discussion took place over several months involving considerable mapping and planning by groups of scientists and engineers. With no consensus visible on the horizon, Jeremy flew to Spacelabs to meet with Glen. A few days later, he announced that he was convinced of the threat to the spacecraft if they attempted to conduct one set of observations before the coming winter and resolved to put off those observations until after the threatening season had passed.

Francis, then newly appointed project scientist on Helen, happened to be visiting during the Paris meeting that was interrupted by the musicians and expressed surprise. A unilateral decision by a scientist without voting, discussion, or even contestation from the audience, would be unthinkable on Helen! It also jarred with what Francis had heard about Paris's devotion to consensus, even if it aligned with how he had heard that other PI-led teams invoked bureaucratic-hierarchical authority in decision-making. But the scientists I spoke to afterward were not concerned; they explained that Jeremy had done the right thing in the midst of a tough call. He had, after all, listened to and tried to accommodate everyone's requests by granting three days for continued investigations before trying to get to the next site. In the latter discussion, only one scientist voiced a complaint in the teleconference meeting by saying he felt "railroaded into" the decision. But the outcome was not persistent ill will toward Jeremy, the engineers involved, or to the scientist who spoke up. Team members after the fact discussed this as a regrettable tough call, saying that their PI had listened well and had done the right thing for the spacecraft and the team.

These instances are instructive in comparison with the Helen case where Everett adjudicated between sides in the debates between scientists and engineers. Everett's authority is bureaucratic-hierarchical. He has the authority to make decisions ex officio by virtue of occupying a position at the top of one of Helen's hierarchical axes, just as the interdisciplinary scientists have authority invested in their position at the top of the science discipline axis, or like the Helen PIs have authority over their instrument teams. The PI on Paris is technically in a hierarchical position on the organization chart—he is the leader of the science team after all. Yet in decision-making on Paris, Jeremy's authority is not imposed through top-down edicts but through subtle, bottom-up achievements. He presides over and institutes opportunities for his team members to be heard, to listen to each other, and to produce consensual outcomes; he conversationally moves them away from coming to loggerheads over existing fracture points; he demonstrates through talk and example the locally assumed relationship between team solidarity, robotic safety, and scientific excellence. In this, the charismatic authority at play within the Paris collective echoes the guidance-based model of participatory decision-making familiar in activist groups (Polletta 2002). In the very few cases where the PI is required to "call it," it might seem like a sudden a switch to the leader's authority overrides and undermines the collective. But such moments, especially as they elicit minimal concern among the group, reveal that charisma is always already a factor in the organization's social order. Team members hold in the balance a notion of consensus as both collective and charismatic achievement.

Further, on Paris, the lead scientist initiates a process that requires both scientists and engineers to speak up and craft a mutually "happy" solution. When those processes break down, as they threatened to do in the All Hands meeting, Jeremy steered the conversation away from threats to that unified social order and oriented the team toward shared processes and practices—ones that did not offer the possibility that sacrificing science observations would relieve workforce constraints. In the Helen case, the "fair trade" that resolved the Enceladus flyby was one of cutting science observations so that the spacecraft could be "quiescent," mitigating engineering workforce concerns. This trade was frequently positioned as a potential solution to the back and forth between scientists and engineers on Helen, including as the team moved into their solstice mission when the workforce was so reduced that the common turn of phrase was that the scientists would have to accept doing less science "otherwise the engineers will die." This social order formed part of what the Helen collaboration considered to be good relations between the scientists and engineers. In the integrating organization, then, trading observational requirements between scientists and engineers was said to keep the spacecraft safe; in the organization dominated by collectivity, keeping the spacecraft safe meant *not* deleting or trading out observations—this would introduce risks due to complacency. These phenomena are linguistically associated as part of team norms and comprise important local rationales for the management of scientific and technical concerns. They are also firmly in line with each organization's orientation toward collaboration, through either integration or consensus.

That said, I met no one on the mission who was more thoroughly committed to the cause of consensus than Jeremy. More than one team member recalled for me how, in the early days of the mission when it was impossible to tend to both Paris robots without working twenty-four-hour days, he attended both robots' TacPlan meetings. When the meeting chair attempted to hold a vote, these scientists recalled, Jeremy interrupted, "We don't vote here!" For him, voting marginalized disagreement and created factions instead of allowing people to move forward together. This story circulated among the team post hoc as an explanation for why getting everyone's commitment through consensus was so important.

It was also upheld as an example of Jeremy's superhuman ability to go without sleep.

Conclusion

What form does *organized science* take on the Paris mission? A small, PI-led organization, the team upholds practices of decision-making that reinforce collectivism and flat hierarchy. This includes ritualized statements of commitment, collective practices of trimming and mutual accommodation, and the crafting of multi-instrument campaigns that answer "two for one" science questions—keeping, as Jeremy often puts it with his characteristic emphasis, *"ev-er-y-body"* in the loop. When this is achieved, the team calls this practice "consensus." The PI himself also plays a central role, although not necessarily through adjudicating in moments of conflict or directing activity. Comparing his conversational shaping and reinforcing of collectivist values to Everett's adjudication ritual, for instance, or even to Malcolm's push for decision-making within the matrix, the charismatic nature of the Paris collective is different altogether. Yet it is another way of enacting decision-making authority in an otherwise lateral, multidisciplinary organization.

There are many relevant details to keep in mind that underlie this comparison. Paris is smaller than Helen. The periodicity of its planning cycles are tight and fast paced—daily, as opposed to in multiweek segments—and it does have the technical ability to sit still if necessary to accommodate observational requests. Its international component is tiny by comparison, although similarly subject to local social forces. While on Helen, the participating nations are highly visible and autonomous partners, on Paris, the Germans and Danes, as well as students of many nationalities, are largely subsumed into the collective order. Alongside the group's effacement of distinctions and status comes a kind of interchangeability reminiscent of Durkheim's mechanical solidarity (Durkheim 1893). This form of tight social solidarity is also cemented through ritual "effervescent assemblies" (Durkheim 1912), such as the emotional energy generated in the TacPlan meetings with repeated statements of happiness. The campaigns also focus on highly visible processes and products of exchange, such that specialization through expertise is contributed to the whole, not fragmented into groups. Observing how Paris scientists and operators interact on the phone line, it is often difficult to tell the difference between them in anything other than the register of their voice, so similarly did they act, speak, and interact in the framework of mission meetings.

Comparing *consensus* with *integration* as practiced on these two missions, then, gives us a starting point from which to examine the effects of this first aspect of *organized science*: that science teams are organizations with their own interactional norms, rituals, and organizational orders. In both cases, local forms of talk and shared patterns of sense-making reinforce and reify these two orientations, visible in moments of decision-making about their robot's prospective activities. Both cases take complex, interdisciplinary scientific work and difficult tasks of prioritization or conflict resolution and resolve local issues with a turn either toward collectivism and consensus or toward autonomy and fairness. On Paris, charisma plays a role in

FIVE

The Environment

Beyond the technical complications of successfully launching and operating a vehicle in space, planetary scientists face the challenge of multiple lines of accountability. Participants are distributed across institutions, each with very different practices, which threatens fragmentation due to a lack of shared assumptions, identity, and lines of communication (Hinds and Bailey 2003; Hinds and Mortensen 2005). They must coordinate across states and borders, funded by distinct national space agencies and subject to restrictive regulations, to which they are all held accountable (Dove et al. 2016; Sheehan 2007; Watson, Kumar, and Michaelsen 1993). While planetary scientists attend certain field conferences together, they also work with diverse communities of astrophysicists, biologists, oceanographers, or economic geologists, courting frictions associated with interdisciplinary teaming (Cummings and Kiesler 2005; Fiore 2008; Ribes and Bowker 2008; Star and Griesemer 1989). Teams work together for decades, but these timescales do not overlap neatly with either congressional funding cycles or even lifetimes, straining both resources and knowledge management (Borgman et al. 2014; Cohn 2016; Edwards 2010; Salonius 2008). In addition, the time cycles of proposals and funding mean that individuals often work on multiple missions at the same time, even proposing competing ventures in collaboration or competition with their current mission colleagues.

Planetary scientists had several ways of referring to these persistent institutional challenges. Perhaps for my benefit, some described them as "the sociology," such as when they informed me that "there's gonna be a lot of sociology at this meeting." Others referred to these issues as "politics" or "bureaucracy." For instance, when I first visited Sam's lab, he warned me, "I'm not going to be doing much science this week; I'm just going be to doing bureaucracy." Such vocabulary expresses the sense that these social facts of scientific life are due to external forces beyond their control.

Even though NASA projects are engaged in a national agenda of space exploration, the day-to-day mission work does not reveal the force of national directives so much as the locally felt need to draw boundaries around mission work, to protect team members from the slings and arrows of an externalized environment with its own accountabilities and pressures. As such, the activities, routines, decision-making, and conflict resolution rituals that teams establish as their organizational order both establish common ground and balance conflicting accountabilities that could otherwise (and frequently do) tear missions apart. Part of "doing the mission," then, is enacting locally shared solutions to challenges in the field associated with multiple contributing disciplines, countries, institutes, and even missions.[1] A focus on how local organizational orders address these larger problems demonstrates the value of each team's interactional practices and norms and its unique arrangements of people and resources. Without this frame of reference, Paris's self-abnegation and Helen's battle talk might otherwise seem like strange ways of running a scientific team.

This may seem a very instrumentalist approach to organizations, one in which cultural practices are simply a rational resource for problem solving in a complex social environment. But if problem solving, environmental fit, legitimacy concerns, or even rational choice were underlying causes, we might expect to see more homogeneity in teams' solutions to the challenges these overarching conditions pose or convergence on best practices. Instead, teams maintain divergence despite their shared institutional contexts and political pressures and even cleave instead to their local collaborative styles. Of course, local solutions are not total: members of the same mission team may have different experiences, sometimes producing a sense of internal contradiction. But for participants, the success of their mission and the proper attainment of scientific goals is predicated upon adopting the team's locally developed solutions to these challenges, not an externally approved institutional style or a standard of conduct (compare with Daston 1995; Daston and Galison 2007; Kohler 1994; Merton 1942). Hence teams do not converge on singular solutions to shared problems or even reorganize based on emergent practices but instead buffer and defend their local organizational orders with great passion as guarantors of their success. Small wonder, as these local solutions are part of how mission team members draw boundaries around their collaboration, experience time, reproduce power relations, manage accountabilities, and even understand the planets themselves.

Institutional Jostling

All spacecraft missions take place against the background of considerable institutional heterogeneity. While Spacelabs operates the two probes that I studied, the scientists who participate on the missions and their engineering and instrument operations staff are located across North America and Europe. Like building fighter jets or other large-scale national projects, these missions distribute federal resources among as wide a group of professionals across as many institutions as possible to develop talent and encourage scientific and technical work. Each place has their own imperatives, accountabilities, routines, and local ways of getting things done.

Of course, this is a source of strength for these missions as it allows them to tap a wide range of experts, contribute to many local economies, and develop a geographically diverse technical workforce, supporting arguments for funding. Yet bridging across these divides also poses challenges not only due to the difficulties of intercultural communication. Studies of contract workers elucidate the difficulties of managing employer-employee relations without the typical boundaries of a firm (Kunda and Barley 2006; Mahler 2016). Sociologists also point to issues associated with divergent "institutional logics"—forms of talk and managing accountability that are incommensurate with participation in multiple institutions (Thornton, Ocasio, and Lounsbury 2012). And because these multiple ties are bridged through communication technologies, considerable time and energy are now spent managing work through devices that are enrolled in other social relations and workplace accountabilities (Mazmanian, Orlikowski, and Yates 2013). Even as distributed teams rely on shared practices to smooth communication channels, trust, and conflict across considerable distances (Hinds and Mortensen 2005; Nohria and Berkley 1994), these are as subject to break down as the fragile technologies that support them.

Distributed work is also a useful lens through which to witness how teams produce and manage power relations in interinstitutional work (Marcus 1995; Tsing 2005). Within planetary science, stiff competition for resources often pits various contributing institutions against each other, including Spacelabs. For instance, several scientists explained to me that a local explanation at Spacelabs for the loss of several missions in the 1990s was not the impossibility of achieving "faster, better, cheaper" (as many scientists put it, "Pick two!"). Instead, the lab blamed it on too much power sharing between contracting centers, producing a lack of oversight and quality control (Conway 2015). Given their considerable responsibilities to NASA to manage mission grants responsibly, Spacelabs worked hard to control coordination between participating centers. Employees I spoke to at the lab described this as an obvious imperative for mission success and risk management. People at other institutions described it as a "land grab" and attempted to resist or negotiate with the lab to protect their local autonomy.

Even the digital systems that facilitate distributed mission work—what Paris calls "remote operations" and what Helen calls its "virtual organization"—were part of managing this interinstitutional tug-of-war. Paris's system hosts instrument operations groups at several participating scientists' institutions where they may employ research staff, graduate students, or undergraduates to involve them in the work of space exploration. They use videoconferencing tools at several major sites, decentering Spacelabs as a managing partner. The Helen project also placed an instrument operations center at the home institution of each of the PIs so that each PI maintains responsibility over commanding their instrument and managing its data. An instrument operator coordinates between Spacelabs and their local institution: this is a technical position responsible for delivering the right code for the instrument to the spacecraft office and bridging any interinstitutional misalignments. Distributing operations centers across the United States and Europe allows people to communicate and manage workflows effectively. How each system is set up locally resolves interinstitutional tensions, allowing Helenites to achieve a local balance of autonomy with integration and Parisians to reproduce collectivity, in line with their unique organizational concerns.

Spacelabs was the de facto "center of calculation" or "obligatory passage point" (Latour 1990) for sociotechnical coordination: both between people and their spacecraft and as a political intermediary between institutions.[2] It was the only location during my fieldwork that was neither "remote" nor "virtual" in actors' talk of distributed operations. Receiving interoperable commands from other institutions often inspired a predilection toward control and oversight, some of which was resisted by partner sites. This was evident in the fate of the computer workstations that Spacelabs provided to remote instrument operators. Operators at each location are responsible for using instrument-specific software to produce the code for their instrument's commands, but they do not upload this directly to the spacecraft. On Helen, once a spacecraft activity plan has been negotiated, kernel files containing spacecraft positioning information are distributed to the instrument sites, where technical teams plan their instrument commands on their own software to fit into the timeline. They then use software and hardware provided by Spacelabs to upload local commands back to the laboratory for review before they are integrated with the commands from other teams and from the engineers to send to the spacecraft. On Paris, spacecraft code is also prepared by "remote" workers to be merged at the end of each day, when it is reviewed by instrument operators, scientists, lab engineers, and software secretaries in a teleconferenced meeting before it is uploaded to the spacecraft.

While certain files like kernels or planning templates circulate as boundary objects between communities (Star and Griesemer 1989), their movements are not seamless. In the matrix organization, changes made in one institutional location may be opaque to another, who must respond without understanding why such a change was made. I once witnessed Helen's Spacelabs-based navigation team cancel a maneuver to change trajectory, something that their team considers a job well done as it spares the spacecraft from strain. They generated a new kernel file that showed the spacecraft's new location to be used in planning. But when an instrument representative logged in to her computer at a different NASA center and loaded the new kernel, all her instrument scientist's high-priority requested observations were off target. In an emergency meeting that afternoon the navigation team, the science planners, the representative, and the scientist all phoned in to resolve the problem. After an hour of discussion, they realized that she had imported an incorrect file so there was no problem after all. Still, watching people at three locations share information across these gaps revealed the complexities of coordination across a team that was not only matrixed but interinstitutional as well.[3]

We might view these sorts of meso-politics as a question of control exerted by a dominant over a subordinate center. Postcolonial studies, however, suggest adopting a multisited view, destabilizing the ideas of center and periphery. This allows us to observe the difficulties associated with establishing and maintaining centrality as well as the frictions associated with the flow of resources and people across locations (Marcus 1995; Tsing 2005). This perspective reveals that maintaining control as the "center" requires considerable work on behalf of Spacelabs employees. After all, each mission's project office at Spacelabs is responsible for reporting the responsible distribution and review of funds to NASA and for demonstrating that the mission is achieving its scientific objectives. Spacelabs employees' interest in extending the lab's purview beyond its walls to manage the work at other locations is, to them, a question of doing their job and doing it well.

It takes work to contest this centrality too. I frequently found that Spacelabs-issued computers sat in the corner or in a disused room at remote sites. At one NASA center that housed a Helen instrument, people accessed the machine remotely from their own desks to send or receive information to and from Spacelabs. One scientist used a Biblical reference to describe this attitude as "deliver to Caesar what is Caesar's"—an approach that animates collaborators' attention to which server one is logged into and where one's data are kept across a variety of virtual machines. On Paris, remote collaboration tools were used to flatten and even the playing field. At two participating university sites I visited, a machine or a hired student accessed Spacelabs' network daily to download spacecraft data to the local network. This allowed graduate students to work on the materials and participate in mission science regardless of nationality or security clearance status. Spacelabs, the PI's institution, and the deputy PI's institution all regularly used parallel videoconferencing units to dial in to their meetings, giving each the same level of visibility and the same status as central nodes. Collaboration tools were therefore enrolled in the management of power and visibility between sites. Their specific arrangement resolved the key issues associated with communicating across institutional boundaries in a creative, locally understood fashion, accommodating varying lines of accountability while managing power lines in ways concordant with their collaboration's organizational order.

Multimission Competition

Interinstitutional jockeying aside, NASA's institutional politics were most visible in each collaboration as a question of balancing needs with other missions. The long duration of mission work and unavailability of promotion combined with the timing of next flight opportunities meant that individuals often ended up working on more than one project at a time. Paris team members I visited were hard at work on a concurrent lander and orbiter and an upcoming rover. When I joined Helen, team members were busy launching a new Jupiter probe, planning an encounter at Pluto, and proposing competing missions to Titan and Europa. A handful of scientists served on both Paris and Helen. Teams had to step away from targeting or TacPlan meetings while other missions launched or landed. They also experienced reduced data downlink and operating times when other missions were at critical junctures due to increased demands on the shared network of satellite dishes that relay commands and data between Earth and NASA's assets in space.

This situation presents challenges for personnel time management, but it also has implications for the sociology of knowledge—what science is done and what is not—and for social relations. Multimission membership threatens to splinter teams into factions, especially when upcoming missions are "the only game in town." As people rely on existing collaborations to forge the connections that establish the roster for upcoming proposals, and as they hope for tantalizing results to promote follow-up investigations, tensions over other missions are never far below the surface—even during conversations about spacecraft planning, strategy, and science. Each team's organizational orders therefore also included individuals' active production of boundaries between missions against this backdrop of competing pressures. Shared practices were thus called upon to enact "a mission" and its shifting borders in a locally accountable way.

Multimission tensions were especially visible during a public debate about the existence of ice volcanoes on Titan, staged at a press conference during a primary conference for the field in 2008. The abstract opened with a provocative question (for planetary scientists): *Are ice volcanoes oozing from Titan and replenishing its atmosphere with methane? Or are these flow-like features the icy-debris that have been lubricated by rain and collapsed into sinuous piles like mudflows?* Helen scientists Arthur, Mickey, and Marina were on the panel along with Sam, an old friend of Francis, and a member of Paris who was an outsider to Helen. Earlier in the day, Sam had invited me to the panel, insinuating that it was something I wouldn't want to miss.

It was certainly the strangest press conference I ever witnessed. Interdisciplinary scientist Arthur, infrared scientist Mickey, and radar scientist Marina presented the results from their recent work announcing the discovery of ice volcanoes on Titan, one by one. As active members of the Titan group and radar team members or affiliates, each one showed images from their instruments and explained to the assembled crowd that what they were seeing was evidence for these "cryovolcanos." But they hedged their presentations with words like "possible" and "ambiguous." Then Sam spoke up to, as he put it, "inject a level of caution." He countered that those same features were the result of running liquid on the surface, which could just as easily occur because of rain instead of icy eruptions. As he showed images of Mars taken by spacecraft in the 1960s to emphasize the pitfalls of overinterpretation,[4] he repeated that because there was as yet no evidence for any interior processes that would cause volcanic eruptions, the team should constrain their interpretations to surface processes alone. According to Sam, Titan was geologically inactive, like Jupiter's rocky moon "Callisto—with weather."

Members of the press were confused. Was there a discovery or not? Were there cryovolcanos or weren't there? A Helen team member in the audience leapt to his colleagues' defense, saying, "I'm very impressed with the work that was done to establish that there are changes going on on the surface." A reporter from *Nature* pressed Sam on what might be resupplying the atmosphere at Titan if not volcanism, while a *BBC Science* journalist asked simply if the group could "paint us a picture of a cryovolcano." Whether there was a discovery or not, there could at least be a good image to accompany the story.

Each side of the debate focused on the robustness of the scientific evidence for or against the volcanoes. But there were undeniably political tensions underfoot. Fostering continued interest in Titan versus denying that the moon was geophysically active could play a role in determining whether it was sufficiently interesting to sustain future exploration, keeping Helen scientists busy past the end of their mission. If Titan was not interesting, that left the door open for Sam and Francis's competing mission to Europa. Since missions bring with them resources for laboratories, students, and investigations, more was at stake in this announcement of discovery than whether there was such a thing, empirically speaking, as an ice volcano.

Cost overruns and decisions on one mission also impacted other missions in development or in operations. During my fieldwork, this was especially the case with a new Mars mission, which became a community pariah.[5] It did not meet its target launch date and was rescheduled to the next launch window two years later, incurring billions of dollars of cost overruns. The delayed project was accused of "stealing" personnel as planners from Paris and Helen were reassigned to the nascent mission, leaving the teams with less long-term expertise and local crises of knowledge management. Jokes at Spacelabs poked fun at the siphoning of money and manpower from existing projects, with one Helen member lamenting in a meeting that "if I could just give them one hundred million bucks they would stop bothering me!" Even NASA Headquarters personnel divided the period into "pre-" and "post-" mission days to delineate a period of austerity due to the mission's extravagances. Meanwhile, planetary scientists described the mission as "eating everyone's lunch" with "a voracious appetite." No aspect of the mission was ridiculed so much as its controversial landing system. Scientists were utterly pessimistic about its chances, colloquially using its name to denote a cockamamie scheme that robbed managers of their good sense and dollars. This was not just bad blood over competition: it reflected real concern as budgets and personnel were slashed to make room for a new mission in development in the wake of the global financial crisis of 2008.[6]

This conflict had implications for knowledge work too. Paris scientists often requested observations from existing Mars orbiters to complement their groundwork, plan their drives, or analyze surrounding terrain. In such moments, it appeared that the Paris team was a subset of a larger community operating several spacecraft at a time on Mars, each coordinating observations and involved in a wider strategic conversation about the planet's exploration. It was also concordant with the Paris team's collectivist orientation and their interest in "two for one science." But when it came to the new vehicle, tensions ran high. At one Paris meeting, the PI even suggested that his team constrain their plans for systematic studies on approach to a crater, in favor of getting to its rim and its clay-rich targets quickly, before the next mission landed a few months later. That way, Paris could be the very first to explore clay deposits on Mars—the unpopular new mission's primary goal, responsible for its choice of landing site thousands of miles away.

Such examples show how other missions are in the room with both Helen and Paris scientists as they make immediate decisions about their mission's operations and scientific findings. This has implications for which missions fly, which data are collected, and which questions are investigated. Yet individuals also use multimission opportunities to draw boundaries around their own groups and establish local mission solidarity. These lines shift and sometimes even conflict, especially in moments when their interests collide. After all, Helen includes scientists working on both Titan and Europa; and Paris's scientists occupy the new Mars mission's roster as well. As scientists switched back and forth from "doing" Helen, Titan, or Europa to "doing" Paris or its successor project, embracing each mission's planning rituals, forms of talk, and social solidarities, this did the work of enacting, producing, and sometimes contesting mission boundaries from the bottom up.

In some instances, mission boundaries were invoked to resolve local conflicts and assert membership in unexpected ways. Recall, for instance, the moment of PSG tour selection when Francis took the bold—and diplomatic—move of asserting that Titan mission priorities *should* be part of the Helen tour consideration. This not only put the Titan agenda on the table and made it a shared concern but also drew the boundaries of the Helen mission to *include* Francis, instead of positioning him as outsider, challenger, and leader of the competing future mission. When members of both Paris and Helen poked fun at successor projects and framed them as an aggressive "other," this fortified their own social boundaries and affirmed their local allegiance in the moment, much like Helenites' way of managing conflict due to "multiple hats." Such organizationally situated practices therefore do the work of establishing each mission as a relevant social unit with particular, shared matters of concern, amid the backdrop of competitive heterogeneity.

International Collaboration

Although NASA's early charge was a response to Cold War and nuclear competition, by the early twenty-first century, the agency had settled into long-standing collaborations with space agencies around the globe, including their former adversaries.[7] Both teams I studied had international components. The Paris team includes a Danish group working on atmospheric dust detection and a German group operating a spectrometer; meanwhile, the Helen team's foundational and extensive international partnership includes scientists across European Space Agency member countries.

While studies of international collaboration highlight problems with intercultural communication, I did not witness these issues during my study. Instead, scientists on both teams praised the strong collaboration between Europeans and Americans as examples of the importance of international collaboration in science. Helenites in particular were especially proud of the international legacy of their project and concerned as to how to continue such collaborations among the next generation of scientists. This speaks to the faith that teammates placed in the mission as an organizational unit to bridge considerable divides. It is, after all, by no means taken for granted that national identities should be subsumed to mission ones. I heard about many other missions while in the field whose instrument team boundaries echoed, reified, and ultimately exacerbated national boundaries instead.

Helenites' way of surmounting the institutional, network, and temporal barriers associated with international collaboration was through nurturing long-standing relationships and regularly celebrating international friendships as central to their mission's success. In interviews, site visits, and sidebar conversations,[8] those elder European scientists involved in establishing this partnership were effusive in impressing upon me the extraordinary nature of the collaboration between these agencies represented by Helen. For them, the mission represented an opportunity to do something unprecedented scientifically, technically, and in terms of the development of their next generation of scientists. It also represented long-lasting friendship, hard work, and commitment on behalf of so many scientists, engineers, and managers. This was especially the case as, as François put it in an interview, NASA was the "big guy" in the partnership. It was therefore easy to make ESA's role "invisible." The success of the collaboration and the full respect awarded to ESA in the mission represented what he called "the richness" of the collaboration.[9]

On the NASA side, these relationships and their benefits were narrated in economic terms. Many planetary scientists operate their labs on "soft money," meaning that they are constantly writing and reviewing proposals for funding to maintain their labs. But European scientists are paid salaries by their home institution, so their participation on a mission is not tied to the shifting whims of review boards. Further, ESA missions are funded up front at full cost for the duration of the mission, while NASA scientists must return to Congress yearly to set their operating budget (Zabusky 1995). Thus as one American exuberantly put it, having ESA-funded scientists on board "doesn't cost us a dime!" A European scientist used much the same terminology, saying, "NASA can get ESA scientists for free! And in return, we get to work on these amazing missions."[10] This sort of conversation also centered on support for nascent missions. When Helen was at risk of cancelation in the early 1990s due to a national recession, its ESA partners wrote a letter to Vice-President Al Gore reminding him of the memorandum of understanding between partner states. With diplomatic connections on the line, Helen survived.

These friendships were nurtured and regularly fêted at ritual meetings as the ties that bound the Helen family. Not limited to the mission's development phase, I observed strong international ties during operations too. American scientists described their European counterparts as among their closest colleagues and collaborators and vice versa. Scientists in Germany, the UK, and France visited institutions in the United States, while American team members traveled yearly to Europe for yearly ESA-hosted PSG meetings featuring local food, drink, and cultural excursions. Legendary on the mission was the strong transnational friendship among the founders of Helen who had met during the prior mission to Saturn—an American, a Chinese, and a French scientist. The latter repeatedly discussed with me the importance of this camaraderie that strengthened through their mission and the opportunities that this afforded for European science and for young scientists in particular. Amid the senior generation of Helenites that included French, Italian, British, and German scientists, I heard nothing but praise for this transnational partnership and the respectful relationships that cemented it on Helen. National politics, regulation, and the vagaries of funding challenge all missions, but Helenites' experience dictated that strong personal relationships conquered all.

Still, a considerable amount of work goes into buffering the tensions of nationalistic collaboration patterns through social relations, especially as these relations are not evenly distributed. At a Titan meeting in Barcelona in January of 2010, in the shadow of the financial crisis but before budget cuts were announced, a scientist described that day as "the mother" of the European side of the Helen collaboration expressed frustration to her American companions as they discussed their plans for future mission proposals over lunch. She decried being shut out of several proposals, frustrated that NASA had all but excluded ESA coinvestigators from its upcoming call for missions. When she insisted that international collaboration "had to start from the beginning," her senior male American colleague described international cooperation as "one of the complications" in the process: it was a "sociological fact," he said, that people assembled their prospective mission teams not out of consideration of their friends but out of "what might make them win." She disagreed and throughout the meeting reminded her American counterparts that the scientific community was "bigger than just America," telling them of ESA's upcoming mission proposal deadlines and plans.[11]

Such cases explain why team members I spoke to were adamant about the importance of transnational ties to their mission but also aware of and constantly pointing to their fragility. After all, these relationships were subject to institutional and international forces beyond mission team members' control. Europeans, for instance, were concerned about how much time their American counterparts spent writing grants and reviewing proposals instead of doing science. And although Helen's international partnerships saved the mission during a financial crisis in the 1990s, this was not an obvious or predetermined outcome (Westwick 2007, 50). While I was studying these missions, US and European scientists were busy planning a joint Mars rover and a parallel flight to Jupiter's moons. Despite the strong heritage of international collaboration on both Helen and Paris that animated these mission plans, NASA withdrew from both collaborations in the aftermath of the 2008 crash, leaving the Europeans seeking another partner to mitigate the resulting threat of cost overruns and launch delays. The Russian space agency stepped in.[12]

Younger European scientists also described a citation gap between American and European national scientific communities. At lunch with a group of French, British, and German scientists, a young Frenchman who had worked in Britain, France, and the US told a story about a colleague preparing a paper for publication: "He gave it to someone to look over, and he said this looks really good, now let me just help you with your references. And he took out all the European papers from the reference list!" People around the table nodded, indicating that this was a common affair. His advisor followed up with the following: "The Americans never cite the European work in their papers. . . . I wrote a paper on this; it is the most important paper on this in the field. And they [Americans] do not cite it. And it's published in *Nature*! It's not even in a foreign journal or any such thing!" Gesturing to a poster on display at the meeting he later described the importance of distinct national networks—British and French in this case—on citation and collaborative practices:

> This is a study done in the UK by UK collaborators. . . . This poster mentions a study done by my student [a Frenchman]. This was the founding study in the field. It mentions it in the text. Do you see it in the references? No. All of these references are British. That's what I'm talking about.

In the opinion of these assembled French early-career scientists, citing one's network was a professional courtesy born of institutional pressures. Because American salaries depended on grants, which in turn depended on publications, and British universities were being restructured toward frequent reviews, it was perhaps an unspoken agreement to cite your peers for the sake of their careers and yours. As one explained it, "You can't cite yourself but you cite your friend and your friend cites you and, you know, it is very close that way." Another scientist suggested that because Americans were always traveling to meetings together, they saw each other and knew their work, but it was too expensive and far for Europeans to come frequently. But there was also clearly a power discrepancy inherent in these citation patterns. As this scientist put it, "For Americans, you can be in good faith and neglect to cite a French study. In France, it can't be in good faith not to cite an American study—that's where most of the science is done."[13]

Citation counts aside, transnational politics were most visible in these collaborations as questions of export control. First adopted by the United States government shortly after World War II to preserve national defense secrets, the International Traffic in Arms Regulations (ITAR) also restricts foreign nationals' involvement with spacecraft and satellite technology. ITAR violations provoke millions of dollars of fines, a situation that tightened with the Patriot Act after September 11, 2001. Yet because ITAR regulations present challenges in interpretation, institutions interpreted them (and were accused of misinterpreting them) in ways that benefitted their own local goals, needs, and narratives. For instance, a scientist at a competing center calmly explained to me that Spacelabs "loves ITAR because they use it as a weapon to keep things proprietary"[14]—pointing to ITAR's enrollment in interinstitutional politicking.

As an international mission from the outset, Helen put many people and systems in place to mitigate the effects of ITAR on their science and operations. It was more of an institutional hassle than an ever-present, exhausting challenge. It was even subject to the mission's style of affectively ambiguous humor as the topic of a Helen Virtual Choir Christmas carol. The local virtues of independence and autonomy on Helen also offered some protection from ITAR concerns. For instance, Everett described how, when integrating the European probe with the spacecraft bus, only the points of *interface* were cleared to share with foreign nationals. The European engineers also flew to Spacelabs to attach and test their probe under the project manager's purview. Such practices were in line with authority and autonomy practices on the mission I have described, as well as with export control restrictions.

Meanwhile, ITAR persistently posed a threat to Paris's collectivist environment. Non-American members were denied access to key systems, documents, and even their own technical plans for troubleshooting their instruments under US law. During the early operations phase, the French and German spectrometer scientists reported being cordoned off into a walled office away from the rest of the team. Because the lab did not give them printer access, they tried to use an old commercial printer that sat abandoned in the room but were chastised for using a piece of equipment that was not cleared for export control. When their own instrument malfunctioned, the German team that built it was required to stand outside the room while lab engineers debated how to fix it.[15] One scientist reported that a foreign student who wrote an instrument's software suite was restricted from debugging his own code once it had been delivered to NASA.

Most foreign collaborators adopted a frustrated but resigned stance toward ITAR, but many American team members on Paris were visibly angry about it and complained to their international teammates and their congressmen. Although they did not suffer professionally from its restrictions, the regulations undermined their team's organizational imperative toward collectivism that the team associated with their success. I observed this in action when Jonas, a young scientist who worked on Paris for several years as a postdoc at an American university, returned to his home country to assume a faculty position. His mission privileges were revoked due to changes in his institutional status. But while Jonas's official membership status had shifted, his membership in the *social* category of the team had not. This created the discomfiting sense of conflicting requirements that balanced rigorously patrolled exclusion from mission activity on the one hand, with the team's expected practices of earnest participation on the other. The tension over who dictates inclusion in the collective—the team or the country—caused anxiety for Jonas and his teammates as well.

I have described above how local mission practices enact each mission as a discrete organizational unit, setting it in relief against its backdrop of inter-institutional politics and national border crossings. It is worth noting, then, that even a law such as ITAR was not set in stone despite its heavy-handed penalties. It was everywhere enforced in American labs but always enacted differently, in line with different institutional or organizational norms. In some places, I was not allowed to view any hardware, while in other places I was shown flight instruments; certain meetings were restricted in one place but not in another; and in all cases, I was assured that this activity was compliant with the letter of the law.[16] Open questions of interpretation were resolved in line with local norms. For instance, if a line of software code is said to be ITAR-sensitive, what should be restricted: the single line of code, the software package, the computer it is running on, the building that the computer is in, or the entire institution in which the computer running the code is located? What if the room with that computer in it is connected to a multiparty teleconference line—can a foreign national be in the remote room as well? To answer that ambiguous question, Paris team members maintained two phone lines. When the TacPlan meeting was over, team members on operations duty hung up and phoned into the line connected to the engineering workroom, while foreigners at participating university sites left the local teleconnected room. This allowed team members to use their remote tools to support their flattened collectivist environment in the context of science planning, if not in spacecraft operations. Even the team's work with and around national regulations was part of the local instantiation of *organized science.*

Temporality and Accountability

Temporality might seem to be outside of political concerns, yet sociological studies of time and the technologies that reproduce them suggest that our notions of time are organizationally contingent as well (Wajcman 2015; Zerubavel 1982). For instance, mission timescales provided an anchor or a reference point for biographies (Cohn 2016). Being selected to work on a spacecraft mission is described as "an opportunity of a lifetime," but planetary scientists joke that it is a "life sentence" too. This is especially the case when there are long transit times to get to the planets under study. Gwen and Victoria both recalled when they had children or bought their first house with reference to an earlier mission's flybys of Jupiter, Saturn, Uranus, and Neptune. Helen's team selection took place in 1989, so by the time the spacecraft arrived at Saturn people had already been working together for fifteen years. By end of mission, this was easily the duration of a productive scientific career. As Jake, a senior Helenite put it, working on outer planets missions was unique compared to Mars because such work took place over a "life cycle":

> It's not like Mars, where you build it and send it and bam! It's there and you get your science. Outer planets, it's a life cycle. Births, deaths, marriages.[17]

Even though Helenites drew a contrast between their work and the work of their Mars-oriented colleagues, biographical intertwinement became part of the texture of life on both missions, especially as the Paris robots' lives extended beyond their initial mission plan to the same duration as the long-lived Helen. As undergraduate students became graduate students and then faculty and team members got promoted, married, or had children and grandchildren, the passing of robotic milestones on Mars marked biographical milestones on Earth as well.

The various scales of mission work produced different embodied experiences of temporality as well, with associated tensions of accountability. Although a career may elapse over a mission's duration, positions are granted by NASA at the mission's outset, so there is limited opportunity for organizational mobility. Individuals therefore work together *in the same organizational positions* for up to three decades. Many younger scientists described for me the tensions associated with these long *durées* as their careers progressed but their status on a mission did not. Those who began working with an instrument coinvestigator or team member as a graduate student had to find ways of staying connected as a postdoctoral scholar or researcher. Taking a faculty or independent research position, although a marker of external status, did not necessarily guarantee continued access to mission planning and its rich streams of data. These scientists were often in limbo, either working as research staff with a mission coinvestigator or taking up a position at a mission center to maintain team affiliation. Even when I arrived on Helen, many key players in science operations were not recognized in official rosters posted on NASA's websites. As two midcareer scientists explained it to me over lunch at my first PSG meeting,

> All the work is being done by these team associates [junior, unofficial members], and you have all these team members who are busy being PIs on their own missions, and they're not there, and they're not doing the work. But then you get to a team meeting and they all go into executive session and [the associates] get kicked out of the room, but there's no way for them to move up in the ladder.[18]

Both teams had to figure out ways of managing the competing temporalities of mission and careers. One official possibility was to open up a call for participating scientists, a position funded by NASA to join a mission after the spacecraft was in flight; a more informal option was to create "team affiliate" positions to enable newcomers to gain recognition for their role and maintain data access rights. Still, this did not amount to moving up to a PI level on mission teams. When one Helen PI stepped away due to a major promotion and handed the reins over to his young deputy, she took on all major responsibilities related to the instrument's development and application, but it was many years before she was allowed to formally drop "deputy" from her title. On Paris too, a decade of operations transpired before Jeremy appointed scientists who had been added to the team early in their careers to take on a deputy-PI role. In neither case was this due to negligence or power plays: there was simply no plan or precedent for mobility.

Moving out was also an issue on the other side of the temporal equation: visible when Roger attempted to retire from his job at a private-sector research facility. In an interview, he explained that it was time to hand things over to the next generation—to Daniel, a midcareer scientist located at the same institution—and that he wanted to enjoy his retirement with travel and hiking in exotic locales. Daniel played an active role in his instrument's operations and was well known to many Helenites as a representative to several targeting teams. It took months of negotiation with NASA to change the contract so that Daniel could take over as PI. Weeks after the announcement was finally made, an instrument malfunction brought Roger out of retirement to negotiate with the project office, who planned to shut the instrument down. A few months later, Daniel left the institution where the instrument was housed, and the contract was not rewritten to follow him. NASA handed the reins back to a very frustrated Roger, now called back from retirement for the second time. By 2013, his instrument was turned off, Roger had left the country for somewhere with no email access for several months, and his facility handed the now-silent instrument over to another collocated team leader in his absence.

These tensions were not the only ones associated with mission temporal scales: another pressure was the challenge of making team members' work legible for evaluation. On Paris, it was common to reuse images from prior TacPlan slides and science presentations, sometimes for several days, and eventually to place these into scientific publications. Although interpretations were always evolving, scientists took each other to task if they felt the annotations did not align with their observations, expectations, or the consensus arising from discussion. When a senior scientist on the team was challenged about his interpretations as labeled on the slide, his graduate student explained to me why accuracy was so important: "If you label the slide, it just snowballs . . . planning leads reuse the slide, and it becomes part of the lexicon."[19] As such, Paris scientists sought to hold each other accountable to their future selves and mission evaluators.

To meet such temporal pressures, each team developed methods that served to both satisfy internal accountabilities and assuage external accountability requirements as well. On Helen, the science team pieced together an elaborate spreadsheet indicating their scientific goals in order to request an additional round of mission funding from NASA to operate from 2010 to 2017. Once this funding was approved, the team assigned priorities to each goal and gave that version of the spreadsheet to the tour designers, specialist navigational experts on their mission. The same sheet was used to evaluate those proposed tours, described in chapter 2. At the PSG meeting in June 2009 in London, the same spreadsheet got a new identity. To a crowded lecture hall full of scientists recently returned from their tea-and-biscuits break, project scientist Francis introduced the meeting's task: to assemble "a traceability matrix commensurate with the level of knowledge about the mission."

The purpose of such a "grand traceability matrix" was to give NASA a measuring stick against which their collective productivity could reliably be judged in upcoming reviews. A NASA representative I spoke to described it as an accountability tool that also offered protection from devastating budget cuts. To craft it, project science leadership began with the spreadsheet that assembled objectives for their extended mission, recently used for tour planning. They then hoped to record "the way we achieve those objectives . . . [as in] we need this many flybys to achieve this observation, how many measurements need to be made, et cetera." Although I had not heard the words "traceability matrix" pronounced at a Helen meeting before, NASA had apparently been surprised in Helen's most recent senior review not to see one included among the team's materials. Kenneth, the representative from Headquarters in the room, insisted that the team must have had one somewhere: "Otherwise, how could they set up their tour?" He then requested that the team show it to NASA's panel at the mission's next yearly review.

Apparently blindsided by this new requirement, the team wondered if what they had already wasn't enough to satisfy the agency. Kenneth replied, "The senior review did see a lot of . . . color-coded charts, and I don't feel like that's what they felt would satisfy their value of a traceability matrix." When interdisciplinary scientist Arthur protested that such a chart could lead to NASA "micro-manag[ing] the observations," Francis counseled that all that was needed was a way for NASA to understand how many flybys or observations were needed to satisfy a prior stated goal.

The team latched onto the new vocabulary, seeing the traceability matrix as a potential solution to local problems of accountability. Even the scientist in charge of shepherding data into a centralized repository wondered if they should include a traceability matrix to facilitate locating online records. But the group was also concerned about how the matrix would fit into local notions of fairness. Would NASA use the matrix against the Helen team, to argue that instead of seven years' worth of funding they could get by with only two? This was especially worrisome as their extended mission examined for seasonal change and variation in the Saturn system, a process that required observations over many years. Engaged in this long-term project, when the mission team went into their first senior review to demonstrate that they were on track, very few boxes on the traceability matrix could be checked off, such that it looked like Helen hadn't done anything at all to answer its top priority science questions! Around this time, certain Helenites began to refer to the matrix as "Francis's spreadsheet"—consistent with instrumental metonymy and demonstrating that it was associated with management. Frustrated, Francis struggled to remind team members that it had a heritage long before he had arrived on the team, saying, "It's *your* spreadsheet!" Difficulties therefore arose as the context around the spreadsheet changed from external solicitation, to tour planning, to internal evaluation, to external accountability.

Like the Paris slide labels, team members are here engaged in managing mission temporality alongside external relations in ways consistent with their local orders. Annotated slides and traceability matrices record a past negotiation during which team members have deployed organizational resources and practices to reconcile differences, needs, and conflicts or to agree upon an interpretation. As such documents "feed forward," the entries in them are considered hard and fast. But should the data in the document later be interpreted in a different context due to inevitable accountability shifts over mission time scales, then this too may cause conflict. This was the case on Helen when the Satellites group realized that their priority threes on the solstice mission planning spreadsheet could not be changed to priority ones when the same document was later used to allocate PIEs. To reopen these allocations to negotiation would leave the group constantly open to rehashing old arguments, recrafting "fairness" anew, and having to retrace its steps. Still, if such documents cannot go back or be reopened for debate, they can go forward to new contexts. When budgets were cut in 2011 and 2012, the traceability matrix appeared yet again. This time, scientists in their working groups had to show which observations would be cut if funding for the mission decreased, by coloring its boxes in red, yellow, and green.

These examples of "traceability" and "feeding forward" illuminate how organizations produce internal temporal alignments, in this case, conducting future work to bring everyone onto the same organizational page. These organizational practices shaped participants' perceptions of their planets as well. This is evident in the way in which temporal challenges were confronted and resolved on both missions, solutions that were bound up in each organization's perception of planetary time. This may seem counterintuitive given that the planets themselves have different rotational periods and orbital dynamics: indeed, team members usually pointed to this fact as the explanation for many of their local temporal practices. But members also knew whether and when questions of seasonality on Saturn, daytime on Mars, or career longevity on Earth were relevant or irrelevant. As certain planetary temporalities were conscripted into local organizational practices, this naturalized each organizational temporal rhythm as "just how one does science" with *this* spacecraft or on *that* planet.

For instance, working on Mars required that individuals internalize Mars twenty-four-and-a-half-hour day (Mirmalek 2020)—at least until the mission's primary funding phase ended. For the remaining decade or more, meetings were determined by a combination of American time zones, weekends and holidays, Mars's rotation, and budget cuts.[20] Saturn's seasonal phases were also conscripted into science questions on Helen when they became convenient markers for mission phases determined by funding cycles on Earth, intertwining organizational concerns with planetary cadences. The daily iterative planning process on Paris and the five-week (later increased to ten due to falling budgets) planning cycles on Helen were always linked to the experience of days and nights on Mars or the fast-moving orbital trajectory at Saturn in actors' accounts.

Like other examples of the reciprocal relationship between instrumentation and scientific framings of time (Canales 2011; Galison 2004; Schaffer 1988), these particular elements of planetary time are reinforced and made perspicuous and relevant through organizational practices, such as observation planning or addressing external accountabilities such as NASA reviews. Even the pace of TacPlans, PSGs, targeting meetings, and sequences cements an organizational perception of planetary time, incorporated into the embodied, lived experiences of team members' work on both worlds (Cohn 2016; Hochschild 2000; Mirmalek 2020; Shih 2004; Steinhardt and Jackson 2014; Wajcman 2015; Zerubavel 1982; Zerubavel 2004). This is consistent with how Sharon Traweek describes the progression of physicists' careers as intertwined with their particle a beam accelerator (1988) and Steve Jackson, Stephanie Steinhardt, and colleagues illuminate the alignment of "collaborative rhythms" essential to scientific group investigations (Jackson et al. 2011; Steinhardt and Jackson 2014). Thus mission-mediated experiences of temporality enroll embodied time with machine time, career time, and future time (Cohn 2016; Ferguson 2013; Schivelbusch 1986). The result is a hybrid and sometimes shifting sense of temporality and its constraints on each planet: a flexible resource that can also be called upon to resolve the many challenges of temporality that confront spacecraft mission teams.

What Challenges Do Spacecraft Teams Face?

Spacecraft teams have to manage their own internal micropolitics. They are set against a shifting background of institutional alliances and competition, of temporal continuities and multimission complexity, of fragmented international expectations and demanding agency accountabilities. That the missions have to solve for so many of these complexities at once allows each team to arrive at local solutions for these challenges, each of which become part of the warp and weft of their organization. Organizational practices from shared rituals to communications tools and common narratives enable groups to meet those challenges and enroll them into the experience of their collaboration, always in locally accountable ways. Such organizational practices also draw a boundary around these persistent challenges as external to their mission's control, designating them external functions of the environment.

As bespoke methods of addressing these challenges arise from each group's organizational context, no two missions are exactly alike in their solutions. Individuals do bring forward "lessons learned" and "best practices" from one mission to another, but there is otherwise considerable heterogeneity as members seek to reconcile so many fields, accountabilities, logics, and careers. Optimizing for all scenarios is impossible, and there are no perfect solutions. Solving for one tension point may exacerbate another, while certain points of strain present a priority for resolution to one team but not another. Each group defends its local practices as otherwise-fragile social orders set against a complex environment on Earth and in the heavens. As each team innovates its own organizational answers to these challenges, these everyday practices are locally legible as ways of "doing the mission"—and, for their members, doing good science besides. This foreshadows how such organizational practices produce, enact, and enroll knowledge, technics, data, and biographies—the second part of the framework of *organized science* and our focus in part 2.

PART TWO

Outcomes

SIX

The Science

Posters line the exhibition hall at the hotel in Puerto Rico where the American Astronomical Society's Division of Planetary Sciences is hosting its annual conference in October 2009. There are colorful images of Saturn's rings taken through thermal filters, radar pictures of Titan's terrain arguing for surface feature interpretation, and camera images of spokes in Saturn's rings, each depicting recent science results and stamped with familiar names. On the other side of the room are the Mars posters. Many focus on specific features on the planet with data drawn from camera, thermal, and particle spectrometry, while others bring together orbital maps, images, and spectral data with data from the robots on the surface.

In adjacent halls, the paper presentations follow this trend. The majority of Mars science talks use data from many different instruments to make a claim about a region, even if they hold one instrumental dataset as their primary focus. The inverse is the case with Helen. Talks largely center on presenting findings based on single instrument views to produce claims about the planet, its rings, and its moons. In cases where multiple instruments are involved, these frequently invoke imaging data already released to NASA's public repository or prior datasets using the instrument's precursor—or they are PAM presentations that deploy the integrated physics suite. The differences are dramatic and cannot merely be attributed to orbital versus ground datasets. Not only are there different norms of operation and interaction on these missions, but *the kinds of science questions being asked and answered are different as a result.*

Prior chapters described the interaction rituals, forms of talk, and decision-making strategies involved in spacecraft observation planning. These processes determine the specific plans for acquiring scientific data such that the organizational *processes* involved in scientific collaboration ultimately shape the *products* of this scientific work. This is the second principle of *organized science*: since scientific organizations shape scientific outcomes, the organization of a science collaboration is consequential for its scientific findings. This perspective is in line with prior studies of factories and software companies that argue for isomorphism between process and product (Burawoy 1979; Conway 1968; Lee 1999; Stark 2009). It is also in line with a long history of science studies scholarship that argues for symmetry between natural order, our understanding of the world, and social order, our social arrangements on Earth (Jasanoff 2004; Knorr-Cetina 1999; Lynch 1994; Pinch 1986; Shapin and Schaffer 1985).

Paris and Helen are both sophisticated and capable missions that bring multiple instruments to bear on the scientific phenomena of interest to the scientists on their teams. In doing so, they have evolved organizational environments that enable them to manage and leverage the expertise of these different scientists and their specialties to better understand the planets. But the organizational context of each mission establishes the conditions under which these different people and instruments work together to produce scientific knowledge. Further, the parallels between scientific and social knowledge suggest that where and how we choose to seek phenomena have a strong impact on what we find. This is not to say that science does not produce surprises; rather, like searching for keys lost in the dark under a streetlamp, we can only seek knowledge based on the tools and principles we have ready to hand. In this way, the overarching logics of integration versus consensus, as they are enacted on both teams, come to matter in the production of datasets and scientific papers that respond to and permit certain types of scientific questions and conclusions—but not others.

This chapter, then, begins the switch in focus in this book from order to outcome, from process to product, showing how the organizational context of mission work plays a shaping role in the scientific outcomes of each mission and our vision of the planets as a result. To demonstrate this claim, I will describe generally observed methods of collaboration around data by both teams, illuminated even more clearly by the inconsistencies that arise when individuals on more than one mission move from one to the other. I then describe two unusual situations, moments of breach to the social order (Garfinkel 1967); as individuals attempt to either establish new working relationships or return to the local order, these moments demonstrate an upset of taken-for-granted expectations of how to collaborate and do science on each mission. Finally, I will describe two different philosophies of science that circulate in planetary science, showing how these are mobilized as narratives to accompany and naturalize the collaborations' local context of work.

Work Patterns and Research Products

Despite their collectivist orientation, all scientists on Paris have datasets, topics, and instruments that they prefer and with which they have specific training and expertise. These are frequently brought together in investigative enterprises or in cross-training to flesh out answers to problems at hand. I have already discussed the role of campaigns like the crater dust streaks that establish shared multi-instrumental goals and then implement those observations over a period of time. These campaigns invoke "hitting Mars with everything we've got" and ask questions that can only be answered with more than one instrument. Some scientists even cross-train on multiple instruments in order to gain familiarity with different scales, visual modes, and spectral readings. For instance, a spectroscopist I visited described how she spent several weeks one summer at the camera image processing center in order to learn how best to combine her spectrometry skills with knowledge of an entirely different instrument altogether.

Even for those with a preferred line of investigation, within the consensus environment, it is essential to get others to agree that an observation is a shared priority. Paris scientists therefore frequently appeal to each other to work on joint projects or campaigns or to align with strategic goals that the entire team establishes in long-term planning discussions. In this way, their work is not seen as a pet project but as a generous contribution of their unique skills to the mission. For example, when Ben was interested in finding out whether the rocks surrounding a local crater were ejecta from its interior, he led a series of requests for multispectral camera observations of them to find out. Ross also worked on these images, and the two of them shared notes and ideas. The photographs were one part of a suite of observations of these rocks, and Ben and Ross led the charge to select which ones would be subject to further investigation using the robot's drill, hand lens, and thermal and spectral readings. While Ben and Ross were involved in determining which rocks to image and how to do so, neither of these scientists fostered a sense of ownership over their observations or the resulting data. Instead, they worked with other instruments and scientists to characterize the rocks that everyone chose together—including Nick, who specialized in infrared spectrometry. When they presented preliminary analysis of their visual interpretation at team meetings, scientists across the team jumped in with suggestions for which other readings to acquire and which objects to visit in order to narrow down the group's hypotheses.

In other cases, some work that can only be done in a particular location by a specific scientist is seen as contributing to larger mission goals. GIS maps, dust measurement, and atmospheric readings are repeatedly performed by singular groups on the mission affiliated with specific individuals or institutions. Their ongoing results loop directly into planning or contribute to long-term monitoring and understandings of Mars, all of which are discussed and shared at the weekly science meetings. Other readings may correlate to measurements taken from orbit, providing an ongoing "ground truth" for observations that generate goodwill among team members' colleagues on other missions. Tony, Paris's deputy PI, was especially involved in establishing these relationships between ground and orbital readings. Also a member of another lander mission and an orbital satellite team, he used Paris data to support these missions and vice versa. Even during science theme group gatherings, as in the campaign description, scientists propose questions and determine which observations using which instruments will answer them. So when a theme group representative speaks up in a TacPlan meeting with a series of requests, these typically involve multiple instruments that address a broader set of goals that the group has agreed upon. This form of multi-instrumental collectivism is a guiding principle in the kinds of scientific investigations that the Paris team crafts and conducts on Mars.

The weekly science team meetings that provide first reports of findings or investigations in progress typify this approach in analysis. As I described in the case of the wind streak campaign, such presentations were often designed to bring in new perspectives, often by building up a coauthorship list that included scientists with a wide range of specialties. For instance, following praise of a graduate student's early stratographic (image) analysis of a cliff face, scientists said that "[they] owe it to [themselves]" to acquire microscopic imagery as well as spectroscopy measurements for mineralogical content, adding that they might be able to get such data "for free" as part of two-for-one observation planning. The student replied, "I'm perfectly happy to add more [spectroscopy] to the campaign."[1] At this point, his observations became a shared concern among the group: campaign science. While this could be considered a problematic incursion on a student's project, such multi-instrumental work was instead framed as a scientific and pedagogical opportunity. Similarly, a postdoc initiated her discussion of meteorites that relied on her chemistry background with the following: "I've been trying to figure out how this complex minerology and geochemistry is related to . . . this layer-cake kind of stratigraphy there." Her slides showed mineral abundance diagrams from the spectroscopy suite of instruments alongside images from the region: targets she had requested in concert with Ben's ongoing image analysis.[2]

This was consistent with scientific problem solving across the team. Mineralogy scientists displayed graphs of chemical abundances correlated with in situ spectrometer readings, and scientists with expertise in remote sensing offered their view of the surfaces under study to help identify other similar objects nearby. When a thermal scientist presented a "field guide" in progress to the region, he compiled it based on the thermal and multispectral properties of the rocks he was most interested in, and the mineralogists chimed in. Campaigns that had already been developed as multi-instrumental investigations continued that way as the group formulated analysis, results, presentations, and ultimately papers—as well as further robotic investigations to test their hypotheses. Graduate students that I met at the camera and thermal headquarters were studying how best to combine and compare readings from more than one instrument, while those at the deputy PI's university were even working to correlate ground-based readings with those from orbiters overhead. Scientific questions, answers, and processes were multi-instrumental at almost every turn, with little sense of individual ownership over results.

This multi-instrumental collectivism endures well into the team's publications of scientific findings. With the exception of early papers that lay out calibration procedures at the mission's outset, single-instrument papers from Paris are rare. Published accounts more typically combine instrumental results to make a claim about the planet's environment. Atmospheres are characterized (Lemmon et al. 2004), rock classes are cataloged (Farrand et al. 2006; Squyres et al. 2006), and soils are investigated (Weitz et al. 2006) using a combination of in situ mineralogical instruments and microscope and remote sensing tools such as the cameras and the thermal spectrometer. These papers describe the properties of the objects they investigate through multiple lenses. Indeed, many of the proposed experiments were written so as to *require* more than one instrument, such as investigation into the magnetic properties of the planet's dust or the qualities of rock weathering (Hurowitz et al. 2006; Leer et al. 2011). We therefore learn from Paris that the atmosphere has *both* optical *and* thermal characteristics, that certain kinds of rocks that look the same through the camera's near-infrared lens have different mineralogical compositions, or that soil grain sizes visible in the microscope correlate with thermal and spectral properties and that these correspond to different terrains and regions around the robot.

Even papers that rely more extensively on one instrument make reference to results from others. The cameras are both a spatial and a multispectral tool, but authors bring in results from other instruments too to understand their interpretations (Farrand et al. 2006; Farrand et al. 2016; Kinch et al. 2007). The mineralogical instruments also publish detailed accounts of their investigations while referring to spatial and spectral data acquired from remote sensing tools such as the camera and thermal spectrometer to do so (i.e., Clark et al. 2005; figure 7). Papers that simply lay out how an instrument works also make mention of more than one instrument in combination to constrain interpretations. For instance, several camera specialists worked hard to publish models for how most effectively to combine the camera's multispectral readings with data from the three other spectrometers on board. One paper begins with the claim that it is "desirable to seek relationships between remotely sensed multispectral imaging data and in situ elemental and mineralogical results"—that is, several robotic instruments—in order to identify materials on the surface (Anderson and Bell 2013).

I choose these examples because they are common cases, not outliers. They are the sort of work that is typically done and published on the Paris

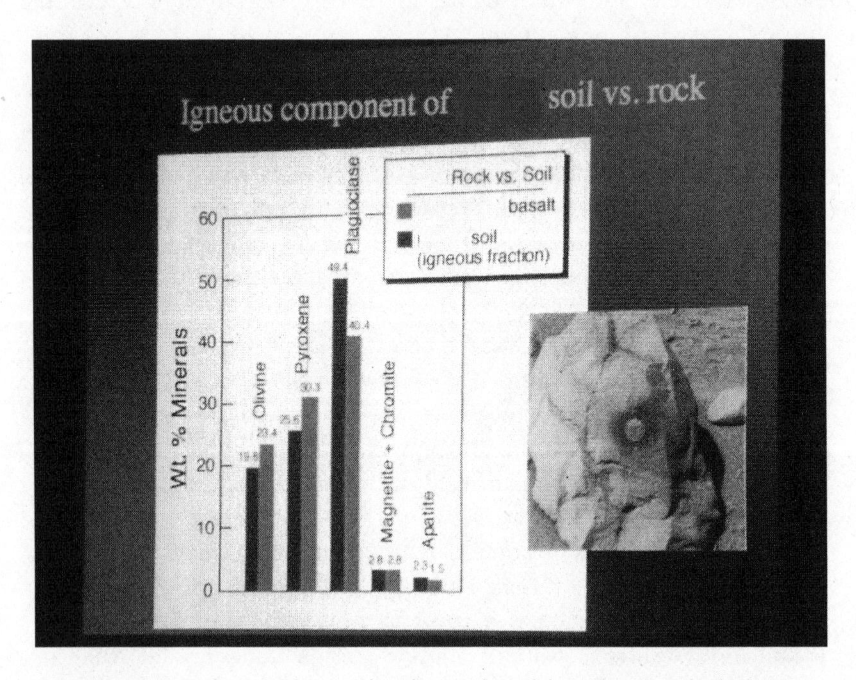

7. Bringing together spatial, spectral, and mineralogical data on Paris. Author's photo.

mission and of the sorts of investigations I witnessed in progress while attending the team's weekly work-in-progress science meetings. This approach held for smaller-scale investigations conducted among the ranks, as well as overview or significant discovery announcement papers. It held for graduate student work and for senior scholars' papers as well. Authorship lists ran long and often included more contributors than just those in a specific research subgroup. Highly specialist subjournals might include papers using a single instrument alone and a smaller group of scientists, but these were less common. They were also cited by others in the collaboration to assemble a multi-instrument publication investigating a single phenomenon using many contributing tools.

In contrast, scientific work on Helen principally took place at the intersection of individuals' multiple hats: at the juncture between discipline and instrument. For instance, at a thermal team meeting in Oxford, thermal scientists Victoria and George discussed thermal data of the rings, Titan thermal scientist Finn described thermal observations of Titan, and others spoke about Titan's atmosphere using thermal data. A notable presentation at this meeting by a promising junior scientist working with Edward to produce thermal maps of the icy satellites used visual data from the cameras as the background for her maps. However, the maps were based on public repository data and were not images collected jointly to solve a shared problem at hand. A few months later, I attended the infrared team meeting (a facility instrument, not PI-led, like the thermal instrument), where I again witnessed presentations divided along working group lines. Scientists gave talks about Titan using infrared data, talks about the rings using infrared data, talks about the atmosphere using infrared, and talks about the moons using infrared as well. Presentations by interdisciplinary scientists combined datasets: for instance, showing infrared and radar scans of Titan together or using thermal or mass spectrometry data together with infrared to understand the planetary atmosphere. On average, between 10 and 15 percent of presentations used more than one instrument to solve a problem.

This was not limited to "front stage" performances such as meeting presentations: work in progress fell along these lines as well. When I visited a group of graduate students at a university in Rome hard at work on the radio dataset, one was using it to speak to a question about Saturn, another about Titan. When I visited the few scientists who were appointed to two teams at once, such as Rod, I was keen to see if their work practices intersected in any way. Rod first showed me his work using the camera images, then opened entirely separate files and software tools related to his ultraviolet instrument work. Across the team, the majority of scientific work I witnessed combined deep instrumental and topical expertise at each intersection of the matrix between instrument (thermal, ultraviolet, imaging, infrared, dust, etc.) and working group (Titan, Rings, Satellites, Atmospheres).

If at the instrument team meetings, individuals presented analysis of the five different working group topics using their shared instrumental lens, at the working group meetings, the matrix pivoted to reveal different aspects of the topic viewed through the lens of one instrument at a time. At Titan group meetings, there was typically a presentation about dunes or lakes on Titan using the radar instrument, another about specular reflectance off the lakes using infrared data, another about clouds using the camera, and another about clouds using the thermal spectrometer. There was also discussion of radio experiments to determine the moon's gravity and whether it had an internal ocean based on radio occultation experiments. While everyone offered assistance in interpretation, everyone knew which scientist was associated with which investigation. Further, each scientist was typically affiliated with the instrument whose tools they brought to bear on the moon. Their studies iterated over time, providing a deeper sense of the import of that particular area of study (i.e., dunes, lakes, gravity, clouds) especially as it related to Titan's variability throughout the seasons. As the mission progressed, they also cited published work by their colleagues to provide a time-lapse view or an integrated perspective as a second-order step in their instrumental interpretations. Views of Titan's lakeshores might combine a camera and a radar view taken a season apart, for instance, and the probe's descent path was eventually captured by both infrared and radar, more than thirty Titan flybys later.

I noted a similar trend a few days after the Helen PSG meeting in London, when the Satellites group convened a meeting featuring one instrument talk after another. We heard about plasma observations of Enceladus's plumes, then mass spectrometer and then ultraviolet observations of the same phenomenon respectively. Each paper was presented by a different scientist who was affiliated with the instrument team whose data they described. There was infrared photometry of the moons followed up with infrared compositional mapping of the same moons by a different author altogether. There was another paper on infrared spectral properties; then there were camera studies of Enceladus or of Iapetus or Mimas. Flipping through the agenda, only one in ten papers were multi-instrumental. Those were typically physics presentations, consistent with the PAM group's style, or papers that drew on data already in the public repository and not acquired specifically for that investigation, such as camera mapping data. It was only at PAM meetings where I more frequently saw many instrumental capabilities brought together in the raw data and planning phase to address particle flow and flux.

Helen's research papers therefore themselves coalesced into "a grand traceability matrix" of scientific inquiry wherein papers, like scientists, emerged at the juncture between instrument and working group. In the scientific results published by various team members, the majority of author lists were exclusive to instrument team affiliation. Such publications regularly made reference to alternate instrumental lenses in support of interpretation but rarely incorporated such data at the first order of analysis. For instance, while I was immersed with the team, published work on the rings emerged from groups of dust scientists (Postberg et al. 2009), camera scientists (Beurle et al. 2010; Hahn et al 2009; Hedman et al. 2009; Tiscareno et al 2008), radio scientists (French et al. 2010; Rappaport et al. 2009), or thermal scientists (Fletcher et al. 2009; Leyrat et al. 2008). Studies of Saturn reported its atmospheric properties (Fletcher et al. 2009), radio experiments (Gurnett et al 2010), or magnetic field detection (Burton et al. 2010). Papers about atmospheric or surface features on Titan drew from radar, thermal, or spectral team affiliates (i.e., Janssen et al. 2009; Nixon et al. 2009; Robertson et al. 2009), while studies of Enceladus's plume published standalone papers that reported results from infrared, ultraviolet, or the mass spectrometer (i.e., Coates et al. 2010; Cravens 2009; Hedman et al. 2009; Hendrix et al., 2010). The thermal, dust, ultraviolet, radio, and imaging teams each circulated their own beautiful maps of Enceladus's surface or of Saturn's rings; the next order of work was to attempt to correlate between these distinct documents to see where the overlaps might be. If it was uncommon to see a phenomenon through two instrumental views at once, it was more likely to see the phenomenon observed through that same lens over time: radar analysis of how the dunes changed with the seasons on Titan, for instance, or how the thermal properties of the rings changed over the Saturnian seasons (figure 8).

There were a few exceptions to this trend. Papers that combined two fresh datasets typically included interdisciplinary scientists on the author list as the brokerage point (Burt 2004) between different teams (i.e., Cuzzi et al. 2010; Soderblom et al. 2009; Waite et al. 2009). The discovery of cryovolcanos is one example, combining both radar affiliates and infrared scientists along with an interdisciplinary scientist as listed authors (Nelson et al. 2009). A few other cross-team collaborations emerged from brokerage within the ranks. A paper about Saturn's emissions included two scientists appointed to two different instrument teams, thereby bringing together a list of authors across several constitutive groups (Li et al, 2010);

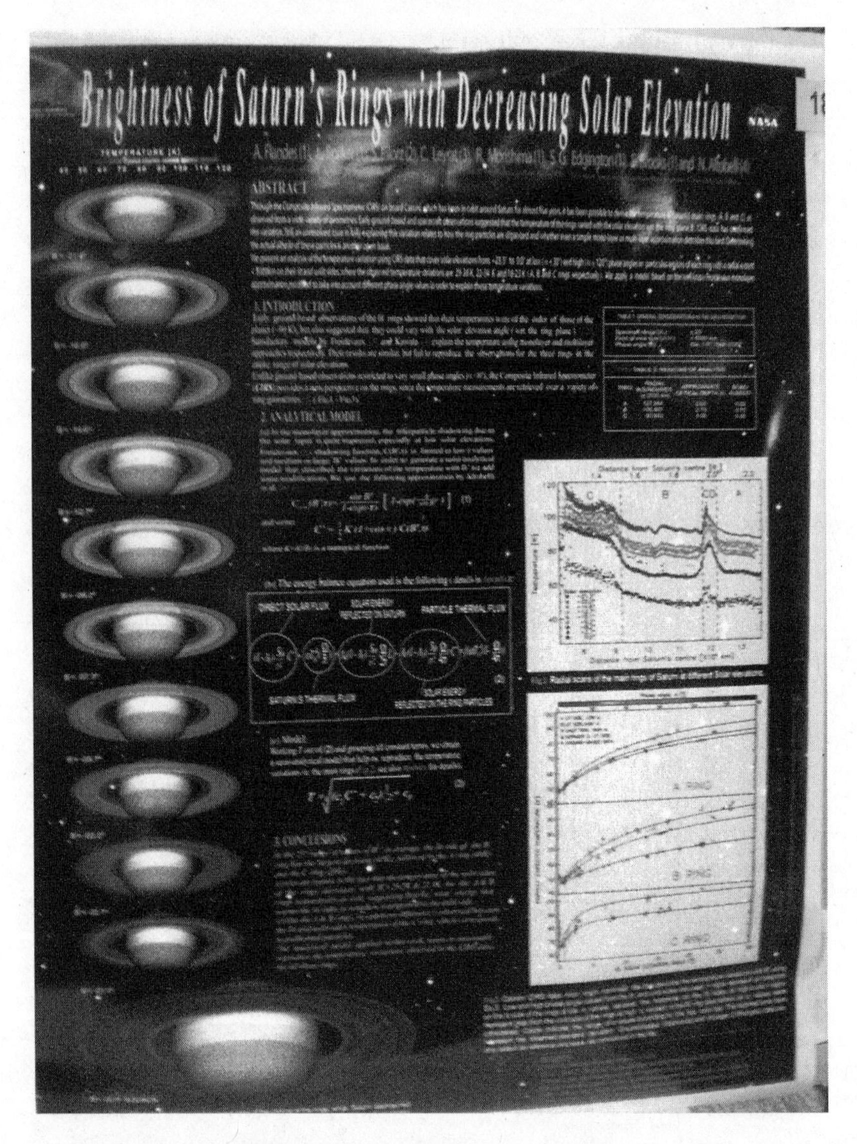

8. Rings analysis of thermal data acquired over time on Helen. Author's photo.

and a few early publications on the aurora emerged that combined optical and physics scientists (i.e., Dyudina et al. 2010). In a few cases, already-published data was used to support an emerging hypothesis, with credit to the original author. This was especially the case on Titan, where scientists built up explanations over instrumental views acquired sequentially over multiple flybys (i.e., Tosi et al. 2010). PAM papers were largely exempt from the single-instrument trend. While there were papers in the PAM team that focused more or less on one instrument—plasma, mass spectrometry, magnetic field, for instance—most others deployed the PAM data exchange platform to use information from across the physics instrument suite to make a claim, typically with that instrument PI's knowledge and often with

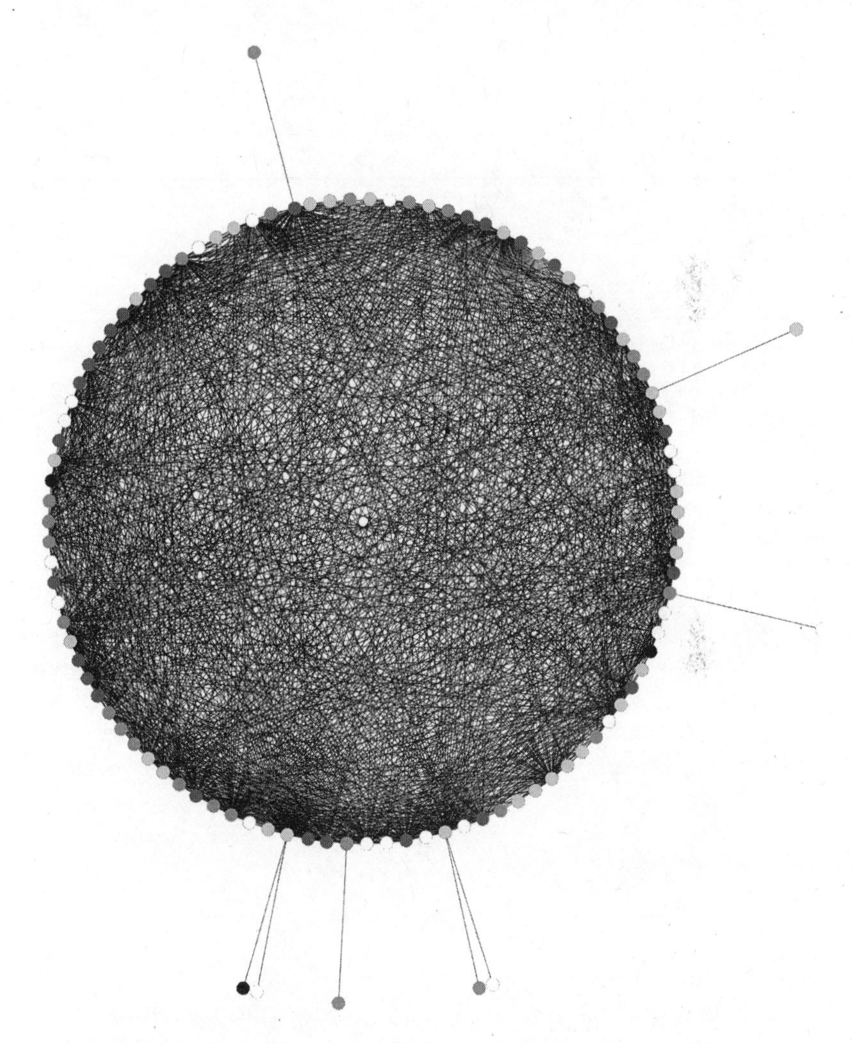

9. Dense, interconnected ties feature in the Paris coauthorship network, visible here at a threshold of seven copublications for team publications to 2011. This single component endures to a threshold of thirty-five ties. With small numbers of weakly connected authors, most authors are less than two steps away from any other author for the majority of measured tie strengths. Nodes are individual authors colored by instrument team.

them listed as a coauthor but not necessarily involving an interdisciplinary scientist (i.e., Jones et al. 2009; Smith et al. 2008).

Such observations of how scientific questions were framed and answered were also visible in traces of team coauthorship.[3] Figures 9 and 10 display copublication networks for Paris and Helen at a threshold of 7 coauthorship ties, for authors who published at least five times about the mission data

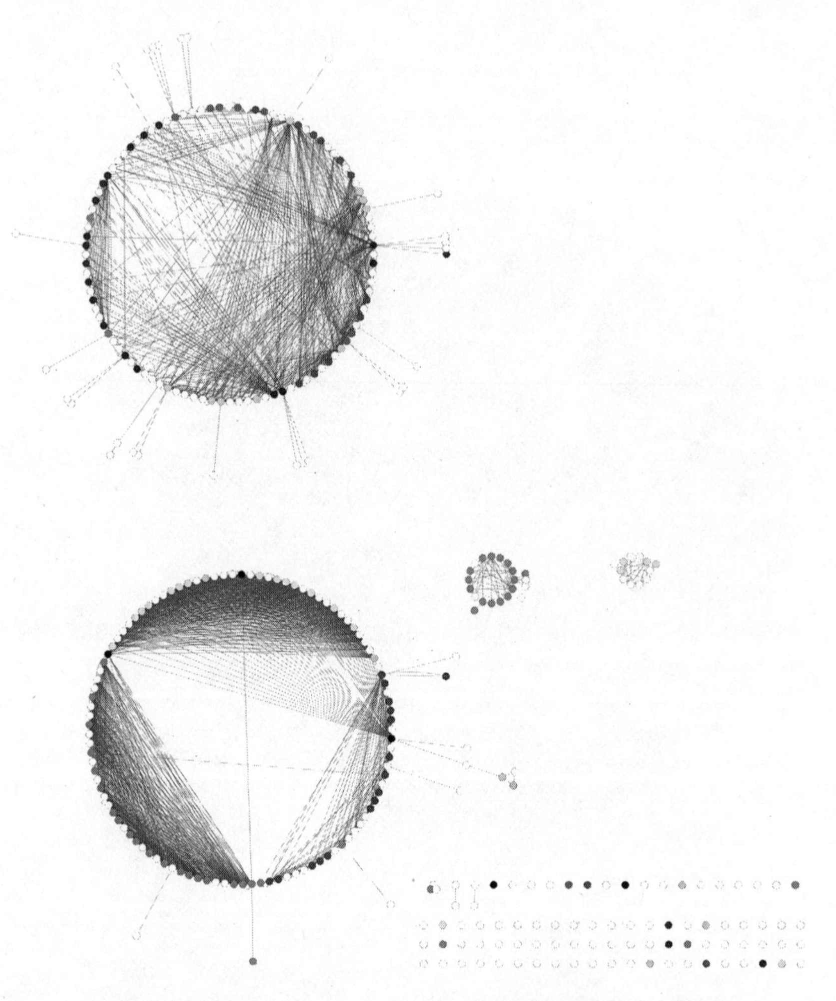

10. At a threshold of seven copublications, the coauthorship network of all Helen collaboration publications as of 2011 splits into a PAM component featuring many cross-instrument ties (top), an optical and remote sensing component with visual, infrared, and ultraviolet instrument publications held together by interdisciplinary scientists (bottom), and separate components for radar, thermal, and other instrumentation. Nodes are individual authors colored by instrument team, with IDS's in black.

between their start in 2004 until the close of my observations in 2010. This tie strength eliminates yearly joint-team summary publications to determine robust partnerships. These structures persist as the tie strength threshold increases.[4] The Paris graph of 436 author nodes and 319 publications appears as a single unified component with many linkages between members of the community. This copublication cluster remains robust until a threshold of 35 ties, when it resolves into a spectroscopy cluster and optical cluster. There are no discernable patterns based on instrumental preference (pictured) or institutional membership.

Helen is of course much larger, with 1,700 authors and more than 1,000 unique publications. Distinct components take shape on Helen: coloring in authors with their instrument affiliations detects a clear pattern. Three instrument teams are immediately visible—radar with its strong internal cohesiveness, the dust analyzer, radio, the thermal instrument. A weakly linked component on the bottom left is the optical instruments—the camera, ultraviolet, and infrared spectrometers—held together by the interdisciplinary scientists of these groups, noted in black.[5] The large component at top left, similar to the Paris component in terms of its number of cross ties, is the PAM group, whose Paris-like interactional norms in terms of data sharing and operational considerations are clearly visible in their coauthorship ties.

Both graphs neatly demonstrate this component of *organized science* by depicting the outcomes of local collaborative practices oriented toward integration and autonomy versus collectivity and consensus. Additional elements are worth noting. Scientists across institutions, nations, funding sources, and space agencies are distributed across these graphs in ways more consistent with their mission status than other institutional ties. This is especially notable on Helen where international investigators are fully integrated within instrument teams as equal coauthors. We also detect no evident clustering among members of the same institution. Thus the missions have resolved these challenges to great effect. It is not that distance ceases to matter but rather that the overarching mission organizational context plays a crucial role in such interactions.[6] Further, due to their positioning on the spacecraft, radar, radio, and the dust analyzer typically cannot operate at the same time or with the same field of view as other instruments, making their data more difficult to combine. Thus the spacecraft's body is also visible as either an integrated (Paris or PAM) or a modular (Helen) entity, according to its enactment on each team.

As a Helen scientist once insisted in a meeting, copublication does not capture all collaborative interactions. Yet the fact that these two collaborations *in the same field* (planetary science) involving many of the same people and institutions have such different coauthorship patterns indicates that organizational elements play an important role in shaping local epistemic cultures—at least as important a role as disciplinary training. As instrument and authorial connections represent forms of shared questions and answers, the graphs demonstrate how the science can look quite different on two overlapping yet organizationally distinctive teams—a compelling validation of ethnographic intuition.

This is not to suggest that the individuals on one or another mission are predisposed to certain types of collaboration but not others. Instead, the collaborative practices that are the norm within the organization shape not only expected collaborative interactions but also the stuff of scientific work—the data itself. For instance, when I presented an early version of this work at a planetary science institute, a scientist in the audience who was on neither mission spoke up. He explained that outsiders to the missions found the Paris dataset to be "broad" but "incomplete," while the Helen dataset was "deep" but was difficult to coordinate—that is, with Paris data, it was easy to find five or six different instrumental readings of a single rock or phenomenon on Mars, but the dataset was far from encyclopedic. You could not, for example, get spectral readings of all the rocks in an area and decide based on that what you, a scientist coming later to the dataset, wanted to look at in more detail. The opposite was true for the Helen data, he explained. It was relatively easy to find a standardized observation taken of a similar phenomenon over extended periods of time but difficult to overlay observations of the same phenomenon upon each other. Such data might not even be available as the result of the integration of different needs and requirements from autonomous teams over a period of time.

This is on the one hand a function of the interoperabilities of the respective spacecraft and their instruments. Radar and the camera cannot observe Titan at the same time due to pointing constraints, the result of the loss of the scan platform and the articulated antenna. But the availability of data in the archive is also contingent on how observational resources are divided up by each group. A spectroscopist on Paris would have to collaborate closely with her colleagues to get her requests for a spectral observation through the process, making a request for ownership over the robot's timeline while in the region impossible. The same scientist on Helen could argue for periodic ownership over the spacecraft's timeline in order to complete an instrumental suite of observations, but she would be at a loss (unless she were an interdisciplinary scientist) to request observations from more than one instrument at once. Thus the data and scientific results that emerge from these missions bear an indelible stamp from each team's local organizational order.

You Can Change Out the People . . .

There is no clearer indication of the gravitational pull toward these organized ways of doing science than those individuals who serve on both Paris and Helen at once. These scientists must contend with the differences in local ways of collaborating or interacting. They may do so by avoiding one mission or the other in alignment with their preferences, but there are other examples of scientists who came often to meetings and participated actively in mission planning and scientific writing. In one of the first working group meetings that I attended on Helen, James chaired the session. I had observed him chair Paris TacPlan meetings, and in a prior interview, he spoke at length about the value of listening to others who bring ideas to the table such that all feel "empowered" about the process and its results. But in a Helen working group meeting when the camera leader, Isabelle, brought up an idea, James shut it down quickly and decisively. Later on, when she didn't have a set of observations to present to the group, he intoned sharply, "Well, if you don't have anything prepared, then you shouldn't be speaking at the meeting." This contrasted with my observations of his leadership on Paris but was consistent with the autonomous operating procedures within distinct groups as well as the affective ambiguity on Helen.

Although he is no "cultural dope" (Garfinkel 1967), James's interactional norms and patterns shifted from one mission to another in line with local expectations. He described this as a question of the imposed structure of each mission's "personality" upon its participants, asserting, "You just take the same people out of this mission [Helen], stuff them in another mission, and they don't behave that way." He offered another flagship as a counterexample, which involved "half these people . . . back twenty or thirty years ago":

> [E]ven though many of [the Helen team members] had changed and forgotten [the past], the ingrained personality of the structure of this social monster [Helen] has not changed a bit. I mean, you could keep swapping the people out and it had no effect.

Several scientists reiterated some version of James's comments in terms of the power of a mission's collective "personality" over individuals in its midst. However, David and Theodore both provided a useful counterpoint to those who felt the force of mission structure upon their actions.[7] Theodore was a veteran of many missions and frequently spoke up on Paris to request a single type of observation. His requests often came at unusual times, such as in the middle of a TacPlan report or when a plan was clearly already oversubscribed. Paris team members politely accommodated his requests but privately wondered why after so many years their colleague still hadn't picked up on the order of operations. Of course, Theodore's interactions were entirely consistent with his work on Helen and flagships prior, where advocating for one's science required making pitches alone for spacecraft time against other engineering and scientific requests.

David, on the other hand, trained first on Paris and then started working with Helen. In his Helen instrument team meeting, I witnessed him present late-breaking results in an early phase of investigation to the team with the intention of proposing a follow-up observation to test a developing hypothesis, which he offered to "put on the table and let the team shoot this down." Such an approach was the norm for Paris science meetings and long-term planning discussions but not on Helen. When David tried to identify a moment where the spacecraft could potentially do the follow-up observation of Titan, his colleague Edgar was incredulous. "This is all deals that have been made," Edgar stated firmly—that is, since the jumpstart meeting had already taken place, these observational opportunities were already fairly allocated through negotiation. While applying the lessons from Paris to his Helen interactions might have been to David's disadvantage in terms of understanding how to get observations, members of his instrument team on Helen described him as very cooperative, a team player who worked well with others.

Breaches and Transitions

The above examples display how scientific collaborations and multi-instrument observations occur in typical cases, but there are moments that depart from these norms. In such cases, tempers or frustration may flare, indicating that something is amiss. Sociologist Harald Garfinkel argues that moments of breach are instructive as they reveal the underlying social order: expectations of how action and interaction are supposed to unfold between members (Garfinkel 1967). In this way, these two naturally occurring breaches provide a window into the expected patterns of asking and answering scientific questions using observational data and requests. I emphasize that while in the prior section I described examples of interactions that *typified* collaborative exchange, here, I will describe in depth two contradictory moments that are relatively *unusual* examples. On the Paris team, this involved a team member's persistent interest in single instrument

assemble a catalog of thermal readings of all the rocks in the region. During one meeting, he proposed "offering up" other thermal observations planned by atmospheric scientists in favor of his thermal observations of the nodules, prompting the Atmospheres representative that I was in the room with to chime in quickly on the line. Off mic, he was notably ruffled ("Nick! He's offering up our observations!") and my field notes recorded a lesson for the day: "Don't volunteer someone else's observation!" Atmospheric scientists were not the only ones at risk; when David tried to defend stratigraphic observations in a plan, Nick called David's request a question of "minutiae." Ironically, this was how many Paris scientists were starting to think about Nick's work. How many more thermal observations would be necessary? Again, acting as TacPlan chair, Tony tried to defuse the situation and return to operating principles, even shutting Nick down in indication of a breach of interactional norms on the team. When Nick tried stocking even more thermal observations on a day where memory was tight, Tony cut him off: "Alright, Nick, that's it for today, dude." And when he tried to suggest moving the robot to a new and difficult area for still more observations, insisting that he had done the necessary image analysis necessary to determine that the area was accessible, an engineer on duty spoke up politely: "If I could just interject, I think it might be wisest to leave the science to the scientists and leave the engineering to the [engineers]." Nick protested, but this time Tony cut him off with a plaintive: "No, no, Nick. *Please.*"

In each of these moments, sitting in the dark teleconference room at a mission-affiliated university, I too felt taken aback. I had never before seen a scientist pit his observations against other groups' requests or against engineering expertise, a Paris team taboo, and I had never heard a TacPlan chair call out a colleague or curtail their voice on the line. Other team members' startled laughter and shocked expressions registered that this was indeed a breach. But soon thereafter Nick began to explain his observational requests as part of a larger series of questions, a strategic objective that presented compelling science for the team. He described how what he was seeing was a unique spectral signature and that his request for more infrared observations was to determine where else this signature showed up. He also demonstrated that the spectral signature on Earth was related to one that was often produced by biological processes: a possible sign of life. This now qualified as a potentially important discovery. Nick also showed how it was part of a larger series of questions that scientists were asking in the same region, related to the dispersion of materials from an ancient source of water.

As the conversation progressed into the summer, other scientists began to see Nick's observations as a shared priority across the team. They commanded the robot to find, crush, and analyze rocks in the region to find out more about this material. Other instruments were brought to bear on the shared problem as a "campaign." By now, the thermal instrument's capabilities and Nick's own unique expertise became the key to a puzzle that many team members agreed was mission critical. Team members thereafter invoked Nick's expertise and the instrument's observational capacities whenever possible as part of a campaign in the region. When the thermal instrument later stopped working due to dust contamination, the team even continued taking observations with the instrument, in the hopes that in future they might be able to calibrate its results and correlate it with their complementary observations. What began as a breach of the social order of collectivity was brought into line with the practices and expectations of the team.

If Paris scientists had difficulty pushing through individual projects, then Helenites experienced breach when individuals who were not sanctioned to use more than one instrumental dataset—such as interdisciplinary scientists—attempted to do so. Some scientists were even upset when other team members used their instrument's data following its archiving to the public repository, the Planetary Data System (PDS). In one case, an attempt to produce a joint press release of the plume region on Enceladus using early data from both the cameras and the thermal instrument required several months to resolve. In another, a young scientist who joined Helen under a new data analysis scheme described being reduced to tears at her first-ever PSG when the PI of the instrument whose data she had proposed to use berated her for not asking permission in advance (If she did not have permission, she wondered, then why was her proposal selected?). Even well-known early-career scientists like Drew and Kevin, who had cobbled together funding from Helen coinvestigators on two different instrument teams as postdocs, published about each dataset independently. Multi-instrumental collaborations among the rank and file were patrolled and constrained.

A transitional moment on Helen demonstrated a move away from this approach toward a multi-instrument paradigm that conceived of joint observations from the outset, instead of relying upon prior downlinked data crafted for other observational requests. When the spacecraft flew past the planet's pole, it observed an aurora borealis, an electromagnetic phenomenon. The physics instruments recorded its presence, but the camera team also collated an impressive multiframe sequence of images into a video they placed online, in which the aurora snaked around the pole of the planet. It was rumored that infrared and ultraviolet had collected observations of the feature too. Many of these scientists had already discussed atmospheric effects with each other at the Atmospheres working group meetings and their subgroup investigating Saturn's rotation, led by an interdisciplinary scientist called Henry, but with no associated targeting team for observation planning. These results especially fascinated the physicists in the PAM working group, who saw potential benefits in being able to align their data with optical, infrared and ultraviolet observations taken at the same time as their "always-on," synergized tools.

Following a cross-team workshop on the aurora at a PSG in late 2009, the PAM group began a discussion about how not just to share data after the fact but to actively coordinate observations of the aurora. As Gabor, the PAM interdisciplinary scientist, put it,

> Unless we share data at an earlier stage, it will impact the science outcome. So the question is . . . How can the project balance intrainstrumental priorities with Helen-wide priorities? In other words, each instrument also has a service role, not only their own science role. They provide a service for other investigations that go beyond that single instrument.[8]

This statement was met with disagreements over details and confusion over purpose, but the PAM group in conjunction with the project science office eventually decided to host another aurora discussion at the following PSG meeting. They invited members of the camera, ultraviolet, thermal, and infrared teams to join them. Gabor hosted the meeting, which he opened by calling himself the "sacrificial lamb" in trying to establish a new method of collaborating. He began by bemoaning that "multi-instrumental observations are really a major political problem," even as he described them as "important to achieve the next level of science." He described his proposal: "to put a group together and to . . . come up with the information, observations, and access to data that are needed to do the best science for the aurora."

The newly added scientists—including interdisciplinary scientists such as Henry, instrument leader Leon, and science planners such as Max—were game to participate but were unclear as to what was requested of them other than simply showing what data they had acquired about the aurora. They were similarly unclear about what they could expect in return. In many cases, they explained that they were unable to give their data to other Helen team members because of their teams' local restrictions around data circulation. For instance, when camera scientist Georgina showed the video to the group, she did not have the authority to give that data away. It was also an essential part of her research program as an early-career scientist. Leon reminded the group of a paper he had worked on with Henry and Georgina, stating, "I hear about issues, but I don't see them in practice." In case they were being accused of not sharing, Georgina, Henry, Leon, and Max reminded the PAM colleagues of their openness to collaboration as evidenced by their presence at the meeting and indicated that they would be willing to accommodate other instrumental requests favorably in observation planning if it came to that. But when Gabor pushed for the capability to not only use instrumental data post hoc but also participate in *planning* and acquiring data, this was met with head shaking and evident confusion.

Meanwhile, members of the PAM group insisted that they already had a strong record of multi-instrument collaboration. Roger accused Gabor of "fixing things that aren't broken!" Others expressed concern about goals and process. Daniel wondered aloud, "I'm still not clear *what* we're trying to do and *how* we're trying to do it, and *how* won't make sense unless I understand *what*." While Gabor was enthusiastic about initiating a "new process" that would produce data that the group considered to be shared from the bottom up, the assembled group quickly pointed out that his "new system" violated too many of their existing rules of engagement, from incommensurate data sharing rules to processes of data acquisition and management. The thermal PI also reminded Gabor that despite his repeated invitations, the interdisciplinary scientist had not yet shown up at a thermal instrument meeting. ("This is what I mean: the sacrificial lamb!" Gabor laughed.) After almost two hours of going in conversational circles, Henry suddenly and loudly broke out into song: "Kumbaya, my Lord!" Gabor begged Henry not to quit his day job as the group dissolved into laughter. But when they broke for lunch a few minutes later, Daniel pointedly stated, "We just aren't willing to put work into fixing something that isn't broken. . . . Don't expect a lot of work out of this, Gabor." The tension in the room suddenly thickened. "I never expected anything," he replied.[9]

Such incomprehension is typical for nascent collaborations in the sciences and elsewhere. Studies of groups collaborating across boundaries reveal that developing shared vocabulary and new processes takes time and effort (Centellas, Smardon, and Fifield 2014; Galison 1998; Kellogg, Orlikowski, and Yates 2006; Paradis, Elise and Albert 2013). This was no less the case for the Aurora group. The group met twice more to share scientific results and after a few months, began to develop a shared vocabulary and fluency with each other's' work. While no "new processes" were explicitly worked out and no top-down notion of required sharing was established, a few bottom-up initiatives started to take shape. For example, an ultraviolet and an infrared scientist decided to work together to figure out how to coordinate observations between their instruments. Each instrument pointed in different directions and required the spacecraft to behave differently when taking observations, but as Max, who was involved in the discussion, explained it to me afterward: "What's ideal for them [ultraviolet] turned out to be not very good for us [infrared], but what's fairly good for them turned out to be fairly good for us."[10] Thus the two teams each accepted a "fair" downgrade to their data in the interest of an alternative approach to data collection. The paper with joint results was published in 2011.[11]

The time and effort it took to get the Aurora group off the ground are indicative of the novelty of its approach as well as the challenges associated with introducing a new mode of work. Changing the style of asking and answering science questions on the mission required adjusting the sociotechnical system, realigning team practices on Earth and the spacecraft in space. It introduced a new operating procedure in the joint observation between infrared and ultraviolet instruments. It also formulated an institutional structure for emergent attempts at this new form of collaboration. The group had ripple effects across Helen, as multi-instrument investigations spread gradually to other domains on the team. Following the aurora observations, for instance, the PAM team extended an invitation to the Satellites group for a joint workshop on magnetospheric-satellite interactions, especially those produced by Enceladus's plume. A few years later, partly influenced by budget cuts that reduced workforce capacity and therefore observational time, the Helen team initiated a new "sharing working group" to plan for observations that would "share" spacecraft time (and the work time that it took to plan for spacecraft activity) as the craft moved into its final weeks of spiraling through Saturn's rings. I will return to these moments and their implications for organizational change in a later chapter.

Local Philosophies of Science

Sociologists of labor describe how different organizations of work reflect and reinforce different ideological orientations and theories of value (Burawoy 1979; Lee 1999; Stark 2009). Similarly, these contrasting scientific contexts of work on Paris and on Helen, with their members of the same scientific community, are also associated with different ideological orientations with respect to scientific inquiry. This was especially evident during a Helen infrared instrument team meeting, held in Colorado on a chilly September day, when team leader Leon invited infrared scientist Mickey to speak to the group about the discovery of cryovolcanoes. Mickey was an active participant in the Titan community with its relatively strong sense of social solidarity, and he had been featured on stage during the cryovolcano press conference announcement alongside Francis's skeptical coauthor, Sam. Fielding a few questions after his slides, Mickey explained that he was attempting to mollify his language to "avoid the kind of loggerhead conflict" that had plagued the discovery announcement thus far. Leon replied firmly:

> But that's science. Science is a process of successive approximation and even when you achieve a consensus, you're never guaranteed to be right . . . eventually if there's some truth in this [discovery], it will come out. . . . But I don't think the process of science is well served if a bunch of people get into a room and say, "Let's agree not to disagree.'"

Mickey disagreed. "Science is a consensus process," he argued. "I'd rather see if there's a way to present these arguments as a consensus view." But Leon was emphatic:

> I think there are a lot of people in this room who do not agree with this case, and as scientists do, they present them to their colleagues, and in enough time, a hypothesis will come to light. . . . I don't think the search for truth is served by a bunch of people [who] get into a room and say. "Let's come to consensus on this." I think some of the best science comes about vehemently or quietly or whatever mode . . . I think people do not agree on this and are going to take [their concerns] to the literature, which is the place for this [argument] to take place.

Mickey tried a different tack, imploring, "We have a resource with which we've been trusted, and we have a responsibility to present [results] to the community." But Leon countered: "Look, it's obvious people don't agree. . . . If you'd told me twenty years ago [e.g., when Helen started] we'd all agree"—his voice trailed off, and he shrugged as if to say he wouldn't have believed it. "People just don't agree. And that's science."[12]

The difference between Mickey's and Leon's point of view reveals a distinction among members of the planetary science community that is brought to bear upon their scientific work. For Mickey, a member of the close-knit Titan community, science is "a consensus process," one in which his presentation of results in a meeting room is aimed at building his argument, considering new points of view that can be incorporated into his interpretation, and mustering allies (as anthropologist Bruno Latour would put it) to support his scientific discovery. I frequently saw radar scientists work together on problems, and their eventual publications bore the names of all team members—not because of a team requirement but because of group participation. At their instrument team meeting, members crowded around screens and printouts to peer at recent downlinked data and build up their interpretations together. Mickey's philosophy would sit equally well with Paris team members. Recall the discursive alignment between robot health and safety, team consensus, and "good science." Under their local philosophy of science, in alignment with Robert Merton's norm of scientific communalism (Merton 1942), collective agreement is the sign that observational interpretations—of Mars or of Titan, for Mickey—are true and that the group is, in fact, doing good science.

Leon's perspective, on the other hand, recalls the Mertonian norm of organized skepticism, a different demonstration that one is doing science correctly (Merton 1942). It is also reminiscent of Karl Popper, champion of logical positivism, who famously argued that science proceeds through "bold conjecture and refutation" (1962)—that is, it is not through agreement but through *disagreement* that science proves anything. This echoes in Leon's repeated statements that the process of science is not best served by groups assembling and "agreeing not to disagree," but rather through "vehemently or quietly" taking their vocal disagreement to conference presentations and to the literature, demonstrating their disinterestedness through focusing on the purported facts of the case alone. As the leader of a facility instrument with little control over his teammates' investigations or conclusions, Leon saw disagreement as a fact of scientific life. Still, in Mickey and Leon's exchange, we see exactly what is at stake for these scientists in adopting, supporting, or contesting one method versus another. The voice for consensus and the voice for disagreement are not arguments about whether one likes ones' colleagues: they are about the correct process of science, the credibility of claims, and the ensuing truth about the planets themselves.

Conclusion

A key tenet of the Strong Programme in the sociology of scientific knowledge (Bloor 1991; Collins 1985; Shapin and Schaffer 1985) is that to better understand science as a social process, one must hold all sides in any scientific debate equal. As such I refrain from asking or attempting to answer the question of which team orientation generates empirically better science. Instead, this chapter describes *two types of scientific organizations that are better at different things.* One is optimized for synthesizing multi-instrumental views of singular objects, the other for collecting encyclopedic or temporally sequenced observations. Neither approach is monolithic: multi-instrument science occurs on the matrix team, just as single instrument science can also take place in the collective, although this is less common. As we examine the different structures' associated outcomes, we can also observe which organizational resources individuals bring to bear on their scientific work. Here, I have described the different philosophies of science in operation, as well as the interactional norms that are policed in case of breach. In later chapters, I will describe how technological resources, data files, and scientists participate in maintaining these orders too.

Before proceeding, however, we must note that the type of knowledge produced by Helen and by Paris about their respective planets is *qualitatively different.* Through Helen's eyes, we see planetary changes over time, craft longitudinal surveys, observe deeply through single instrumental lenses, and produce global maps of magnetospheres and moons. With the Paris robots, we see specific cobbles, rocks, or other phenomena through many different lenses with many instruments, each contributing to a fuller picture of the object under study. The planets participate in this knowledge work too. There are, of course, no rocks to observe on Saturn and no rings on Mars, and different phenomena offer themselves for investigation on each planet. However, the way in which investigated objects are constituted on these two planets differs, in alignment with their investigating team's organization. Helen's autonomous teams and practices of integration collude to produce neatly matrixed results located at the intersection of instrument and working group. Meanwhile, Paris publications are imprinted by their collectivist practices, multi-instrumental work, and their synergized payload—a feature shared within PAM on Helen. It is possible on occasion to use the spacecraft differently, as the cobbles and aurora examples show. But the vast majority of presentations, publications, and dissertations I observed during my time on the missions took this approach, constructing each planet along organizational lines.

Mars and Saturn are both multifaceted, dynamic places, and these missions offer complex solutions to the challenges of bringing disparate groups together to study them. Like any organization, each approach has its advantages and drawbacks, and holding both styles in comparative view may suggest alternative ways to solve a particular scientific problem. But the comparison especially illuminates the importance of the organizational context to scientific work. If the kinds of observations that get approved through the meeting process on Paris are those that engage multiple instruments and team members in answering shared questions to achieve collective consensus, this team will almost certainly produce the multi-instrumental, synergized scientific questions and answers of the type that animated Paris mission posters and presentations. This makes sense of why Nick's attempts to produce an encyclopedic scan of cobbles were unsuccessful until he combined his interest with those of other scientists in an investigation involving more than one instrument. Conversely, under autonomous integration on Helen, single-instrument observations are not only the norm but emergent and selected for in the planning process. The exception of the PAM team and the Aurora working group demonstrates just how much new ground was broken to encourage other ways to collaborate between instruments.

This chapter, then, begins to develop the second principle of *organized science*—that *scientific organizations shape scientific outcomes*. The outputs of the collaboration are visible and measurable as scientific papers, publications, and presentations that answer the very questions that drove observation planning in the first place. As a result, Saturn and Mars look like quite different planets—not only because of the *tools* but especially because of the *organizations* that we have brought to bear upon their investigation. These differences are not due to geophysical or astronomical properties and are certainly not because Mars is a more synergistic place than Saturn, whatever that might mean. Rather, the teams' different organizational orders as they approach both targets in our solar system produce very different kinds of knowledge—and planets—as well.

SEVEN

The Spacecraft

So far, I have described the people and practices that compose each mission organization, and the tensions they face in their work. But perhaps these two organizational environments are simply the result of how the spacecraft are constructed. This was certainly an explanation I heard frequently from planetary scientists, who offered that it was simply more difficult to relate to an orbital craft, with its weeks-long segments of activity, than to one that wheeled daily on the ground. In this chapter, then, I discuss the importance of the spacecraft themselves. Science studies scholars commonly weave discussion of technical apparatus through their analysis from the outset, so it is unusual to leave such a conversation to this late in the book. However, viewing the organizational context on Earth first not only mirrors a fieldworker's experience of the site, with the team's object of commands millions of miles away; it also offers a useful perspective upon how each group's organizational social relations are inexorably enmeshed and intertwined with the spacecraft themselves.

The idea that the technologies we use *demand* certain organizational patterns of use and interaction, can disrupt organizations, and enforce cultural change is known as technological determinism, and it has been repeatedly debunked by a long list of scholars who study technology and society. A different way of looking at these missions is to begin from the assumption that both are *sociotechnical organizations*: complex systems that interweave software and hardware with organizational practices and interactions. Actors' descriptions, then, that their interactions are natural or necessary for their particular spacecraft are the *outcome* of such sociotechnical interweaving, not its cause. This perspective is in line with contemporary studies of scientific and enterprise communities whose work is suffused with technical systems. Studies of new technologies entering organizations show that the ways in which the tools become "imbricated" (Leonardi 2012a) with work practices typically reifies that organization's power structures and existing practices and that such technologies may become incorporated into local power struggles over resources and authority (Kling 1991). These studies have also introduced vocabulary for describing the intermeshing of the social and technical. Some draw on sociologist Anthony Giddens's work to describe a method of "structuration," whereby technologies are one of a set of tools that individuals use to structure their work and through that, the organization (Barley 1996; Orlikowski and Yates 1994). Others do not see such a ready distinction between the social and technical, using the language of "sociomateriality" and "configuration" to describe how objects do not stand alone but through social processes come to incorporate the very human context that gives them meaning (Mazmanian, Cohn, and Dourish 2014; Orlikowski 2010). Regardless, because the spacecraft and the tools used to command and work with them were constructed by overlapping groups of people, it is unlikely that they challenge the organizational assumptions of NASA writ large (Pinch and Bijker 1987). Further, as I have described, vehicles like Paris have now flown with Helen-style matrix organizations, while orbiters like Helen host Paris-like teams. So while the technical systems may *reinforce* certain patterns of organizational interaction, they do not necessarily *require* them—even if these modes of interaction are so well established on each team as to feel perfectly natural.

A useful middle ground is suggested by semiotician Morana Alač. In her studies of robots in groups, Alač convincingly describes how people build up a concept of their robot's potential and possibilities through interaction with it, creating what she calls a "body-in-interaction" (Alač 2009). This fuses together social expectations of interaction on the one hand, with the limitations of robotic hardware on the other, making the two categories impossible to discern in practice. In line with this work, and that of anthropologist Lucy Suchman (2011), I will show how each spacecraft participates in producing and reproducing their respective missions' organizational orders: not simply by virtue of their design but through how mission participants configure their bodies-in-interaction. The organizational practices, interactions, and forms of talk I have described in earlier chapters establish the interactional context in which the spacecraft acquires its form and meaning. Taking this argument a step further, I will show how the spacecraft participate in reinforcing local team solidarities and distinctions, in an effect I call "technosolidarity."

Technosolidarity recalls founding sociologist Émile Durkheim's stated connection between a society's social form and the way that group members feel toward one another, or as he put it, "The form of a social system's organization dictates the degree and type of its solidarity" (Durkheim 1893). According to Durkheim, groups with limited or no division of labor experience tight social relations among all members of the group, cemented through ritual assemblies that generate "collective effervescence," and totem objects that they associate with their groups' fate. But as societies grow and individuals specialize and establish systems of exchange among different groups of experts, Durkheim argues, this primary loyalty migrates to one's subgroup first, and secondarily to the society at large. Durkheim associated this shift with a transition from "primitive" to "advanced" societies—a perspective that is out of step with contemporary sociology and further does not suggest why advanced technical teams might still experience these social effects. Technosolidarity therefore expands Durkheim's definition to include the *sociotechnical* relations common on today's technical teams. In line with Durkheim's definition, I will describe how mission team members experience solidarity relations with their teammates that are consistent with their organization: Paris's collective "happiness" and interchangeable social roles versus Helen's allegiances and fairness in intergroup negotiation recall Durkheim's mechanical versus organic or even industrial social forms. Further, I argue that mission members also experience such relations *with and through their machines*—that is, the way that the spacecraft are interpreted and used and the unique configurations of collaborative software established under distinct organizational conditions ultimately support each team's local solidarity relations and social structures. In this way, continued interaction with the spacecraft and the tools of its command reproduces each mission's organizational commitments as well.

This may be especially visible on spacecraft teams because the technologies involved in these collaborations are so very far away. As a result, considerable interactional work goes into interpreting their location and status and planning for their every move. This is observable in the copious meetings, software suites, and gestures that account for the spacecraft's current and future movements. I will begin by surveying the software tools, both bespoke and commercial, that each team uses to both establish the spacecraft's possibilities and to structure their team's conversations. Then I will turn to the bodily techniques (Helmreich 2007; Myers 2008) that scientists use to understand their spacecraft. Throughout, the intermingling of hardware, software, and human bodies is consistent with and reinforces each team's cultures of *integration* or *consensus*, producing technosolidarity relations that differ for each organizational form.

Documenting Entrainment

On mission teams, the electronic files, visualizations, and documents that are part and parcel of each team's software suites are not only the *result* of work but also the *sites* of work. They are active focal points for collective attention and the place where work is done.[1] As such, they feature in the organizational landscape as one of the means of producing local solidarity relations. For instance, meetings on both teams are suffused with documents, whether projected on a shared screen or emailed in advance to get everyone on the same page. Target teams or TacPlan meetings proceed with the document as the focus of "mutual entrainment," a shared focal point for collaborative work that gives all participants a sense of belonging to the group involved (Collins 2004). As evolving and live documents, they are manipulated and changed as part of the meeting itself. For this reason, perhaps unsurprisingly, each spacecraft team prefers different software with different visual metaphors for capturing various aspects of their work.

Some of this software is off the shelf, mundane, and downright quotidian, produced by the Microsoft Corporation and bundled into its Office productivity suite. PowerPoint slides were displayed at every TacPlan meeting on Paris, while Helenites preferred Excel as a planning tool. So central were these tools to determining spacecraft plans that team members kept these applications running continuously on their computers, often with several files open at a time. Organizational ethnographers have already noted the importance of PowerPoint in corporations as "part of the machinery that produces knowledge" (Kaplan 2011, 321), as a genre that "shapes the ongoing work of organizational actors" (Yates and Orlikowski 2007), or as "embedded in communicative action" (Knoblauch 2013). Although these studies are based in an enterprise context where certain presentational styles were common, Paris scientists did not use the familiar bullet-point-and-header-slide-deck templates built into the PowerPoint software nor did Helenites use the number-crunching functions of Excel. Instead, a particular set of conventions for software interactions and display developed in line with technosocial relations on both teams. Incorporated into each meeting, these support the team's collectivist or integrative goals.[2]

For instance, the typical TacPlan meeting opens with a slide deck assembled by the Paris long-term planning lead. The first slide lists the date on Mars and on Earth and the primary occupants of key meeting roles for the day, followed by several full-page images taken by the spacecraft and annotated to establish its location. Charts follow that relay information about onboard power and memory status, constraining the meeting's conversation

11. PowerPoint slide used in a TacPlan meeting on Paris. With permission.

with an overall "bit count." The information in these slides and transitions between them structure the meeting ritual, pace the conversation, and get everyone on the team "on the same page." Everyone follows along as the leaders of the team go through the deck one slide at a time, their speech punctuated with "next slide . . . next slide." The slide deck's stable format and rhythmic presentation lend a ritual quality to each individual meeting, no matter the specific issues of the day (figure 11).

In addition to this ritual atmosphere, the slides are chock full of high-resolution images that situate the viewer in the body of the robot, looking out over the terrain or sometimes looking down from an orbiter overhead. These are decorated with text, lines, circles, arrows, or dots to indicate prospective plans, areas around the robot, or other agreed-upon interpretations of the terrain around the robot. Planetary geology has a long history of annotating maps to make features of the terrain perspicuous to view.[3] But in this case, the markups served an essential organizational function: to gather everyone "on the same page" on the way toward consensus by generating a shared sense of place and common ground for a collectivist discussion. Such techniques are not impossible with the business flow-chart images built into the software, but these tools are clumsy and strained the capacities of the software. As they struggled with auto-place features and enormous file sizes, scientists also frequently eyeballed locations for dots or lines or used Comic Sans, WordArt, or different sized fonts overlaid on the image. So central to decision-making were these documents that they were even subject to humor. When the spacecraft was about to reach a ten thousand meter driving milestone, a team member took to Photoshop to digitally place a four-digit odometer stuck at 9997 on the robot's body, dangerously close to rolling over to 0000. The image was shown in the PowerPoint deck as a visual joke and remained there for up to a week, generating chuckles from team members each time it was displayed.[4]

Similar documents were on display as planners hosted strategic discussions about scientific goals. These included large maps based on an orbital image, annotated with the robot's current position, potential drive directions, and observational opportunities. These pictures served a structural role for consensus-oriented conversation by grounding team members in prior decisions, albeit without constraining them too tightly. For example, as the spacecraft explored a specific region, the team repeatedly updated a series of images with prospective and past drives. The result of such discussion was frequently a new map, marked up to indicate the team's recent decision and displayed at the next day's planning meeting. Returning to these maps repeatedly over time not only kept track of planned objectives but also opened up the possibility of adding new goals to the map view. In this way, discussion could indeed become "endless" (Polletta 2002) as it was drawn out over considerable time, and each return to the map afforded an opportunity to revisit current conditions and concerns. In some cases, a long-term goal remained permanently on the horizon as the team continually added new considerations to be achieved in the area before moving forward. Ultimately these maps provide a series of snapshots of where the team was in their planning process at a particular point in time in an evolving conversation oriented continually toward consensus.

As Paris scientists used PowerPoint, Helenites used Excel. Whether PSG, target team, or PIE meeting, at the outset, the group is presented with a spreadsheet, usually already familiar to them through prior iterations. The spreadsheet is projected onto a shared screen in the room and circulated by email for remote participants to examine during the conversation. They may modify the spreadsheet or create a new one based on what came up in discussion. That spreadsheet is carried forward to the next task, generating new spreadsheets and so on. In my years on the Helen mission, I never saw a spreadsheet that tabulated numbers, invoked formulas, or generated graphs. Instead, Excel provided an environment for collaborative text entry in a tabular, gridded environment best suited to categorization, priority assignment, textual elaboration, and color coding. This was put to use to support the mission's task of integration by giving each part of the organization its proper line, column, or box for self-expression. Recall during tour selection at the PSG in chapter 2, when the Titan group filled in the tour selection spreadsheet in pinks and oranges alongside red, yellow, and green. Spreadsheet work was so common that it was also subject to humor. For instance, when PAM scientist Daniel was asked if his instrument team had any input prepared at the start of the Titan jumpstart, he quipped straight-faced, "No, but it could be a colorful Excel spreadsheet, so we have to do it."[5]

The spreadsheet also draws boundaries around who gets to speak for which cell based on group or instrument membership, consistent with and reinforcing Helen solidarity relations. In the tour selection case, each instrument had a box in which they could make their judgment call about the tour. Thus members responsible for the radar instrument, for instance, got to speak for and fill in the radar cells—but not the other cells. The spreadsheet as a whole gives a sense of the overall fairness of resource allocation, even as it visually displays the wearing of multiple hats on the team. Indeed, it is something of a visual metaphor for the matrix organization, with individuals located at the intersection of instruments and working groups just as their inputs are directed to a particular intersection of rows and columns.

Excel facilitated integration but also demonstrated its challenges, such as when teams had to aggregate across the many columns and rows to make a final decision. While the software is good at computing averages or other numerical functions, calculating based on text or color is not straightforward. The Titan group confronted this issue during tour selection when they set about trying to relay their carefully delineated instrumental needs in the rainbow spreadsheet into the three colors required to report back to the PSG. The group met the challenge with creativity and no shortage of Helen's affectively ambiguous humor. First Cliff, the interdisciplinary scientist working on the spreadsheet at the front of the room, suggested assigning numbers to the boxes. This would allow them to compute averages and determine which recommendations should be red, yellow, or green. He chose values on a scale of one to one hundred to correspond to each color and began inputting them into the cells. But others asked, how were these numbers chosen? What did it mean to give eighty-five to lemon-limes? Should all lemon-limes have eighty-fives or did the color represent a range? One scientist suggested using "a continuous color scale." His colleague corrected, "We should use *numbers,* which *are* a continuous scale," and another quipped, "Can we use irrational numbers?" When the averages were computed and one column got a score of 83.6, Daniel asked wryly if

command requests are often displayed in a list, spreadsheet-style form. Any team member could use the tool in real time. Paris scientists loaded Conduct! during the TacPlan meeting following the initial PowerPoint slides, when the software secretary inputted observations into the tool as they were proposed and approved through conversation in the TacPlan.[7] As each observation appears on screen, it can be dragged into temporal order and modified to show an updated count of bits and power requirements. Members on the line along with the TacPlan chair recommend moving such observational blocks around to accommodate the day's memory or power constraints. Thus the group follows along on screen as they collectively edit each observation until consensus is achieved, reflecting and reproducing the team's consensus orientation.

Helenites also developed a database for science planning into which any mission scientist, using a web interface, could input an observational request. One output of this database software also looks much like a spreadsheet, listing observations in terms of start time and coloring each line according to its responsible instrument or organizational unit (such as navigation or spacecraft office). The database can also print out data in a spacecraft timeline. In the background of these visualizations, gray and white vertical stripes represent an hour of spacecraft time. Running horizontally across colored bars representing instrument requests: thermal in pink, infrared in yellow, cameras in gray, dust analyzer in orange, and so on. The timeline includes other essential elements to bear in mind while planning, such as moments when the spacecraft needs to turn in order to send data back to Earth (figure 12).

Rings scientist Walter has a knack for reading spacecraft timelines. To him, the colors that streak across the page represent not only instruments and requests but the spacecraft's movements and location in the Saturn system. Sitting in a bare-walled office at a college in London, where he was a visiting scholar away from his prestigious American university, he led me through a visualization line by line:

> That's [*points to a line across the bottom*] the gravity low-frequency radio. Mag [*points to a long red bar*] is the magnetometer [observations] and [*points to another colored bar*] is the instrument that does plasma [observations]. And these tend to be operating continuously. These are the requests from the optical and remote sensing instruments [e.g., the cameras and visual spectrometers]. The next day is the day of periapse [the closest approach to Saturn], which is indicated by this red dotted line. . . . And not surprisingly . . . this is where . . . all the requests get concentrated.[8]

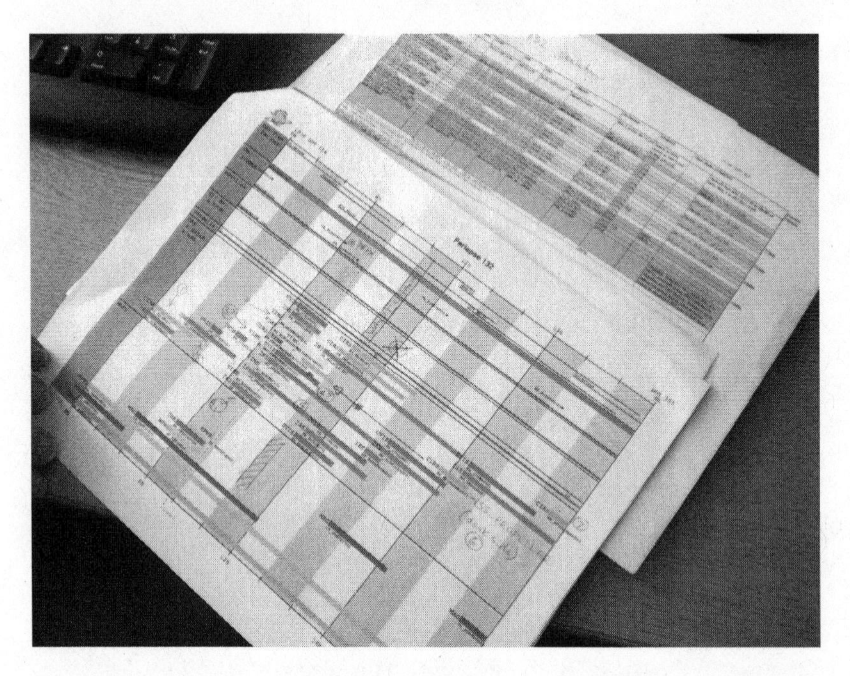

12. Helen spacecraft timeline in Walter's notebook. Author's photo.

Gesturing to each colored bar, Walter describes an association between instruments and observations for a particular purpose. Even in the software and in these visualizations, it seems, the spacecraft is partitioned into organizational units. He also talks through the visual representation of conflict when a cluster of instrument requests around the red dotted line indicate that a fraught conversation is likely to erupt at the next planning meeting. When I notice that many of the conflicting lines are of the same color, Walter explains that this is because people on the same instrument team do not necessarily talk to each other first about instrumental priorities but rather place their requests directly into the database to discuss later. As he put it, "There's so much planning to go on here that essentially none of the [instrument] teams ever get to the point of looking at all of their competing requests and deciding as a team which one is more important: Is it more important to do an Enceladus observation or a Rings stellar occultation at this time?"

Walter keeps another set of colored printouts from the database taped into a notebook interspersed with timeline printouts.[9] In these charts, each line depicts a spacecraft orbital period as a whole, and colored sections along the line represent not divisions among instrument teams but among the disciplines. This view represents another slice through the matrix organization by delineating which sections of activity will get planned by which targeting teams: purple for Rings, blue for Atmospheres, green for PAM. At the center is a red line indicating the craft's closest approach to the planet, while at the edges of each line are always Titan flybys—moments when, Walter reminds me, the spacecraft makes orbital adjustments to set up its next orbital segment. The occasional touch of orange indicates that the segment includes flybys of other small moons, accommodating the Satellites group's requests. A quick scan over the page permits an appraisal of the "fair" allocation of observational time, prompting Walter to explain that the orbital groups' sections were smaller because they were planning targeted flybys. At the same time, this representation also indicates where the spacecraft will be during each part of its tour through the system, writing the organization's distinctions onto the spacecraft's journey and its potential observations.

These bespoke database tools also assist in the process of decision-making and record meeting outcomes in ways consistent with the local organizational culture. One Helenite described how placing observational requests in the system before the targeting team meeting is "almost like a legal document in the process," placing a stake in the ground to indicate where an observation is requested. But another noted that this practice could make certain people distrustful of the software and of the multiple hats people wore during input because "people . . . putting the [observation] designs into the database really want to have their designs in the database because they want to have their observations." Further, each targeting team used the tool differently: sometimes as the starting point for negotiation, elsewhere as the record of the completed integration process. On Paris, on the other hand, oversubscription in advance was the start of a process of collective trimming toward consensus and locating "two-for-one" opportunities. It was, therefore, not necessarily the functional elements of the software but their organizational contexts of use that were oriented toward group-wide consensus or multigroup integration. In this way, both bespoke and commercial software tools were enrolled in the local challenges of science planning in ways consistent with and productive of solidarity relations on each team. And as both sets of tools were premised upon understanding, situating and negotiating around spacecraft activity, they projected organizational divisions onto the spacecraft, drawing its contours and potential in line with solidarity relations on Earth.

Gesture and Interaction

This construction of the spacecraft is especially evident in the "body-work" (Myers 2008) used to make the robot present in the room as an interactional element in planning. Bringing their hands up to either side of their face and twisting slowly from the waist, Paris scientists imitate the cameras atop their robot's mast in a physical reminder of their robot's wide-set robotic eyes. Splaying their arms to their sides and tilting forward from the waist, they shuffle their feet or roll their wheelie chairs to show the robot's drive path. So frequent and unconscious are these movements that they are often performed alone, not as communicative gestures but as ways of working through the robots' experiences. "Get your camera," a scientist called out to me at a team meeting, a few hours after I had presented some work in progress to the team about what I playfully termed "the Paris dance." He was bent over with his arms thrown out behind him, in the midst of talking his colleagues through the robotic maneuver that got them the images he was analyzing. "I'm doing it; I'm doing the dance!" Everyone in the room laughed.

Paris team members articulated the robot's body in very specific ways. During her shift, camera operator Liz explained her way of moving the arm like the robots. She first picked her left arm up and held it in line with her torso, elbow cocked and forearm swinging as if from a hinge, saying, "There's this bit." Then she stopped, arm still held in midair, peered at an image of the robot on her screen, then used her right hand to maneuver her left arm, holding it just below the shoulder. Still looking at the screen she said again, "There's this bit and it goes like *that*" and rocked her arm up and down in the elbow socket. "But it also goes like *this* I think," she continued, rocking it from back to front.[10] This imitation of the degrees of freedom of the robot's arm relied upon but also challenged the physical capabilities of the human body. The awkwardness of the gesture expresses the differences between the robot and the human even as it permits individuals to step inside the robot's body and experience an alien planet from its frame of reference. As such, the body of the robot—more specifically, shared local ways of embodying the robot—is an important resource for scientists pursuing the work of unilateral agreement (Vertesi 2015).

The spacecraft's body-in-interaction—as it is imagined, invoked, and gestured into the room at a distance—is a hybrid object, a mixture of human projection and machine parts (Haraway 1991, 2007; Suchman 2011). It is also a key site around which group solidarity relations are performed. Much like the tribal totems that Durkheim describes among collectivist groups, the body of the spacecraft is the locus for expression of the unifying social relations between team members on Earth. No wonder the Paris robot's safety and science depended, in team members' accounts, upon maintaining their unified stance. I also heard about and witnessed illnesses or injuries that occurred at the same time as a catastrophic event that the robots endured on Mars. Visiting the engineers who were working to free their robot from a sand trap, I noted a Paris driver had a bandaged knee. She explained that she got injured while salsa dancing one evening, around the same time that the robot got stuck. Favoring her good leg as she stood, her stance echoed that of the model robot behind her, listing to one side with its wheel sunk into the flour that a young scientist had sprinkled in the sandpit to approximate the texture of the soil.

There is embodied knowledge on Helen too, which also expresses local social relations. People often explained their instruments to me using their hands: the ultraviolet PI set his hands apart a few inches with palms and fingers rigid and flat to demonstrate that instrument's field of view or the rotation of different focal planes, while a thermal technician described a similar feel for his instrument's field of view with his hands. They also had an embodied sensibility to their instrument's requirements, range of motion, or sensitivities. When Rod, the member of two instrument teams, asserted that he had to fight with himself to establish observational priorities to group laughter, George followed up with "What did that look like?" Rod responded by associating a feeling for each instrument with feelings in his body:

> For example, with [the cameras], I'm more sensitive to the sun than I am with [the spectrometer], so I had to decide whether to make it [an easy] rider or a [rough] rider. So at the moment, I'm making it [an easy (i.e., to ease the requirement that the camera look at the sun)].

Rod's decisions about the spacecraft's potential activity rest on his camera-like sun sensitivity to where the instrument will be facing, as well as his knowledge of where each instrument is located on the spacecraft with its associated field of view. He describes an embodied process of understanding how the spacecraft will have to juggle those competing requirements, leading to his decision for a particular class of observational rider.[11] His elision of self with each instrument also contributes to his instrumental metonymy.

Even as Helen scientists had embodied knowledge of their instruments, they frequently had to reconcile different instrumental tendencies and

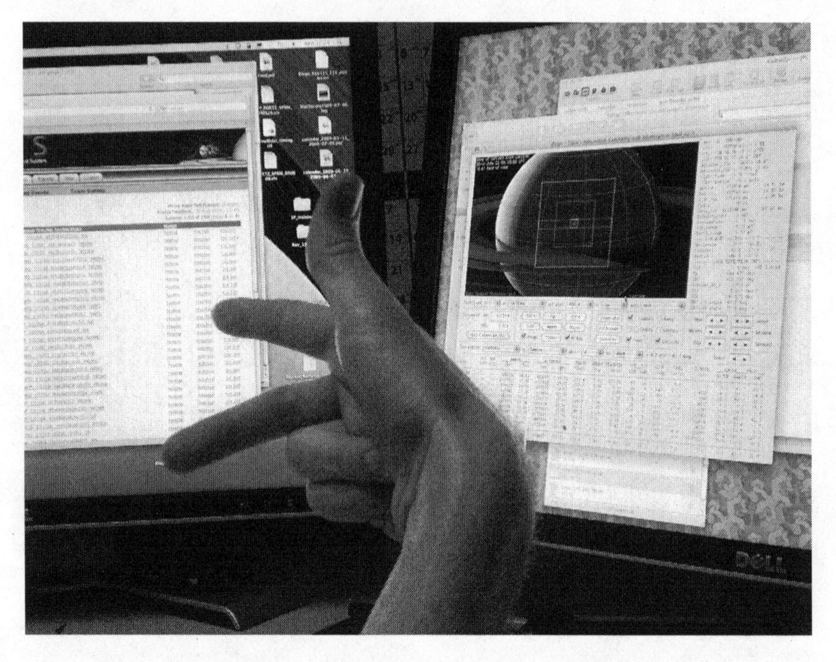

13. Noel enacts the spacecraft with the right-hand rule. Author's photo.

sensibilities without sharing that same intimacy of understanding between teams. To do this, they used a shared way of approximating the spacecraft's pointing—its attitude, twists, and turns in space—by using the right-hand rule, an instructional technique used in physics classes. Holding their right hand out at arm's length from their bodies, the elbow slightly cocked to best observe their hand's full range of motion, they extended thumb, index, and middle finger to create the three spacecraft axes of x, y, and z and twisted from the elbow accordingly (figure 13). Team members used this as a sort of shorthand much like Paris scientists used their own arms and hands to embody their robots: as an effective way to coordinate and calculate between sociotechnical relations on Earth and in space at the same time. On Helen, however, these techniques reflected the principles of integration.

Planning for Helen requires working with pointing parameters. To develop an observational request, the team must command the spacecraft to point one part of its body toward an object in the Saturn system, and another part along a different axis must point elsewhere to establish the craft's altitude. One engineer even explained to me that her job was to craft "a whole series of commands that say, 'Point here, point here, point here, point here.'" But there are considerations associated with pointing and turning, as certain instruments will be destroyed if they look at the sun for a long period. So to maneuver between instrumental observations, the spacecraft must not only point correctly—to positive or negative x, y, or z—but also rotate in such a way as to avoid harm. The way that Helenites embody their spacecraft with the right-hand rule, then, is a way for scientists to work out the potential requirements and conflicts inherent in their observational requests while avoiding any potential dangers before passing them along to the engineering team.

I recorded an example during the Titan jumpstart, when the group had to figure out how the spacecraft would accommodate a transition between the camera and the spectrometer. Talk, software, and gesture came together in this exchange to establish how the spacecraft might accommodate these requests before they could approve the proposed plan (figure 14):

ELEANOR (CAMERAS): The problem is that we don't need the turn like that after, we need it before closest approach—

LYNN (SCIENCE PLANNER): Yeah we need it on closest approach to get—

FINN (THERMAL): You wanna point the radiators . . . [2 sec]. What about pos[itive] x perpendicular to Titan?

ELEANOR: We need to find out when this turn has to be moved.

JEROME (INFRARED): [*Sticks his arm out with the right-hand rule, index finger pointed down*] Is it possible to have minus y to Titan [*pushes his hand along parallel to the floor*] and turn [*twists at the wrist*] 180 degrees? [3 sec] [*Opens both arms up and shrugs shoulders*]

ELEANOR: [2 sec] Yea.

BETTY: Can we put the radiators up, Howard [MIMI rep]?

ELEANOR: We can do that, but [*puts her hand up in an open cupped hand*] you can't image by doing it [*twists at the wrist in front of her face*] cos you—

JEROME: Ah! [*Sweeps both hands down in front of his body, in a "throw it away" gesture*]

BETTY: Can we put the radiator straight up?

FINN: Yeah, that's what I'm talking about—

ELEANOR: Or wait maybe [*points at Jerome*] you guys *can*, depending on the turn, how big the turn is [*twists wrist*]. But your integrations may be too big—

JEROME: Maybe put [*right-hand rule, index finger down, left hand cupped around right index finger*] minus y to—

HOWARD (PHYSICS): [*Looking at numbers on his screen*] If you put pos x to [the sun], then you will still go through a quick turn that puts the sunlight on

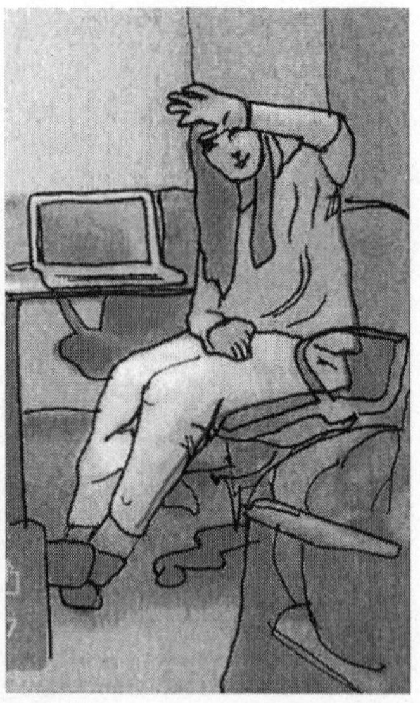

14. Jerome (*left*) and Eleanor (*right*), negotiating observations between their instruments, enact the spacecraft with the right-hand rule. Author's photo.

the radiators, but it will not stay there very long. I don't know what happens to the spacecraft, but it still looks like you're flipping the spacecraft around.

BETTY: I think we need to look at this one in more detail, don't you?

LYNN: Another pirouette.

Eleanor and Jerome both use their right hands to gesture at each other in order to coordinate between their proposed observations. This includes a turn before the approach to Titan, minding not to point the radiators at the sun, not to rotate along a certain axis while taking photographs, and not to rotate in such a way as to restrict the infrared spectrometer. Eleanor forms her hand into an open cup to embody the camera she represents at the meeting, and as she twists her arm, she demonstrates how her instrument is unable to take pictures while rotating 180 degrees. Terms like "you guys" invoke instrumental metonymy. Jerome responds by showing how minus y (where both their instruments are located on the spacecraft) could be a stable pointing position toward Titan with the spacecraft rotating around it as a fulcrum. He gestures this by grasping his right index finger with his left hand to show that that point would be stable while twisting his right wrist as much as possible. Note how this exchange is essential to the process of integration. Each instrument is represented, each describes their needs, and the coordinating device of the right-hand-rule-as-spacecraft enables scientists to work together to align these needs.

Scientists also possess external spacecraft models that they use for calculating pointing. Thermal expert Edward has a cardboard cube on his desk several inches square, the sides labeled positive and negative x, y, and z, on which he drew Helen's instruments facing in different directions. Picking it up, he zoomed it out to arm's length, moving it toward or away from the light in his office to show me how the spacecraft would point toward Enceladus or the sun. A few blocks away at a different institution, an empty plastic juice container on a desk represents the spacecraft's body, with stir sticks punched through it to represent its axes.[12] At Spacelabs, Gwen has a flat circular cardboard disk about a foot in diameter with a hole in the center for a pencil to slide through as the z axis, with x, y, and instrument names labeled around the edge. She rotated the disk around the pen to show how they would manage primary observations by facing particular instruments at Titan while keeping the radiators out of the sun and allowing the particles instruments to collect data too. The Titan target team apparently used this model to plan the entire primary mission, so "this little cardboard spacecraft has done a lot of duty!" The project even issued everyone a coffee cup depicting where the instruments are placed on the spacecraft bus. With characteristic Helen humor, George reminded me that it was useful for *either* representing the spacecraft *or* drinking coffee but not both at the same time—unless you were prepared to spill hot liquid all over your keyboard.

These cardboard models, plastic bottles, and mugs are working objects. Along with the right-hand rule and the Paris dance, they are embodied ways of working out, through attenuation to robotic pointing or mobility, exactly how to produce a sequence of activities on board a distant machine. They thereby produce a "body-in-interaction" (Alač 2009): a way for scientists to intuitively understand, account for, and make decisions about their distant probes. But even as they are used to express technical capabilities and mechanical requirements, they infuse this sensibility with a particular form of social relations between team members on Earth. Paris team members fully embody their robots in such a way as to cement their collectivist organizational order, the robots' totemic qualities expressing the extension of charisma to their machines. Meanwhile, Helenites maintain the instrumental metonymy and distinctions of their matrix, holding their spacecraft away from their bodies to deploy it like a sociotechnical calculator. Each spacecraft's "body-in-interaction" is consistent with their organizational distinctions and social relations on Earth.

Betty drew this contrast in technosolidarity relations as she compared work on Helen to the Mars community's strange "commitment" to their machine:

> The thing that happens on Helen is that the emotional commitment isn't to the machine. The emotional commitment is to the project so, for example, it's not the same thing. The machine doesn't represent what we're doing. To some people, the best representation of the work of Helen is the scientific publications. To some people, the best representation of the work of Helen is the wonderful cooperation between the Europeans and the United States. . . . To some people, [it's] the fact that the spacecraft is still healthy at this stage in the game.

As Betty sees it, the Helen mission does not mean one thing: it means many different things to different people, depending on their organizational hats. Maintaining this sense of different needs and requirements throughout planning is an important feature of achieving fair, successful integration, and is projected onto the many symbolic meanings of the machine. In this way, forms of social organization on Earth—the autonomous groups on Helen or the collective on Paris—are written onto the interactional bodies of the two spacecraft.

An implication of these differing technosolidarity relations has to do with care and responsibility for the craft. On Paris, technosolidarity relations are oriented toward collective responsibility and care: the notion that "*ev-er-y-bo-dy*" (in the PI's words) is responsible for the success of the robot. Meanwhile, the interactional body of the Helen spacecraft reproduces technosolidarity relations consistent with the matrix's sensibility toward fairness in integration. This may be why the only team member I saw who had a "Helen dance" was Connie, the mission's chief engineer, who alone threw her arms wide to bring the whole spacecraft into her body in gesture. This does not mean that team members placed their spacecraft's health and longevity at risk. Helen scientists always made sure to point radiators and instruments away from the sun and to preserve the wheels or save fuel, and they were concerned when a fault occurred or when the spacecraft went into safe mode and lost their data or observational opportunity. But the differences in "degree and type of solidarity" that Durkheim points to remained especially visible in moments of breakdown. When Paris's thermal detector failed, team members repeatedly wailed, "My kingdom for a working [thermal spectrometer]!" Unable to conceptualize the robotic body without its infrared eyes, they continued to use the broken equipment in the hopes that someday they would find an algorithm to retroactively calibrate its data. Meanwhile, when Helen's plasma detector encountered a fault and the project decided to turn it off, the only ones at a loss were those on the detector team itself. That team's observational time was redistributed to the other instruments, and the Virtual Choir sang about the situation in a Taylor Swift parody: "Now we [the project] don't let them [plasma] play, much to their dismay."

Coda: PAM's Immersive View

It could be that these observations are simply due to different technical layouts in the spacecraft involved, not their organizational orders at all. After all, Helenites frequently explained to me that they had a different relationship with their spacecraft because it was so large and flew through space, as opposed to the Paris robots, which were more "human scale" and "cute" and rolled on the ground. Of course, from the Paris point of view, the robots are not humanlike and their bodies are awkward to approximate: the very factors that their gestures attempt to mediate. A useful test of the theory was a visit to the PAM working group: those Helen instruments that examine fields and particles and maintain a more Paris-like organizational orientation.

As I have already noted, consistent with prior studies of particle physics collaborations *and* with the autonomy of different subgroups on Helen, PAM has its own collaborative structure. At a PAM science meeting in Annapolis in 2012, I came to realize just how different their view of Helen and of the Saturn system is from that of the other working groups. Immersed as their instruments are in Saturn's plasma and magnetosphere, every physics reading that Helen takes is a record of where it is *right now*. These "in situ" instruments therefore record the spacecraft's own direct encounter with space. No wonder, then, that PAM members are intensely, acutely aware of the spacecraft's location as a question of time, date, and orientation in space. Of course, remote sensing scientists working with the cameras and spectrometers are concerned for accurate views of their targets, including how different incidence angles for observing these targets will affect light shadowing or radar backscatter. But over three days of PAM presentations, I watched scientists credit their observations to highly specific days of the year and local Saturn times alongside flybys, segments, and spacecraft position. They were also fluent with spacecraft locational "kernel" files from the navigation office, which they used to know *exactly where Helen was* at every point in time.

Think of outer space like an ocean, constantly moving, with waves on the surface and currents below, and some underlying principles directing its ebb and flow. The plasma and magnetic field environment around Saturn are like this ocean: endlessly in flux, with particles of all kinds hitting magnetic field lines and collecting or losing charge as they move. Like the challenges of gathering data in a rushing river or monitoring global climate (Edwards 2010; Ribes and Jackson 2013), Helen could fly through exactly the same coordinates at two different times and pick up two different readings. This means that it is not clear which observed changes are regular and which are irregular. It also means that the fields and particles readings that the spacecraft takes are not building up a single picture, as one might attempt to map Titan with radar coverage, for instance. Instead, they must up readings of continual fluctuations over time.

Like Paris team members, I noted how members of the PAM team used their hands and bodies to inhabit the whole spacecraft as their sensory apparatus, gesturing to show the phenomena they are flying through and detecting. For instance, an instrument's deputy PI, Howard, and his postdoc sat together at a break during the meeting, looking at a plot of energetic ions detected by Howard's team's instrument. He pointed to his screen on occasion to bring out certain features in the plot (as in Alač 2011). But then he used his hands to "be" the electrons or the field-aligned currents and put his hand in the right-hand rule position to "be" the spacecraft, showing the latter's position and orientation with respect to the currents' location and direction. A few minutes later, a scientist walked up to his colleague at my table and opened his discussion with "Helen was over *here* and Saturn was over *here*," using his hands to reflect the relative positions of the spacecraft and planet in order to understand the data he was seeing from the spacecraft's position at a particular moment in space and time. This gestural version of the spacecraft, the hand with its outstretched three fingers, was not body-work to assist in negotiating twists and turns used to plan data acquisition or handoffs between instruments. Instead, it was a mechanism through which PAM team members achieved immersion in the dynamic Saturn system, connection with the spacecraft, and their instrument suites' sensibility to the shifting environment outside.[13]

The experience reminded me of how anthropologist of science Stefan Helmreich uses the concept of transduction, a term deployed by cultural studies scholar Adrian MacKenzie to describe the embodied relationships between humans and machines (Mackenzie 2006). In his ethnographic writings about the Alvin submarine, Helmreich describes the sensation of immersion in the sub, the water, and the cultural environment: "a sense of being at once emplaced in space and at times, porously continuous with it" (Helmreich 2010, 10; 2009). He attributes this quality of immersion to transduction, the work of creating an immersive environment that in this case brings elements of the alien ocean outside into the pings and sounds that are sensible to the observers in the Alvin sub, such that the scientists on board have the collective sensation of "merging with our data" (Helmreich 2007). Helen is also immersed and turning in space while particles swirl and churn around it bolstered by the planet's shifting magnetic field. Even the spacecraft's own electromagnetic signature affects the system: it must be calibrated out of the data or shaken off the spacecraft through rolls that detect its own signal amid the flow of materials outside. Sitting in the PAM meeting room, I suddenly became aware of a vivid, bodily sensibility to the transductive work of Helen, its sensors, and its science team, an immersive sense of rolling in the shifting magnetic environment around Saturn. In three years of attending Helen meetings across myriad teams and groups, I had never had this sensation before. Consistent with the notion of technosolidarity, Helen's interactional body *can* be experienced holistically—as Durkheim would predict—amid a group with a more collectivist social order.

Technosolidarities in Action

Far from challenging organizational cultures and structures, software and hardware on these spacecraft teams are brought into alignment with these elements and reinforce them continually. Software tools facilitate integration or consensus work, such that images with annotations on Paris or colored spreadsheets on Helen produce both spacecraft plans and organizational order, simultaneously. Bespoke tools are used within planning groups to produce either consensus or integration accordingly. Even the spacecraft's own body is made present through elaborate gestures that either assert a collective connection with a charismatic robot on the one hand or implement a sophisticated sociotechnical calculator on the other.

An important implication of this chapter is that technologies, as they are organizationally experienced and enrolled, play a role in asserting solidarity relations—that is, social relationships on Earth are held together through the projected, interactional body of the hardware itself imagined through software and gesture. On Helen, these ways of representing the spacecraft at a distance reflect the division of responsibilities among team members and the wearing of multiple hats. Individuals have an embodied sensibility for their part of the machine in alignment with their organizational affiliations.

Science and engineering team members on Paris alike, however, display an embodied connection with their robot that exceeds rationality. Totemic social relations make for continued connection between team members, their colleagues, and their robot in space. In this way, organizational elements fuse with the body of the robot, making its interactional body a site that reproduces Earth-bound social distinctions in space.

Importantly, this fusion is naturalized by team members. Members' accounts described these mechanisms in matter-of-fact terms. For them, the spacecraft drives the operational concerns and the organization, not the other way around. In their accounts, this is simply how a flyby probe or a roving vehicle *works*; the other team must work differently because their spacecraft is different. Hence the spacecraft's interactional body—which reflects group processes—is invoked to mediate the very same social relations that frame its interpretation. In other words, what the *interactional* body of the craft could or could not do was called upon to resolve the question of how the spacecraft would or would not operate. This practice further naturalizes the local way of doing science, spacecraft management, and social relations on each team. I will return to this point in the last chapter of this book.

For now, this analysis demonstrates how the specific intimate interconnection between social relations and technical interactions reproduces the organization.[14] After all, when Hẹlenites use PowerPoint, as they do to make science presentations, they do not experience the "collective effervescence" (Collins 2004; Durkheim 1912) of the TacPlan meeting—just as Paris scientists, when they load Excel, do not debate how to color in their charts. The PAM group did not use the right-hand rule to manage trade-offs but developed an immersive sensibility to the spacecraft akin to Paris body-work. Local ways of working with software and hardware are consistent with each mission's organizational norms, interactional regimes, and associated social relations.

As such, instruments, software, and hardware play a prime role in *organized science* too. As the stuff with which the work gets done, the focal point of collective endeavor, or the object under everyone's care, such technologies are roped into local divisions of labor and resource allocation. The ways in which these technologies are locally configured and enacted are consistent with organizational distinctions. They also enact different forms of technosolidarity: team members' relationships with each other and with their machines as mediated through and figured by their technical work. If anything, then, these technologies do not undermine or determine the practices associated with mission work. Instead, they harden, "make durable" (Latour

1991), or otherwise naturalize organizational relations among the mission teams on Earth. For teammates, these practices render the essential role of organizational elements invisible in their technologies and vice versa. The lens of *organized science* reveals instead how the technologies of scientific work, labor, and care essentialize organizational distinctions and solidarities, building them into the collaboration's everyday tools themselves.

EIGHT

The Data

The Paris PI and camera lead still recount the astonished looks on the engineers' faces at Spacelabs when the two of them asked how to make their images instantly available to the public. Usually these engineers had to promise to planetary scientists that their mission's freshest data would be protected from science journalists before a discovery announcement could be made, from academic journals complicit with embargoes, and from competing scientists in the field. Such problems typically plagued visual datasets such as photographs, as outside scientists and amateurs alike readily made claims about pictures based simply on what they could see. Further delight ensued at NASA Headquarters when the two scientists outlined their open image data policy. This would build on the success of a prior Mars mission, whose images enjoyed tremendous public audiences on the early internet. The mission would also commit to the shortest period of data calibration and validation yet, releasing other data to planetary scientists through the agency's data portal, the Planetary Data System (PDS), every three months.

This promise of "open data" is reminiscent of the rise of the open source community in programming, with its values of free information exchange. Scientists deployed the argument that since the instruments were taxpayer funded, it was only right that taxpayers could see for themselves the fruits of their dollars. This contrasted with the prior, deeply held commitment that PIs and their teams worked hard for years to be guaranteed the datasets that would fuel their laboratory work, feed their graduate students, buffer their publication lists, and establish their careers.[1] More important, NASA Headquarters was always looking for ways to engage the public and demonstrate the value of their missions to Congress. They were so pleased by the Paris team's example that they began demanding that other ongoing missions follow suit—including Helen. But data already moved in very different ways on the Helen mission. A series of arguments ensued that would haunt the team for years to come.

To understand how data sharing practices can be so disparate on both missions despite their common scientific field requires that we leave aside the cultural narrative from the turn of the twenty-first century that "information wants to be free." It also requires us to look beyond the local rhetoric that justifies data sharing decisions due to public funding; after all, military technologies are taxpayer funded but are not open access projects. Finally, because of divergent practices, we cannot reach to disciplinary, institutional, or field-wide explanations for data management. Instead, I will advance a local organizational explanation for these divergent "evidential cultures" (Collins 1998), showing that how data circulates is consistent with and reproduces each organization's orientation toward integration or consensus. In this view, data are the stuff with which organizational relationships are produced, maintained, and circumscribed, and the challenges affecting mission participation are mitigated. Participating in locally appropriate data sharing regimes is participating in the organization's local social order.[2]

This explanation builds upon prior studies of biologists, astronomers, and natural philosophers that show how scientists circulate data, specimens, or instruments in accordance with local norms. Sociologist of science Harry Collins ascribes differences in data exchange as evidence of a laboratory's open or closed "evidential culture," dependent upon local institutional reward structures (Collins 1998). Historians of science are likely to describe such cultures as situated within a "moral economy" of science, one in which scientific objects from fruit flies to telescope time circulate in concordant ways with local understandings of what makes a good scientist (Daston 1995; Kohler 1994; McCray 2000). To participate in a scientific moral economy or evidential culture, scientists must build, negotiate, and have access to infrastructures that enable circulation even as they also support their local community of practice (Birnholtz and Bietz 2003). Historian of computing Paul Edwards calls such systems knowledge infrastructures: "robust networks of people, artifacts, and institutions that produce, exchange, and police knowledge about the natural world" (Edwards 2010). Knowledge infrastructures may include refrigeration trucks and units for cold storage and sample handling (Radin 2017), vast server farms for data crunching, or delicate equipment, file cabinets, and databases (Borgman et al. 2014).

In part due to these different moral economies, information infrastructures, and institutional environments, sharing cultures vary widely across scientific disciplines.[3] Physicists share their data broadly, sporting publications with hundreds of authors, while biologists are more selective about with whom, when, and why they share specimens; meanwhile, astronomers place enormous datasets online, using crowd-sourced manpower to locate galaxies but relying upon security through obscurity to protect ongoing investigations. These differences across fields are well known, but the planetary science example demonstrates that there are differences *within* single fields as well. Paris team members share raw data internally across all members of their mission team, post calibrated versions to the web every three months, and see their data as an inherently open resource. Meanwhile, Helenites share data with members of their instrument teams, post calibrated data online a year after its acquisition (raw images are released quickly), and place conditions upon how and when calibrated data may travel further afield.

This should by now be unsurprising as the practices and interactions on these missions diverge in so many ways. But this chapter is not a moral tale. Instead, I will show that data moves in ways concordant not with lofty idealisms but with local norms that underline social relationships—and that the relationships at stake behind these infrastructural transactions and evidential cultures are *organizational* ones. In other words, data follows pathways consistent with organizational orders and attendant social relations. Data mobility is therefore both contingent on the organization's social order and reproduces that same order through its circulation at the same time.[4] For this reason, despite its use among my field site participants, I avoid the language of "data sharing" and prefer instead the less morally tinged language of "data circulation." Moving away from ideological prescriptions about what data *should* do helps us instead see what data *does* do and how its circulation participates in the organization's social order.[5]

The Values of Data

What makes data so valuable? As the stuff of doing science, it is the fuel of scientific discoveries, the basis of analysis and papers, and at the core of scientific careers. Despite the popular "Eureka!" notion of discovery, scientists often know what the outcome of an experiment should be in order to make a discovery in the first place, a phenomenon Harry Collins calls the "experimenter's regress" (Collins 1985, 2004). Even the long-sought Higgs Boson only appeared upon construction of the Large Hadron Collider and the experiment designed to find it! Discoveries are also often collective. To whom should the attribution of credit go in an experiment where tens or even hundreds of individuals are involved? These factors point to an essential organizational element to discovery. What a team discovers with their group-built instrument and how they manage the credit and argue legitimacy for these discoveries is yet another place where we can glimpse the scientific organization's social relations at work.

In planetary science, this is often attributed to a question of expectation and planning. Because missions require so many people, so much time, and so many social resources, the very observations that are planned are those that the team knows (or hopes) will lead to a discovery or further elucidation of the object in question. In some cases, there is simply an unexpected element of obvious importance that stops the team in their tracks; on the Paris team, this potential moment is referred to as finding a "dinosaur bone" (something the team has never discovered). But such moments are rare. Instead, discoveries frequently follow long periods of concerted, collective work on behalf of many team members to go after an expected finding. Even the extraordinary finding of plumes on Enceladus, the legendary "discovery story" of a finding that required turning the Helen spacecraft around, although surprising in its magnitude, was not entirely unexpected. A 1981 *National Geographic* story about an earlier Saturn flyby (involving many eventual members of the Helen team) described the hypothesis that Enceladus was being "recoated from its interior" due to tidal forces and fueling the E-ring, complete with an artist's depiction of geysers on the icy surface (Gore 1981, 15, 24, 30–31). Helen proved this hypothesis to be correct a quarter-century later.

The canonical story among planetary scientists of this often-misunderstood aspect of discovery—that is, the role of expectation and hypothesis-testing in planning—is that of the volcanoes on Jupiter's moon Io. A camera operator was the first to observe the images as they came down from a visiting spacecraft and suspected she might be seeing eruptions on the surface. In public press articles, she was readily credited as the volcanoes' discoverer. But team members explained that they already suspected that the eruptions would be there for the seeing. That is why they asked the spacecraft to observe Io in precisely that way at exactly that time, with the limb of the moon lit just so to reveal the volcanic plumes, if there were any to be seen. Not only had they planned the observation; they fought hard for the time to observe Io using their instrument over the contestations of other teams. In their accounts, the operator was simply the first to confirm what the scientists expected to see in the first place.

There is a related concern about discovery: that of being "scooped" by another member of the team or by an outsider. Scientists also invoked the misplaced discovery narrative in the story of a graduate student who, excited by a photograph of Jupiter released on the front page of the *New York Times*, penned a scientific article about what he saw in the image. The dual violation of writing about a press release image (one that was not yet calibrated and released to the science community) while also not being a member of the instrument team was too great: his immediate career prospects plummeted. Several years later, two female graduate students who were not on a mission team penned a paper about a spacecraft image of one of Jupiter's moons. When the instrument team in question protested, a team member who was in their academic department claimed them as his advisees to protect their paper and their reputations. And a Helen PAM member told me that a prior magnetometer PI had jealously protected his data, to the detriment of the other physics instruments, as an explanation for the Helen group's dedication to data circulation. It was also equally likely that a mission team member could step on an existing investigation on a different side of the mission of which they had no prior knowledge. This was especially possible on large missions such as Helen, where the distance between subsidiary collaborations can be so large that there was little visibility as to what others are working on amid the ranks.

Such problems especially plagued Helen's camera instrument. As the photographic data was mandated by Headquarters to go online quickly after acquisition, this left them open to interpretation by Helenites, external scientists, and amateurs alike, to the distress of its team leader, Isabelle. Isabelle understood the rules of the road as requiring any Helen scientist who wished to use camera imagery to align, situate, or otherwise understand their data to seek a collaborative agreement with the imaging team. She therefore admonished several scientists on Helen for publishing findings using camera data without engaging her or her team. The majority of Helenites I spoke to, however, told me that once the data was in the official repository, they did not need such an agreement to use it and were astonished at Isabelle's admonitions. Amateurs were also in the mix. Two bloggers told me that they confronted Isabelle at a conference with images freshly downloaded from the spacecraft while she had been on a plane making her way to the event, claiming that they had "discovered" ring spokes or moonlets. They were offended when Isabelle flatly denied their discovery claims. After all, her team had fought for the spacecraft time, location, and pointing to command the observation—precisely because they suspected such features were there to be seen in the first place.

In another case that spilled over onto Wikipedia, some team members accredited the sighting of new moonlets to a British scientist who was an associate of the imaging team. Isabelle jumped onto Wikipedia to correct the entry. The credit should go to the imaging team, she claimed. A user pointed to a press release from the scientist's institution highlighting his role in spotting the moons to promote their university, but she argued "press releases are . . . *not* an official statement of discovery," as discovery statements had to be approved by the International Astronomical Union (IAU).[6] It was unreasonable to give an operator credit for discovery, she explained to me later, "when he's discovering it in images that someone else planned and we all knew as a team that was part of our goal." The changes were made, unmade, and remade again until the page was locked by Wikipedia's editors. The whole situation was evidence to Isabelle that "people should not really pay attention to Wikipedia."[7]

I use these examples to show how groups must navigate consistent tensions in data sharing that relate to producing legitimate and community-appropriate discovery claims. For this reason, mission teams enact a local regime of data management to mitigate these concerns, to produce legitimate discovery claims, and to appropriate social relations at the same time—a combination of both formal rules of internal circulation and informal knowledge about how data may travel, both of which uphold organizational relationships specific to the local scientific collaboration.

Internal Circulation: Following the Rules of the Road

To mitigate these tensions, each group pens a series of formal rules to govern data management and publication rights, laid out as a contractual agreement among team members. This document is called "the rules of the road." This is a local, incontrovertible agreement that lays out exactly how data ought to be exchanged and delivered among team members, as well as to the centralized data repository operated by NASA, the Planetary Data System (PDS). "Rules of the road" documents typically indicate whom team members may share with and when and specific roles associated with data access. They stipulate an initial period when the data belongs to the PIs and team members with exclusive publication rights. The aim of this period is to give the instrument leaders time to correct or standardize the data and to protect mission scientists from being "scooped" on their results during the time when they are busy caring and planning for an instrument. Many members of the field impressed upon me that because PIs invest considerable time and resources over their careers to design, build, and manage an instrument, it was only fair that they should reap the first rewards of those resources in publishing about precious information about distant worlds. Another common narrative about this period was that junior and early-career scientists making their way up the ranks benefitted from the freedom to conduct their scientific work without competition in order to build their reputations and careers. The period was therefore variously referred to as the "validation period" by those keen to show they were willing to share or the "proprietary period" by those concerned about data protection.[8] After this point, teams must deliver data to the community at large via delivery to the Planetary Data System, which houses all NASA-funded mission data. The timing and requirements associated with this data release vary.

On Helen, teams may circulate data only among their own instrument team rosters during the initial months after its acquisition.[9] Interdisciplinary scientists may use data from more than one instrument but need to get a letter of approval from the PI or team leader responsible for that instrument to approve data use during that time. At the end of this time, all teams are meant to release their data to the PDS in validated packages to satisfy external community members, although in practice, some teams released on schedule and others took up to a year or more. This is consistent with the organizational norms of the mission: teams operate autonomously, integrating between them through sanctioned channels for multi-instrumental work, with PIs, team leaders, and interdisciplinary scientists in positions of relative authority. Indeed, so consistent are Helen's data sharing norms with their organizational order that during my time on the mission there were many different rules of the road in use. Some instrument teams had their own policies for how publication credit should be assigned or how investigations were to be divided up among members. The PAM group had their own rules of the road to govern their multi-instrument collaboration, permitting downgraded versions of data to be made available to other PAM collaborators to use in their investigations. It was only after I left the mission that the PAM rules of the road for data sharing were expanded mission-wide, after several years of work.

Just as Helen's rules of the road support instrumental autonomy and matrixed hierarchies, practices of enforced sharing, copublishing, and collectivity are enshrined in the Paris rules of the road. The rules describe mission roles such as PI, coinvestigator, student collaborator, or participating scientist, but then state emphatically that "there is no distinction" between any member of these categories. Raw data for all instruments belongs to all team members during their rapid validation period of only three months, with images released immediately to the internet. Further, the rules state that "any Paris data product . . . shall be made available upon request to any Paris team member or collaborator" and that publication and coauthorship on papers are "open to all team members and collaborators" should they ask, be invited, or "make a substantive contribution" to the paper, with "lead authorship . . . established by consensus." These data sharing and publication practices reinforce Paris's collectivism: no hierarchical distinction, the requirement for awareness and visibility team-wide for any ongoing investigation, and openness to all coauthors, from PI to graduate student.[10] As much as these rules demonstrate an approach to an unabashedly "open" data policy internally, then, they do so insofar as this is consistent with the team's organizational orientation. Another way of saying this is that data are a shared resource on Paris not because they are data but because *everything*—instruments, robots, and scientific expertise—is a shared resource. Importantly, this does not apply to external circulation: the Paris rules of the road end with a complete list of all mission team participants with whom such reciprocal obligations are expected. A clear line between mission participants and outsiders facilitates this pattern of resource circulation, as is the case in other collectives.

By describing how data may circulate on and outside the team, each team's rules of the road participate in their local organizational order. Like the missions' heritage lines, these circulation patterns were refracted through the changes to mission structures in the 1990s and the eventual distinction between Outer Planet and Mars missions. A mid-1970s flagship rules of the road underlined how data would be drafted, shepherded, and shared under that team's hierarchical social relations.[11] PIs and certain types of scientists were the sole possible sponsors for research projects, and while using unpublished data required the permission of a PI who "may refuse coauthorship but not the use of his data," official instrument team members had the right of first refusal over topics involving their datasets. But organizational and technical changes struck in the mid-1990s, with divergent effects. A follow-up flagship in the late 1990s added a proprietary period, gave the PIs the right to restrict external data circulation, and required instrument PI approval for any scientific analysis involving an instrument's dataset, including keeping "records of such agreements." *After* the proprietary period had ended, scientists were allowed to use data already in the PDS but were still "encouraged to seek the participation of the producer of the data to ensure the proper use and application of the data."

These two missions featured hierarchically organized teams, even if each awarded different responsibilities to their PIs for managing data flow. The tightened restrictions visible in the latter's rules were potentially due to shifts in the contexture (Lynch 1991) of planetary science in the late 1990s. After all, early missions used vidicon tubes to record images, with data played back to be stored on physical tapes. Once, touring Spacelab's image processing facility, I met a lab technician who described hand-delivering spacecraft images to PIs or to the press fresh from their darkroom. There was no Planetary Data System, but a series of libraries placed at key institutions for planetary science work carried official silver plate images and black and white graphs for reference. Beginning in the 1990s, spacecraft flew digital cameras with data computationally accessed, copied, and transferred online. Not only was exchange easier with digital files, but so was the ability to combine readings into something greater than the sum of their parts—and the inability to limit such capacities to the hierarchical roles of PIs and interdisciplinary scientists, upholding the mission's social order. This may have led to stricter rules to maintain the hierarchical structure.

The project scientist and the scientist in charge of the recent flagship's Data Working Group, Jake and Joseph, eventually joined Helen, and these rules and their associated norms for data circulation migrated with them along with the outer planets community. But missions within the Mars Program took a different tack, deploying digital data circulation among their nascent program in the 1990s to create a connected community. The scientist who penned the rules of the road for one of the program's first orbiters eradicated the "proprietary" period, replacing it with a "validation period" for cleaning and calibrating data to release to the PDS in six-month chunks. Instrument team leaders were exhorted to "make their own data available earlier" or even come up with "a number of versions available on a timelier basis than the typical six-month period." This scientist then carried these rules forward to one of the first missions operated under the "faster, better, cheaper" model on Mars, where with the spread of the commercial internet at the time, "a subset of the data are made available [online] in near real time." This cleared the way for live internet release of photographs. This scientist migrated with others to work on Paris, where their close social relations from the smaller mission repeated and the new Discovery-style organizational form took hold. Thus these two styles of data circulation are not due to any inherent tensions associated with digital tools nor with any inherent predilection toward openness or closed cultures. Rather, they traveled with individuals as standard practice in the 1970s and were refracted through the organizational and sociotechnical changes of the 1990s.

External Circulation and Internal Tensions

As data moves from being a relational element among mission members to circulating among a broader community, it must transform to somehow interlock or connect with external knowledge infrastructures despite local idiosyncrasies (Baker and Yarmey 2009). A series of tensions plague the mobility of data from a currency of relational management among team members to a currency of relational management with team outsiders. These tensions are in some cases outlined as imperatives in the rules of the road. But they are also visible as scientists describe certain elements of data work that make it unruly for archiving despite the importance of external social relations to internal mission politics.

This became explicit on Helen at the PSG in January 2009 when a stately senior scientist named Patricia—who had succeeded Jake as project scientist of Helen's flagship predecessor—delivered a hotly debated presentation about improving data delivery to NASA's Planetary Data System (PDS). Instrument teams were already "compliant" with their commitments to NASA, but scientists in the broader community claimed that delivered data suffered from poor "ease of use." She therefore asked "how we can, as a Helen community, be helping in terms of data usability: first in terms of what we can do essentially for free and then what might be harder and demand NASA funds . . . so that any scientist years from now can retrieve the data and use it without any previous software." The responses to her plea for "dialogue" illuminated tensions associated with data mobility. As the scientists described properties of their data that they considered to be intrinsic, they simultaneously described challenges with its social transformation: from a local resource that circumscribes organizational relationships, into something that must circulate externally to navigate other relationships entirely.

For instance, because data are only made visible thanks to a suite of software and hardware tools—and because specialist datasets often rely on contingent, local systems and human relationships—ensuring the viability and visibility of these data into the future was a Sisyphean task. PIs resisted the requirement to deliver specialist software to the system that might assist for good reasons. One PI was juggling four competing calibration scripts, each belonging to a different subgroup in his team: Which one should go online as the definitive version? Another scientist insisted that their full calibration pipeline was already in the PDS but due to software upgrades, "five years from now, it may or may not work." Human relationships mattered too, especially the tacit knowledge associated with working with an instrument team and understanding the instrument's analytical techniques over a long period of time (Collins 1985). Another PI explained how his team went through considerable effort to write an instruction manual and train individuals, but "it still took us a long time to get our own team to use the data." His colleague described how "it's a mentor-student relationship: we know our datasets better than the outside community"; this made it difficult to teach someone how to use the data when they were not available in person. Still another claimed, "It is dependent on my team. People who don't know enough about the data are not likely to succeed." Despite best efforts, "it's impossible for the noninsider to either access or use the data."

These concerns and criticisms are well known to those who study large-scale scientific data systems and repositories. But note how the organizational overtones in such statements, such as insider-outsider status, boundary work, and team autonomy, become conflated with things that data *just can or cannot* be expected to do. For instance, being a member of a team is coextensive with an insider's knowledge of how to use the instrument's data, and data familiarity is something that delineates relationships among team members. Indeed, organizational relationships were quickly conflated with expertise. One PI asked Patricia why NASA didn't just suggest to the community members that "the best way to work with Helen data is to establish a contact with the individual Helen teams." This would also uphold the mission's local norm of autonomy (Patricia quipped, "They couldn't find your phone number"). Organizational relationships also bracketed outsider expertise. When one scientist complained that "I don't know exactly how to tell the nonexpert" how to use their data, Patricia countered that this didn't mean it should be "easy" to do but rather a question of "communicat[ing] to the nonexpert user a little bit better what they need to do." At this, a PI exploded, "I don't think that the PDS or our contracts have ever asked us to do anything so unrealistic as to be easily useful to anyone who is not an expert user." When Patricia explained that the comments arose from scientists in the community (presumably also experts) who were unable to work with Helen data with ease, a wry response from the audience was "In other words, science is hard." The representative from Headquarters clarified: "I'm not all that interested in making it *easy* for the community; I'm interested in making it *possible* for the community."

Even saying "the data" ran the risk of ignoring how heterogeneous mission datasets were. A PAM PI argued that his data required a multidimensional matrix, one sensitive to swirling and changing plasma models, in contrast to what he called the "two-dimensional universe" assumed by remote sensing instruments and Patricia's report. Such heterogeneity mirrored Helen's own organizational tendency toward autonomy, resulting in some datasets that were complete and on schedule, while other PIs were several years behind in posting their data. Concordant with local norms, this heterogeneity needed to be upheld, not ignored. Eventually the conversation turned to the mission's legacy and the need to preserve data usability long after the teams were disbanded. At this, the tone softened as scientists described precious data that was "lost" following missions in the 1970s and 1980s and considered how to translate what one PI described as the "intellectual activity and judgment . . . [and] experience with that data" for future generations. The challenge of temporality could also be a resource in integrating the autonomous team. Eventually, the assembled scientists concluded that the problem was not the data itself in the PDS, but "the expectations of the PDS" and the expectation that the scientists should—in addition to planning operations, mentoring students, attending meetings and conference calls, reviewing grants, and writing papers—meet external data usage expectations as well.[12]

If such concerns plagued a mission with high degrees of autonomy and independence in data production, then other concerns dogged the Paris mission with its continual push toward collectivism. While images circulated freely online, one instrument team leader was apparently vocal about keeping raw data away from external, competing scientists. Further, even the mission with a commitment to "open data" experienced difficulties in communicating this data usage to its broader community audience in planetary science. Because the Paris mission is a narrative mission that has unfolded over time, the robots' daily operations rely on the team members who are in the meeting room and the changing location of the robot on the planet's surface. Thus the "story" behind any particular observation is complex, contingent, local, and evolving. It can be extremely difficult if one is not in the room to know where the robots are, what they are doing, and why.

This caused a degree of anxiety among Paris scientists as to how best to serve their broader community. As one put it,

> I kind of have heard people [outside the mission] complain. But it's hard. It's not the fault of the people who are on the mission. We are too busy, we are doing what we can do . . . as quick as we can. It's very open, it's much better than the Europeans, than Japanese, India, China, but still there's limitations, people only have limited time. . . . [There's] some kind of a feeling. You cannot get it from the outside, you need to be in front of a computer looking at data day-by-day, then when someone mentions a name and you know immediately where that's located and what's around them, but for outsiders, they have to look through lots of images. . . . It's harder because you're not inside. You're not in the field.[13]

Because being on Paris cultivates a situated, embodied, and collectivist approach to data production, stripping the resulting data of this rich contextual information to make it visible, indexical, and accountable to outsiders is a difficult proposition indeed. It is therefore difficult to access data or even know that an observation exists to be found if one is not already a member of the team. Paris mission scientists attempted to address these issues through their own data release portal. An online tool developed and maintained by a university department participating on the Paris mission brings together all images, instrument reports and documentation, site maps, and other location information related to particular sites on Mars. These sites take team members' embodied, situated sensibility for their robots' locations and the temporal unfolding of their mission and transpose this to being a portal for a web-based interface. This is consistent with the methods of operating on the team, using the robot's body on the surface as a rallying totem and orienting position from which to begin the consensus process.

But with so much data to load and visualize over an extended mission timeline, mission sites now number in the hundreds: a vast increase over the system's original intended aims. Thus most team members and outsiders I interviewed admitted that they did not use it, relying instead on the visual memory they had built up of their robots' locations, the stories of backroom discussions, and naming schemes that referenced particular times of year (i.e., a rock called "Wishbone" was investigated around American Thanksgiving). Not being there during the time such observations were acquired presents an outsider searching for data with the problem of finding a needle in a haystack. Taken together, then, the PSG discussion and the online tool show how difficult it is to translate data crafted within an organizational setting for external audiences with other expectations entirely. This is not due to the inherent material qualities of digital or instrumental data but to how such data and their circulation patterns are inscribed and intermeshed with the social relations of each organized team.

Attention to these norms of external circulation can help or hurt internal social relations. For instance, the Helen radio team at the time of my ethnography had already published a highly cited paper in the astrophysics community confirming general relativity, but they were otherwise far behind in placing radio data online and publishing. Because the radio science experiments require many, many passes by a planet or moon in order to get adequate triangulation of instrumental results and eliminate artifacts, it can take several years of flybys in order to calibrate the readings and answer scientific questions. Further, one is not always certain from a single pass that the data are correct. As a radio scientist described it,

> You need to be certain of the data that you are publishing. . . . The question is, do I have adequate data to publish my result and consider my results reliable, or can it be that I am relying on certain regions [that affect the measurement]? Am I mapping the gravity field well enough? . . . I don't want to publish another paper saying . . . sorry, all these [papers are] wrong because the data were inadequate.[14]

So sensitive was this instrument to spacecraft wobbling, however, that several other instruments had to sacrifice their observational opportunities whenever the radio took a measurement. It could therefore be unfavorable for radio to be seen to be noncompliant with release policies when it came to requesting further observational time from the spacecraft, an issue already in the background at the tour selection meeting. This was the threat when project scientist Francis and deputy Victoria visited the radio team meeting at the PSG following Patricia's presentation and tour selection. While it was clear that the team worked very hard, Francis explained that without data released and publications, they would have trouble getting more:

> Publications need to happen in order to justify new data. . . . The goal is not to have you guys have an unfunded mandate [i.e., deny you the ability to do your science]. . . . Experts in the community can't use the raw [instrument] data and can't use the processed data. The team is meeting its mandate by getting the raw data out there, [but] if the community can't use the data, then why are we collecting more?

Francis drew a relationship between previously collected data and future data collection on the upcoming extended mission. He used the taxpayer argument, saying the mission is "taking 80 million dollars a year," and because Helen "[serves] the community" they had a responsibility to release their data in a timely way: "We want your team to shine and for Helen to shine." He also claimed that better data practices would argue for more gravity science on other missions, making radio students the inheritors of Helen's legacy. And he was happy to claim that the data was "preliminary" on the PDS. So even if that data are preliminary, invalidated, or raw, it is worth releasing, but it is not OK for the data to be difficult to work with—despite the fact that the data might be difficult to work with precisely *because* it is preliminary, invalidated, or raw.[15]

The radio case has much in common with knowledge infrastructures such as the PDS and online Paris tool. Translating local, organizationally situated knowledge to broader audiences across spheres of context is always challenging despite best intentions (Baker and Yarmey 2009, 11; Borgman et al. 2014). But it also demonstrates how engaging in circumscribed external relations is essential for managing local relationships too. The pressure to perform in order to participate in Helen's success, argue for follow-up missions, and acquire further observations from competing peers was keenly felt. Within a year, the radio scientists published a much-anticipated paper about Titan's internal structure in *Science*, based on years of flybys, that supported the possibility of cryovolcanoes on the surface. But asking radio scientists to deliver "validated" data to the PDS and to publish before they got the additional readings that might validate that data had consequences for their science. Based on data acquired in later flybys, the team retracted that paper a few years later.

The Power of Data Circulation

Because not all planetary science relationships are organizationally circumscribed, maintaining compliance with the rules of the road is essential to mission social order. Those I spoke to who were graduate students and postdoctoral scholars on missions in the 1990s reported passing printouts of images or floppy disks of data under the table to their institutional peers, sometimes without their advisors' knowledge or an instrument PI's consent. One confessed that the imaging website for a flagship mission's cameras was hacked by an outsider. These stories speak to the many relationships and accountabilities that underlie such missions, as well as to reasons to clamp down on unsanctioned mission data exchange. Far from the mission's echelons of power or central membership group, data circulation can easily follow other lines of social relations, such as those among friends.

But data can also move in ways that protect or prescribe internal boundaries and social relations that are well known to members, if obscured from outsiders. In this case, the rules of the road play a role in policing members' proper participation in the organization. For instance, because data are open on Paris, flare-ups occurred when individuals attempted to withhold data from others or otherwise circumvent the collective. Although I never met him or observed him in action on the team, Owen has a private company making instruments over which he holds patents and data interpretation rights. He resisted making his data into a shared commodity, putting him at odds with the rest of the Paris team. By the time of my observations, he was no longer actively participating on the mission. Team members were unusually strained as they discussed his unwillingness to share but did not remove his name from the team roster.

On Helen, data circulation rules were used to police Isabelle in her frequent interinstitutional tug-of-wars with Spacelabs. The majority of stories about Isabelle centered on her control of the camera data. One story that I heard from several team members was of the moment when NASA mandated that all Helen images be released to the public immediately upon downlink. Isabelle resisted. On a mission oriented toward fairness, why were the *images* the only dataset that was required to be released? The rules of the road protected everyone else's autonomy over their own data. Other instruments could take their time and pore over results and their calibration routines, but forcing her team to release images early would leave the camera team uniquely vulnerable to scoops from scientists and amateurs alike.[16]

Isabelle was prepared for this challenge. Former camera leaders advised her that if Spacelabs ever pressured her to roll over for new requirements, she should take her grievance to NASA Headquarters. A review meeting in Washington therefore placed Isabelle, then-project scientist Malcolm, and project manager Everett in a room with higher-ups at the space agency. Accounts of this meeting by its attendees differ, although its outcome was legendary.[17] Isabelle argued to support her existing data release policy, which protected the scientists on her team from being scooped and respected existing collaborative agreements with other Helenites while making lower resolution images available upon downlink and producing beautiful images for the public. Malcolm thought that Isabelle's push for a slower schedule for release until "after the team was through doing science on them" could produce, as he put it, "a delay of years." The Headquarters attendees were reportedly unsympathetic to Isabelle's view: one attendee I spoke to remembered that at the end of the meeting, "people walked out shaking their heads." The NASA official in charge directed Everett and Malcolm to ensure that the imaging team would not be deprived of resources unfairly compared to the other instruments on Helen but also that they release every full image to the internet upon downlink, as Paris had recently announced. Isabelle described this moment as getting "piled on by everybody. Because everybody had something to gain by looking at our images."

Data circulation violations locally justified project actions to restrict Isabelle's authority, including her ability to see, manage, or control her and her team's budget. For instance, at the outset of the mission, Isabelle planned to send out a press release on the day that Helen arrived at Saturn to generate excitement about the mission and its findings. This would include an image taken as the spacecraft approached the planet. The announcement was planned for 9:00 a.m. (EST), but to avoid asking her team to come in early at her institution in the West, Isabelle made what she later described as "a drastic bad decision" by instructing her staff member to release the text the night before, at 5:00 p.m. local time, under embargo, with a note saying that the photos would go live the following day. She recalls that Spacelabs' public relations officer "read [her] the riot act" on the phone, and she received a letter indicating that her reprimand would be swift and strict. She wrote to a NASA official and offered an apology but later received a letter that she recalled was widely carbon copied accusing her of plagiarizing the press release that she said her own public relations team had written and given to Spacelabs for release. As she put it, "It said [that] from here on in your contract is gonna be monitored, you will not be able to travel without asking approval." She recalled a series of "really punitive steps" including an audit of her spending the following year.[18] During my observational period on Helen, there was another slip-up associated with Isabelle's team making a press release visible online only a short time before it was supposed to go live. Isabelle attributed this mistake to her publishing software mishandling daylight savings time, but this fueled her reputation at Headquarters for not coordinating her press releases with NASA public relations representatives. Soon after, Everett announced that her own scientific budget as well as her team's budget would henceforth be subject to his control.[19] On another occasion, I learned that Everett had approached other camera team members to share the data that Isabelle was reluctant to release. As Isabelle put it,

> They expect me to run my team without really knowing what the budget is or at least who's getting what. . . . They're using the fact that my team members [on the facility instrument team] are under contract to [the agency] and not to me to prevent me from seeing their statement of work. So they have gone out of their way, I mean, they are pulling every single thing they can to undermine my authority with my team members.[20]

The data tensions, fiscal concerns, and discovery anxieties that Isabelle voiced were ones that I heard repeatedly in a decade of ethnographic work in planetary science. Photo interpretation is something that many individuals from armchair enthusiasts to trained planetary scientists feel comfortable doing, a tension I have discussed in detail elsewhere (Vertesi 2015). Even Malcolm explained to me in conversation that the images were relatively easy to interpret, while other instrumental results were "gibberish" until calibration and processing. And although Isabelle reported using her own contacts at NASA headquarters to defend her from budget cuts, Everett controlled the project's purse strings: even the project scientist had to request money from him for scientific purposes. The very fact that violations to data circulation rules—the particulars of data release, the timing of an embargoed post, or the unlicensed sharing of an image—elicited sanctions and interpersonal hostility speaks to their importance to upholding organizational orders on the mission. The way that data are supposed to circulate maintains the mission's matrixed, intergroup relations. Meanwhile, the way in which violations are articulated, produced, and enforced maintains the mission's embedded hierarchy. This latter process also supports and reinforces locally acceptable forms of organizational alignments—and ostracism. After all, many scientists referred to Isabelle's data circulation history as evidence that she was "the singular nonteam player." As such, conformity with data circulation expectations crafted or contested local notions of cooperation and of being a "good" scientist (Daston 1995).[21]

Conclusion

When it comes to data circulation, relationships matter. On planetary missions, as on any scientific collaboration, these relationships are often organizationally circumscribed. The rules of the road outline roles for members alongside the conditions under which data circulate to them. These criteria on each mission establish appropriate participation and legitimate discovery claims, as well as illegitimate use. Data must be prepared for outsiders while managing relationships within the mission as well. As data travels, its path reflects local social relations and reproduces those relations at the same time.

Not only are datasets shaped by the organization to answer specific kinds of questions, as described in the prior chapter, but the way they circulate among that same team also reinforces and reproduces organizational orders. After all, when it comes to "data sharing" discussions, Helenites pointedly state that they *are* already sharing data. This is, of course, true. Data circulates according to instrument team and interdisciplinary scientist contracts; it circulates in a downgraded quick-look form amid the PAM collaboration; instrument teams place it on the PDS as quickly and as well as they can. Shifting from the moralistic tone of "data sharing" to "data circulation" is necessary if we are to include the many knowledge-making endeavors that deploy Helen-style methods of data management as legitimate ways of maintaining an organization, accomplishing scientific excellence, and producing social and natural order. This framing also explains the Paris circulation network, which reifies communalist and collectivist practices internally such as those around scientific questions and instrumentation—and which confronts altogether different challenges when their data travels afield.

Essential to maintaining organizational order is the ability for distinct teams to control their data production and circulation, thereby delineating the contours of their membership as any other collaboration would do.

Examining data in this light moves us away from the simplistic notion that data are shared because they are digital or that datasets should be free because they are publicly funded. NASA missions did not suddenly embrace open data policies because of some inherent property of data or public funding but due to organizational changes. In the opening vignette, Paris proposed an open data regime that was entirely consistent with their social order and historical trajectory through the Mars Program: one that upheld local values of collectivism and consensus and in which a single PI could speak for an entire mission team. Transporting this regime to Helen, with its altogether different rules for effective collaboration and its heritage in earlier flagship models, made for a culture clash in many respects. On a mission in which instrument team autonomies must be upheld for successful expert integration and scientific work, demanding that the borders between teams be loosened through data circulation across boundaries threatens the mission's collaborative endeavors. It deprives Helenites of an essential tool by means of which they assert their social relations and manage their multiple hats. Assigning this imperative to one instrument also breaks the mission's code of fairness. The order to share widely took place within existing strained social relations between the camera team leader and her mission's hierarchical, multi-institutional character. We must therefore see such popular narratives about data sharing as post hoc rationales that members use to explain and provide moral support for their existing data circulation system. This itself supports their local organizational order—and their mission's political position too.

The social fact that local data circulation regimes produce and reproduce social orders illuminates another way in which scientific organizations shape scientific outcomes—the second principle of *organized science*. Adopting this perspective can explain why data sharing cultures and regimes vary so widely across scientific collaborations and why, when conventions change externally, scientists may seem resistant to new norms. These implications echo across many collaborative ventures across the sciences. Yet because data circulation enacts local social relations, it produces and reproduces social orders both among insiders and outsiders—as well as those within the organization who suffer punitive effects. Where data travels, who is permitted to analyze or speak publicly about it, who controls its production and release, and who is credited in discovery narratives demonstrate who it is that is permitted to make knowledge about data and when. Finally,

NINE

The Personalities

Saturn's icy moon Enceladus holds many mysteries. An ocean world surrounded by a layer of ice much like those at the Earth's arctic poles, its orbit around Saturn, like all moons, is affected by the planet's gravitational pull. But Enceladus also orbits in lockstep with fellow moon Dione, in a resonance that pulls it into an eccentric path around Saturn. The contradictory push and pull of these tidal forces strains and fractures the tiny moon's frozen surface, opening enormous gashes around its southern pole, heating the ice and spewing enormous geysers of water into space. These geysers are so significant that they create a disturbance in the Saturn system's magnetic and plasma field, causing auroras to snake around the planet's poles and even forming an additional ring around the planet. The ice, rock, and dust jettisoned from the moon also regularly bombard the planet's other small moons; smearing their faces with ejecta and pockmarks.

The other moons do not erupt like Enceladus does, assaulting its neighbors and affecting the makeup of the planetary system. They fall into different orbital planes and rotate in resonances with each other or with giant Titan, or they may find a different gravitational sweet spot with Saturn. Titan, itself about the size of one of the inner planets, looms large but is far enough away from Saturn so as to maintain its own regularity and pull its own weight in the system. But maneuvered into the very zone where competing tensions are rife, this small moon spews its innards continuously into space, affecting the entire planetary system in its wake.

Like Saturn's many moons, people on planetary missions find harmonies and resonances with each other. They share characteristics in common, shepherd each other onto particular paths, and keep each other in check and in line with recognized orbits. But occasionally, individuals fall between orbital paths. Caught between conflicting resonances and gravitational pulls, like Enceladus, they too erupt violently, sometimes repeatedly. The result of competing frictions and organizational impossibilities, these eruptions can cause chain effects down the line, leading their peers to believe they are difficult, squeaky wheels or simply impossible to work with. Planetary scientists have a word for such people: they call them "personalities."

The idea of a "personality" places the blame for such behavior on innate, psychological characteristics that make these individuals such a challenge. Indeed, at my first Helen social event, two scientists joked that instead of a sociologist, what they really needed was a psychologist. In their accounts, "personalities" just *are* "egotistical," "self-absorbed," or "paranoid," "bitches," or "jerks." Several planetary scientists explained to me that the joy of assembling a PI-led team for a smaller class mission, instead of being on a flagship or a facility instrument team assembled by a review board, was that they could enforce a "no-assholes rule" (Sutton 2010): choosing only people they got along with and therefore assumed were innately easy to work with. Missions are also said to have "personalities," cultural forces, and interactional modes that impose themselves on individuals. Like people, missions simply are—or are not—successful by virtue of this quality. "Personality," for planetary scientists, thus denotes the unchangeable, essential, and impervious that one simply has to either work with or learn to avoid. Although planetary scientists believe that rock formations, planetary atmospheres, and orbital pathways are shaped by complex, dynamic, and interactional processes, they typically take their peers at face value.

Although it seems like a natural fact, planetary scientists' concept of "personality" is what anthropologists would call an *emic* term: a member's category with a definition that is taken for granted by the group under study. The organizational lens allows us to analytically unpack "personalities" and other scientific personas somewhat differently. This chapter traces interactional effects between organizational orders on the one hand and identity characteristics on the other to demonstrate how organizations participate in producing scientific personas from lauded leaders to "real personalities." I demonstrate how the combination of institutional arrangements and locally acceptable identities conspire to develop members' ascriptions of their peers' innate personal characteristics. Taken together, this analysis sheds new light on the biographical outcomes I observed in the field for the participants in each organization.

This is important because we know that inequalities persist in the sciences and elsewhere. Scholars have identified many reasons this may be the case. Situations like "the Matthew Effect" describe how the most senior scholar on a paper accrues accolades while coauthors do not, a situation that worsens demonstrably for female scientists, even when they are the lead author (Lincoln et al. 2012; Merton 1968; see "the Mathilda Effect" in Rossiter 1993). But stories of scientific women whose contributions have been ignored or even sabotaged, like Rosalind Franklin, pepper the archives alongside those like Marie Curie who receive continued praise. Meanwhile, men and women alike who work as laboratory technicians remain largely "invisible" to history (Shapin 1989). Some argue that these discrepancies are due to scholarly networks, which may exclude (Merton, 1965) or, through mentorship, buffer minorities in their careers (Alegria and Branch 2015; Gaughan, Melkers, and Welch 2018; Smith-Doerr 2004; Whittington and Smith-Doerr 2005; Xie and Shauman 2003). Others show how stereotyping matters for who is or is not accepted into a field, for instance, describing the limited identities available to both men and women as they assume the mantle of "scientist" (Dunbar-Hester 2008; Ensmenger 2015; Fara 2002; Milam and Nye 2015; Traweek 1988). We also know that social markers such as these play a central role in hiring, evaluating, and promoting employees in other occupations (Pedulla 2014; Rivera 2012; Rivera and Tilcsik 2016; Schug, Alt, and Klauer 2015). This is especially pronounced in organizations where advancement is based on "merit"—in which gender inequalities paradoxically increase (Castilla 2008; Castilla and Benard 2010). In the sciences, such factors lead to debate over what kinds of organizations best advance minority careers: bureaucratic ones, with their clear criteria for advancement, or flat hierarchies that rely more on existing network ties (Freeman 1972; Smith-Doerr 2004)? Dynamics of inequality are clearly afoot in the sciences, but the causes are complex.

I confronted these paradoxes on the two missions I observed, where both men and women were celebrated for their competence and leadership or denounced for their difficulty to work with. Certain men were continually placed into important roles, while others were eclipsed by their female peers and experienced difficulty getting ahead. On Helen, the majority of the project's senior roles were occupied by women who were praised for their work, while one senior high-performing woman was decried; on Paris, gender and other social markers didn't seem to matter at first blush. Over time, however, I noted that a *combination* of organizational positioning, identity performance, and social networks determined how individuals were perceived by their peers. This evaluation was also tied to a person's interactional behavior: not whether they uphold a code of virtuous scientific conduct (Daston 1995; Merton 1942) but rather whether they adopt an acceptable identity, one that is available and legible to members of the field as a way of being a good scientist or engineer. Adopting one of those available identities brings reputational benefits by virtue of linking individuals to an affinity group: strong, crosscutting friendship networks of scientists that displayed similar attributes, much like clusters of moons in orbital resonance.

Over many years, I observed these affinity networks intersect with organizational environments to produce different "personalities." In Helen's matrixed context of autonomy, multiple hats, and battle talk, individuals who belong to affinity networks are buffered from mission fault lines and interinstitutional misunderstandings and spared accusations of misconduct, even when they have to make difficult calls. On Paris, meanwhile, a charismatic identity was made possible and reinforced in the leadership position, while other scientific identities and their associated networks were subsumed into the collective. Those who fell between institutional cracks without a strong network while running afoul of identity norms became known as "a personality." These dynamics presented implications for team members' careers. Individuals' biographies and peer evaluations are therefore also an outcome—like scientific papers and data—of *organized science.*

This is a difficult chapter to write. Empathy is an essential part of the ethnographer's toolkit, especially in complex organizations where experiences can diverge along several fault lines. Over my many years embedded with planetary scientists, I have come to care for each of these individuals and their stories as I sought out and genuinely attempted to understand their points of view. At the same time, I have observed persistent patterns in the field that are impossible to ignore. Because these patterns give clues as to how inequality persists in the sciences and engineering, as well as the complexities of that inequality in action, and because these dynamics surely recur elsewhere, I believe that this story is crucial to tell. While I attempt to capture individuals' perspectives as faithfully as I can, discussing individuals' career trajectories, alliances, and concerns with analytical distance does not indicate my dispassion. Instead, I hope to generate a degree of generalizability with which we might give name to and understand these dynamics in operation in other workplaces, wherever we may find them.

"The Personality"

I had already heard a lot about Isabelle Bacconi before I even joined the Helen team. It seemed every planetary scientist had an opinion about her, but everyone I interviewed asked me to turn off my recorder before launching into stories about her. There was that time when she'd stormed out of someone's office, when she wanted to charge admission for an award lecture, when she butted in front of someone in a coffee line, when she refused to share camera data, or any of innumerable eruptions between her and other scientists, students, or space enthusiasts. Individuals presented these stories as evidence that she was egotistical, self-centered, and impossible to work with, but I wondered why Isabelle's side of the story rarely circulated. I also faced a clash of narratives as I tracked these stories down. When I spoke to scientists that Isabelle was reported to have slighted, they sometimes did not remember the episode, offered an alternate explanation, or bore her little ill will, but they offered stories of other times when she had caused a colleague injury. In the case of the award lecture, I learned Isabelle wanted to open the audience to a broader public and sought an admission charge to cover the cost of renting a larger lecture hall: the conference organizing committee agreed. And why would someone I knew to be an honest and dedicated colleague hold a grudge against a scientist for beating them to a cup of coffee a decade ago?

I first interviewed Isabelle while I was studying the Paris team. I didn't detect the egotism that others had described. She was generous with her time and discussed her scientific work on Helen with enthusiasm, only alluding to the fact that she had faced many difficulties along the way. When I joined the Helen team, I had the opportunity to observe her interactions with others, collect more stories from Helenites, and hear her side of these stories. Although space precludes full review of the fissures between Isabelle and members of the Helen project, at the core of the conflict was typically a combination of interinstitutional misunderstandings and organizational disempowerment, compiled with gendered expectations for her behavior. Taken together, these were locally coded as being "a personality."

Organizationally speaking, Isabelle was placed into a precarious position. As the only woman on the camera team and one of its youngest members (appointed in her midthirties), Isabelle was put in charge of a roster of senior scholars. Some of her teammates were already famous from their work on prior missions, others—such as Jeremy—went on to illustrious careers as PIs of other teams. A few of them had even applied for the leadership position that Isabelle herself had won over them. Such factors evoked the role incongruity and negative status effects commonly experienced by female and junior leaders and well described in the literature (Anicich et al. 2015; Eagly and Karau 2002; Heilman 2001; Ridgeway 2011; Ridgeway and Correll 2004; Rudman and Fairchild 2004; Rudman and Glick 2001).

As the only woman placed into a high-level decision-making role, with female representation at well under 15 percent of the organization, Isabelle was also subject to the exact dynamics that Harvard Business School professor Rosabeth Kanter attributes to proportionally skewed groups (Kanter 1993).

For instance, when I asked why she got the job as team lead, she said plainly, "The scientist in charge of the judging panel said that I wrote the best proposal," but consistent with Kanter's concept of the *tokenism eclipse*, wherein majority members persistently downplay the minority members' achievements (1993, 216–17), others on the team I spoke to quietly confessed that they thought it was because she was a woman. She suffered the pitfalls of heightened visibility associated with both her position and her media experience (Kanter 1993, 213–14). When she first joined the team a fellow scientist asked pointedly if she planned to hog the media spotlight as she had on a prior mission, and when she was made a science advisor to a *Star Trek* film (she was known to be a huge fan), Spacelabs colleagues made snide comments about it in the hallways—although they were enthusiastic about a male engineer who was an advisor to *Battlestar Galactica*. Such are common examples of how minorities in skewed groups experience organizational isolation through the uniqueness of their position and its inherent visibility.

But Isabelle faced organizational challenges as well. She led Helen's camera team, representing the most public-facing instrument on the mission, whose data was invaluable to many different types of scientists. As a facility instrument, Spacelabs managed the construction of her instrument and the contracts of her team members, whom NASA had selected in an arrangement she called "a shotgun marriage." When managers are given great responsibility but have limited authority, their position is characterized by organizational scholars as one of *structural powerlessness*. Rosabeth Kanter describes how women in such positions are accused of being controlling micromanagers whom their colleagues love to decry, even though this managerial style is the *result* of their structural powerlessness, not the *cause* of their organizational ineffectiveness. Consistent with this finding, the other leaders of facility instrument teams I interviewed expressed frustration with their impossible task (except for one, whose instrument was based at Spacelabs and whom I did not interview), and they were often subject to interinstitutional misunderstandings and labeled "personalities" too. Yet gendered expectations for interaction heightened the backlash, as "powerlessness tends to produce those very characteristics" negatively attributed to female leaders such as bossiness, territoriality, and holding back talented subordinates (Kanter 1993, 202, 249). These same criticisms were frequently repeated to me against Isabelle, who saw her rightful role as more on par with that of PI.

Status effects, the tokenism eclipse, the double-edged sword of visibility, and structural powerlessness are well understood and cataloged in the literature on women in leadership positions. But these compounded in the face of the interinstitutional misunderstandings typical of large-scale scientific collaborations. For instance, when Isabelle interviewed past imaging team leaders for lessons learned, they told her to be wary of Spacelabs trying to take over her instrument. So when her instrument liaison at Spacelabs suggested changing a feature on the camera, she resisted. He recalls that his suggestion was made to "help" her and interpreted her "terse" reply as an example of her wanting "more control" and "micromanaging" things. Meanwhile, Isabelle fought back when Spacelabs announced changes to the cameras, relaying them as examples of *the lab* micromanaging *her* and her instrument team. Describing a related incident, she recounted being shut out of the process ("[The project manager] didn't ask me to join in this decision, he just made this decision"), and she resented the "dictatorial statement" from the project about how to proceed.

As I described in prior chapters, Isabelle's concern for managing the balance of power between her institution and Spacelabs is not unique, nor is her concern for her instrument's production and its data circulation or for protecting the scientific interests of her team. When other scientists espoused these concerns, their behavior was explicable and excusable. But in Isabelle's case, mundane conflicts and misunderstandings across institutional fault lines exploded to become personal and acrimonious. Much like in Calvin Morrill's (1995) study of a conflict-laden firm—where the senior woman on the team was labeled "the princess of power" and assumed to be motivated by greed, revenge, and inherent bitchiness—women in these roles are seen as responsible for their own ill fates instead of responding to insurmountable organizational pressures. Of course, just like Enceladus's pressure cooker orbit, *anyone* placed in this organizational position would erupt, but team members accounted for this organizational outcome as due to Isabelle's personal characteristics.

Interinstitutional conflicts combined with the longevity of mission work and static positions produced a continuous loop for these dynamics. Like the organizational ostracism described by Robinson and colleagues, Helenites were wary of Isabelle and downplayed her concerns and contributions (Robinson, O'Reilly, and Wang 2013). In local accounts at Spacelabs, she needed to be "contained" because she would "not play well with others." This in turn motivated project members' attempts to work around Isabelle's "personality" by attempting to curtail, circumvent, or override her limited authority over her team or her scientific production. As one manager admitted,

> I haven't spent much time trying to understand the why of her behavior. It's all speculative. You can't do much about it. You just try to work with it as best you can. In her case, you had to overpower her by one device or another before she would pay attention.[1]

But the more Spacelabs tried to bring Isabelle into line, the more she resisted, and the more distant the institutional narratives between Spacelabs and the camera command site became. Further, these repeated interactions instilled in Isabelle a sense that the project and its members were persecuting her unfairly for doing her job:

> My job is to . . . present [my team's] case. . . . I feel like I'm not being listened to. I get louder . . . and then pretty soon everyone's looking at me going, "What a bitch!" And so meanwhile, I wanna say, "I'm sorry, but you made me this way. I have to act this way just to make you pay attention to me. . . ." You develop a reputation.[2]

In companies, such dynamics play out over promotions, production, or the bottom line (Acker 1990; Castilla 2008; Correll 2004; Fernandez and Sosa 2005; Kanter 1993; Morrill 1995). In science, they play out over the acquisition of scientific data and the ownership over topics and results. For instance, I described how the targeting teams were Helen's site for decision-making and management, playing an essential role in how and what data are collected. Explaining the origin of this arrangement, an interviewee revealed this structure was a way to avoid allocating the mission's "personalities . . . her own" observations. This workaround was locally interpreted as an imposition that Isabelle *enforced upon Spacelabs*, not as a sanction against Isabelle:

> We [at Spacelabs] ended up having to spend additional money and additional resources in order to build capabilities into the system to say well, because Isabelle did this [last time] . . . we've got to make sure that she doesn't do that for the [next] tour, so let's do *this*.[3]

Like Enceladus's icy plumes that impact other moons, recoat its surface, and cause magnetic disturbances in the Saturn system, such "personalities" are produced by organizations but also impact and disrupt them.

As another example, Malcolm recalled in an interview a moment when Ian, an IDS, did not have a collaborative agreement in place to use the camera for an observation,[4] so he approached Malcolm—then project scientist and a colleague from both a prior mission and graduate school—for help.

Malcolm explained to me that he asked Helen software engineers to create a cloned camera in the project-wide database by means of which he could override Isabelle's control over the instrument and input Ian's observation directly. He called this "the Project Scientist's Discretionary Instrument" and recalled her shock when he explained to her that he could now commandeer her instrument any time he chose.[5] As the frontline brokerage point between the camera team and the project, the relationship between Isabelle and Malcolm was especially acerbic. He blamed her for launching a coup to remove him as project scientist, reporting that he canceled her meeting room reservations to leave her in the lurch. One of his jokes in the office, if he had not heard from Isabelle that day, was to proclaim the day to be "a kosher day—no Bacconi." Meanwhile, Isabelle recalled Malcolm dismissing her invitation to the White House Office of Science and Technology alongside NASA's then-female chief scientist as "a women's coffee clutch" and announcing to the PIs before launch that he would make her "mud-wrestle" with James to resolve observation conflicts.[6]

Yet Isabelle credited the difficulties she faced on Helen largely to interinstitutional dynamics, not institutionalized sexism. As she saw it, the crux of the conflict was Spacelabs' continual interest in controlling the cameras and her triumphs of public engagement with the Helen mission. She described this as motivated by "resentment"—of her position, her authority, her visibility to the public, and her excellence as a scientist. This made sense of the ill will she experienced from other, non-Spacelabs scientists too: some carried over from the prior flagship mission, where she was identified as a rising star and media personality, and some was the result of her appointment to a leadership role over others who felt they deserved the job. It also explained prior camera team leaders' experiences and frustrations, as well as sanctions upon her instrument, its budget, and its data cap, and the "hatred" she described between her and other project leaders. Her continual refusal to "roll over" for Spacelabs was, to her, proof of her strength, resolve, and deserved leadership role in the field. Refusing to be victimized, Isabelle fought back repeatedly when it appeared her team, her authority, or her reputation were under threat.

"Resentment" was an important explanatory mechanism for Isabelle, a form of lay sociologizing that reveals how participants make sense of their social world (Garfinkel 1967). The term belies a fundamental belief in meritocratic advancement in the sciences—a commitment she holds in common with her peers. But this form of talk was as unique to her as her organizational position, speaking to her organizational isolation. As I will describe, many planetary scientists band together in homosocial friendship networks, extending attributions of merit to their peers and, with it, opportunities for advancement. But Isabelle did not extend this credit to others like her in the organization, as there was no one else like her in the organization. She saw her own appointments as meritocratically earned but others' (like Malcolm's, or those who sought to lead her team in her stead) as undeserved. Many of her peers, meanwhile, countered her explanation of "resentment" with egotism. This established a looping dynamic (Goffman 1961a) that maintained her solitary status and ensured that her side of the story rarely circulated with credibility among the team.

Other women in planetary science who were placed in similar positions faced comparable issues in kind, although never in degree. For instance, one scientist I interviewed described being shut out of Paris despite her relevant expertise. She was later informed that her competitor, a male scientist on the review board, had dismissed her application, saying aloud, "I will not work with *that woman.*" She related subsequent difficulties getting mission work, grants, and institutional status. In another case on Helen, I witnessed another mission "personality" caught in a looping situation (Goffman 1961a) where her outbursts, themselves often the frenzied result of being caught between competing tensions, became justification for assuming her "craziness" in the first place and a rationale to further ostracize her from decision-making circles. Another young woman tearfully explained to me that she was outspoken about inequality in the field because she thought it was important for science but now found that her grants were getting turned down and her tenure case was in jeopardy. Female "personalities" not only face limited mobility in their careers due to the reputations they acquire from their frustrated outbursts but are described as "psycho," "nuts," "crazy," "paranoid," or "off their meds." This leads to denial of those career opportunities typically associated with merit in the sciences, such as grants or mission status. Because such individuals come over time to distrust the organization and to adopt a defensive position toward it, newcomers who have not experienced the many decades of such offenses take their first experience of lashing out as evidence of a psychological characteristic instead of as a response to an environmental condition. Since one of the things such "personalities" are fighting for is a place at the decision-making table to begin with, the very label of being "a personality" sets up a vicious cycle of organizational ostracism (Robinson, O'Reilly, and Wang 2013) and heightened outlandish behaviors on behalf of both the personalities involved and the organization attempting to shut them out. Indeed, fighting back only added to the long litany of offenses that marked these people "a real personality" and "difficult to work with."

I never met a man in planetary science who experienced the full organizational backlash associated with Isabelle. But I did meet and interview a handful of men who were also described as "personalities." These men were also placed in untenable positions, sandwiched between institutions with incompatible requirements. When they resorted to outbursts to attempt to solve problems they were labeled "personalities." Leon was a leader of a facility team and contractually unable to exert authority over his instrument or his group and so was often in situations where competing requirements from Spacelabs and his own institution or contract were at odds. He chuckled as he recounted that another team member once called him "an alligator—all teeth and no ears." Owen abandoned Paris entirely in an effort to exert control over his instrument's data, which was against the team rules of the road. His mission peers and many at Spacelabs described him as "a personality"—but also "a genius." On one of my first days in the Helen project office, I sat in on a meeting (the second of three) with senior personnel as they spent a full hour editing an email draft line by line, wordsmithing sentences "to avoid setting off any personalities." While it was taboo to discuss outright, the understanding was that this was code for Isabelle, Leon, and a few other non-Spacelabs leaders.

Organizational factors in the collaboration such as the struggles between institutions, invisibility across sites, and structural powerlessness clearly play a role in attributing eruptions to inherent characteristics instead of to interinstitutional miscommunications. It eventually seemed to me that anyone whose science or instrumentation was involved in a Spacelabs decision yet who was not consulted in the process risked being labeled "a personality" should they stand up for their needs. Yet Leon retained his position as a professor at a high-status institution, Owen continued to win contracts for subsequent missions, and both were labeled "difficult" but never "egotistical." As such, other men subject to these pressures may be called "personalities" but were not subject to the full range of organizational ostracism experienced by Isabelle. This highlights the role that prescriptive gender stereotypes, tokenist dynamics, and status effects play *alongside* the organizational positioning and interinstitutional challenges in planetary scientists' evaluation of their peers. This combination of frictions and conflicts produces both Enceladus-like eruptions and outcomes for each "personality."

The Buffering Role of Affinity Networks

There are plenty of women in planetary science who do not suffer the same backlash as Isabelle and many men who are not described like Leon or

Owen. This might lead us to believe that gender and organizational position do not matter to career outcomes. But another interactional effect between the organizational environment and identity work is at play: this time, in the form of distinct, robust affinity groups that extend across the field. Members of these social networks not only demonstrated strong support for each other and helped each other as part of the same friendship circle. They also adopted largely the same self-presentational style in line with a particular gendered identity in planetary science and, like all social networks, offered information and opportunities to their peers. Because they are drawn together over their similar, gendered identity characteristics and extend social capital to their members, I refer to these affinity groups as *gendered networks*.

Prior work in organizational sociology has noted the importance of homophily (essentially, the principle of "like attracts like") in hiring and promotion decisions within firms (Correll et al. 2017; Rivera 2015). Kanter describes the armies of similarly dressed secretaries, managers, and young superstars, whose similar appearance, interests, and other social cues distinguish rank and aspiration and seem to reproduce themselves across the organization (1977). Studies of hiring also demonstrate an overreliance upon determining similarity or high-status characteristics in merit-based organizations (Castilla and Benard 2010; Castilla, Lan, and Rissing 2013b, 2013a; Rivera 2015). As these networks and associated judgments of fit tend to break down by gender, race, social class, and other markers (Rivera 2012; Rivera and Tilcsik 2016), they afford not only different visibility in terms of hiring opportunities but also different extensions of social capital to their members (Burt 1998; McDonald 2011). This makes individuals in different networks that feature different social markers appear to be better or worse for upcoming jobs or promotions, regardless of their qualifications (Ibarra 1997, 1993; McDonald 2011).

Planetary science is no different. In my time in the field, I observed many groups of long-lasting friendships that reached across mission and institutional divides. These individuals were drawn together around shared interests but also around similar styles of dress and comportment:[7] visual cues that were colloquially associated with their innate personalities and assessments of their scientific work, engineering achievements, or management skill. Unlike mission "personalities," whose outbursts at the crossroads of several institutional tensions were organizationally decried, the women and men of NASA who belonged to one of these networks and displayed that group's accepted style of masculinity or femininity were sheltered from hostile interinstitutional dynamics and even supported in their careers by virtue of these networks. If "the personality's" isolation from these groups reveals the push and pull of competing organizational forces within the collaboration, these affinity networks demonstrate methods for buffering against such forces, limiting negative impacts on individual careers.

For instance, Malcolm, James, Jake, and Ian all went to graduate school in the same department at the same time, gaining their doctorates early in the 1970s while Gideon completed his PhD in an adjacent field. These scientists subsequently worked together on an outer planets flagship mission, establishing their reputations as central players in the field and cementing the social ties and trust between them. When Helen's team roster was announced, Ian and James were appointed interdisciplinary scientists, Gideon was made an instrument team lead (and later, head of Spacelabs), and Jake, busy in a leadership role on another flagship, recommended his friend Malcolm for the job of project scientist. In interviews, these men described the importance of long-lasting friendships, the value of working with people one trusts across projects, and the "fun" of doing so. Jake laughed as he described his ties to these founding members of the mission as "very much like a mafia family deal": because each group came from or directed specific graduate schools, "they all have their own regions of control." He was joking, but some outsiders were serious in describing these ties as constituting "an old boys' club" that left them out of opportunities and created an exclusionary mystique about this group of friends. More than one senior team member recalled for me that following Helen's launch, Malcolm invited select individuals from the top echelon of the project—his graduate school friends chief among them—to celebrate at a private martini party with him, but he did not open the invitation to all Helen's leadership.

Meanwhile, Victoria and Gwen worked on that same prior flagship in support roles, having met as undergraduates attending the same physics program at a local state school. They later undertook their doctoral degrees at another university together and had children at around the same time ("Right after the Uranus flyby!" recalls Gwen), sharing their experiences with daycare or school drop-offs with each other as they began their careers at Spacelabs. They described being each other's support network and trusting each other's managerial capabilities and scientific intuitions and explained that they were devoted equally to their jobs and "equally devoted to having fun," as Gwen put it. Arriving on Helen and placed in various deputy managerial roles to boys' club members, they jokingly began to call themselves the "old girls' network." By the end of my fieldwork, this friendship group had expanded to include a range of mentees such as Karen and Rose and occupied senior managerial positions across the Helen project. Members of this group are admired in the organization for their professionalism, efficiency, judgment, and cool-headedness and are also widely acknowledged to be tremendously "competent."[8]

Such network effects are well known to sociologists of organizations. Studies of boys' clubs, for instance, emphasize exclusivity in interaction as well as a hegemonic control of resources at the top of an organization (McDonald 2011). Members of Malcolm and Jake's friendship network typically occupied positions of power on mission projects, especially during critical moments early on when they were responsible for laying down the principles of resource allocation and decision-making. In the Mars Program, the same roster of senior men led project after project (Conway 2015). These individuals repeatedly appointed each other and their acolytes to powerful roles in successive projects or during different phases of the mission's production over decades. Consistent too with Burt's (1998) and Ibarra's (1997) findings about female networks, the old girls' network offered opportunities for advancement to its members, although typically at a lower institutional level than the boys' club with less social capital to exert beyond immediate projects. Both groups' members could be identified by their form of dress and comportment, which were locally associated with their shared innate characteristics. Powerful men at Spacelabs frequently wore collared shirts and slacks and were on the whole clean-shaven, while their followers wore slacks and thick-weave T-shirts with mission logos on the breast. Meanwhile, members of the old girls' network cultivated a polished appearance in skirts or slacks and sometimes suit jackets, their blonde hair worn long over their shoulders (Herbert 1998; Kvande 1999).[9]

Boys' club and girls' network members populated the top of project hierarchies at Spacelabs, but there are other ways of being a planetary scientist. Francis came from a different crowd. He wore his graying hair long and pulled back into a ponytail; his friend Sam sported an earring, goatee, and sardonic sci-fi T-shirts; friends such as Marvin, Hugo, and Mickey were usually outfitted with beards and beads, sandals, and flowered shirts. Elder members are fans of Stewart Brand and the Grateful Dead, while younger acolytes go to the Burning Man festival.[10] When I asked two of them how they knew each other, they shrugged and said that they gravitated toward each other at conferences because each was similarly "weird," referring to their shared left-wing politics and predilection for skepticism.

Seen alongside the boys' club, this counterculture group reflects a longstanding division between "hippies and suits" in the history of science and engineering.[11] Members of this counterculture group share a strong sense of irony about all things organizational and bureaucratic, especially when it comes to NASA politics. But this skepticism does not extend to the scientific enterprise, where they are most earnest about their science, quick to uphold the scientific method, and publicly outspoken in their critiques of science. Hence Francis, along with Sam, coauthored the paper arguing against cryovolcanoes and included a word of caution about the discovery in press announcements, despite being newly appointed science leader of the Helen mission. Francis also blocked a subsequent press release from a Helen instrument team, believing that it outpaced the group's findings and would step on another team's claims. Being responsible to the science, he asserted, was more important than sensational headlines.[12]

If becoming "a personality" is the result of falling between institutional cracks and being caught up in disempowering organizational dynamics, then affinity networks represent a form of harmonic resonances and alliances that protect individuals from accruing blame for organizational effects. Notably, those women who were mentored within the old girls' network reported being very satisfied with their positions, with little sexism or antagonism toward them in their workplaces based on their gender. They were working on interesting technical, scientific, and managerial challenges; were considered to be "competent" and "very good" by their peers; and were grateful for those women at the top who looked out for them (Fox 2001, 2010). On Helen, I was frequently in meeting rooms where women far outnumbered men. Rose once explained to me that she felt more comfortable on Helen because of the number of women on the team and these experienced role models. Such women distanced themselves from Isabelle's experiences in the organization—and she from theirs. Unlike the lone "personality," the affinity network's buffer allows participants the role distance they need in order to perform their job, such that they can make tough calls without negative attribution to their person (compare with Rudman and Glick 2001).

Affinity networks also presented implications for scientific advancement. Although these friendship groups brought together individuals with homosocial ties, individual scientists in the networks described their preferences for each other firmly as a question of *scientific merit*.[13] This was especially visible in the process of stewardship and curation of the next generation of network members. Across planetary science, I observed as one or two young men were selected from a year's crop of graduate students, taken aside, and groomed by boys' club mentors with influence in the field. They were introduced to "the right people" and invited to "the right" meetings, dinners, and drinks and described as "very good." They ended up in tenure track faculty jobs at high-status universities or in a highly visible position at Spacelabs. This is at the same time an introduction to a particular clique as it is an inauguration into a local power elite. Old girls' network members also mentor "competent" and "good" newcomers and promote other women in the organization with similar self-presentational styles into leadership roles. This protects them from the negative backlash effects associated with female leaders like Isabelle. Their networked ties to senior women also present implications for maintaining the circle of power across generations of project work. Because high-status members of affinity groups are able to promote those with whom they share trusted network ties into positions of authority, this continues their influence over ongoing projects. I can only report on the result of internal hires and not their processes, but team members took note when Victoria was promoted to project scientist and Gwen's mentee Karen was appointed her deputy; following Karen's departure, Victoria's instrument teammate, George, got the job.[14]

Boys' or girls' club members are not malicious in their decisions to include and exclude individuals from their networks and from associated opportunities. Consistent with Emilio Castilla's "paradox of meritocracy," they see themselves as upholding and enforcing a meritocratic order.[15] Even the counterculture group adopts and integrates new members who are similar to them whom they concomitantly see as promising, even if they have relatively less power to share or bequest to their members. They also circulate graduate students and postdocs (Slaughter et al. 2002; Traweek 1988), but their acolytes are more likely to get caught in soft money cycles, in second-tier university positions, or in administrative jobs despite their long lists of scientific publications and considerable project experience. Due to their strong sense of scientific robustness, accuracy, and justice—which, like all affinity networks, they extend to members of their homosocial group—these scientists were often visibly frustrated when their and their friends' efforts did not result in an external reward.

Another way to avoid the personally accruing effects of organizational dynamics is to adopt an accepted identity that outwardly eschews politics and networks altogether. The "geeks" at NASA are one such self-presentational style, with much in common with "geek" identities in computing, gaming, and tinkering (Dunbar-Hester 2008; Ensmenger 2015; Taylor 2012). Men and women alike in this group cultivate a persona that is interested in "just the science" or "just the engineering": a focus on the job at hand with a disregard, obliviousness, or even flagrant antagonism toward workplace politics. During my fieldwork, Spacelabs geeks congregated at a downtempo bar in the nearby town that featured board games and bar trivia nights alongside an encyclopedic beer selection. They attended concerts of touring bands such as They Might Be Giants that play songs about science or technology and participated at DragonCon, a science fiction and fantasy genre convention for dedicated enthusiasts, where they spoke on panels about their work. Geek men and women across the field wore jeans and mission T-shirts, with science fiction or planet-themed shirts and accessories purchased through websites like ThinkGeek or Etsy.com. If they colored their hair, it was typically with a bright hue instead of hiding natural gray.

Women who adopted a geek identity had the reputation of being good engineers and organizationally committed, as well as good at their jobs, allowing them to occupy important positions in the organization without backlash. Older women in this role described themselves as "one of the boys," a phrase I also heard repeated by a pioneering woman in rocket science I interviewed, with similarly close-cropped graying hair and self-presentational style. For instance, Connie, Helen's chief engineer, wears her hair gray and short, always dresses in a mission T-shirt and jeans, and does not describe her experience at the space agency as being dominated by sexism or discrimination. She proudly displays a photo in her office of herself sitting inside the spacecraft, physically connecting the wires that comprised the craft's communications infrastructure, demonstrating both her technical expertise and intimate knowledge of the spacecraft. Betty too in her "uniform" of mission T-shirts and long skirts, kept her own gray hair cropped close. Although I saw Betty and Connie placed in confrontational and difficult situations that could easily accrue them the moniker of a "personality," they were more typically praised for their devotion to the mission. Adopting a geek persona or being "just one of the boys" appeared to shield individuals from accusations of being swayed by mission politics or being too interested in a self-beneficial outcome in the midst of the matrix team's organizational complexity.

Unlike in e-sports or tinkering communities, geeks I met on these missions do not use their technical knowledge to maintain power over others or outsiders (Dunbar-Hester 2008; Taylor 2012) nor are they earnest, dedicated "organization men" (Whyte 1956). Instead, their geek identity insulates them from power dynamics and interinstitutional politics, which they profess to be in the way of their work. They also tend to avoid appointments that place them higher on the organizational ladder and do not seek to climb the hierarchy. One respected Paris engineer proudly reported turning down promotions when they were offered because additional managerial responsibilities were not of interest to him. As such despite the fact that they do much of the grunt work on the missions, geeks have very little power to make change. While they may be the fuel of the organization in terms of producing spacecraft commands and scientific papers, individuals who adopt a geek persona do not steer its direction.

An additional persona was available to men alone that allowed them to step above organizational concerns altogether. Just as the negative qualities of the "personality" are attributed to innate characteristics, these men are ascribed innate qualities that set them apart. Jeremy is a quintessential example, with a self-presentational style that is at the same time laid-back, as if unaffected by politics, and intense, buzzing with internal energy. A university professor, he typically wears cowboy boots and plaid shirts, recalling the large buckles and boot cut jeans I saw worn by men across NASA who are similarly labeled "charismatic" or "good leaders." At Spacelabs, a book review in a major newspaper described a leading mission engineer's cowboy boots, while yet another wears baseball caps as a sign of his gentle irreverence (compare with Law 1994; Morrill 1995). Such singular individuals are upheld with a higher degree of awe, and others ascribe to them intrinsic qualities that simply set them apart from other men. They are clear and powerful orators who speak without notes, and people are inspired and cheered to work alongside them. They are frequently granted positions of prominence and are treated with respect and even honor, although they typically reject being in the spotlight and evade overwhelming praise. They can also buck the rules and may even be rewarded for doing so.

Although these cowboy scientists' and engineers' professional networks are extensive, they largely work alone. The few individuals they do take under their wings are ambitious acolytes who, once they themselves are placed in positions of authority, remember their mentors. This may allow such individuals to enjoy the benefits of being in influential positions and having friends with unquestioned authority without embedding themselves within a single network. Long after their involvements with specific projects end, they continue to be showered with awards and accolades, and those who speak ill of them are quickly shot down by others who leap to their impassioned defense. One counterculture acolyte privately expressed outrage to me that his mission's local cowboy seemed only to "like white boys who are just like him" at the expense of others' (namely, his) careers, constituting a violation of his perspective on merit-based advancement. Otherwise, the near-universal public admiration for successful cowboys is remarkable. At one event I attended, Sam and Francis led a standing ovation for Jake, their former leader on a prior flagship mission, *before* he even began his speech. I had previously assumed Jake's authority among the scientists was because of his "mafia" connections to the boys' club. It was on this occasion, however, that I noticed his bolo tie.

Organizational position determines which cowboys are successful and celebrated for their leadership and which are considered "personalities."

Jeremy's position at the top of his own mission hierarchy with the power to produce change was as much a key to his perceived competence as a leader as structural powerlessness was to Isabelle's or Leon's perceived shortcomings. But men who attempted to deploy leadership tactics while buried in a bureaucracy are at risk of being described as "personalities" or difficult to work with. Jeremy himself stopped attending Helen science meetings, frustrated by his experience on the project. This underscores how labeling such men "charismatic" is itself an organizational accomplishment—that is, organizations that permit charismatic authority allow us to ascribe positive leadership characteristics to their visionary leaders. But it also demonstrates that charisma is not supported among the rank and file. Because efforts at charismatic leadership are not tolerated from among lower echelons, such men's (and they were all men) ability to effectively command a group is contingent on their collaboration's social organization as well.[16]

Cowboys and geeks, the suits and the counterculture, boys' and girls' clubs: these recognizable identities in planetary science resonate with work in science studies examining gender and scientific careers. For instance, Sharon Traweek describes the archetypal stories about what makes a good physicist alongside the career arc of a neophyte as a form of "pilgrim's progress." In her analysis, "affect and gender are significant components in the division of labor in laboratory research, as well as in decision making, dispute making, and leadership styles that are part of the whole realm of power and tradition in scientific research" (Traweek 1988, 105). Historian of computing Nathan Ensmenger too describes the emergence of the "computer bum" and the hacker identity as part and parcel of professionalization strategies among computer scientists, including hiring that favored the "antisocial, mathematically inclined, and male" (Ensmenger 2015, 51). Such examples reveal each field's self-fulfilling prophecies of who is to be successful, trusted, hired, or promoted, echoing studies of hiring practices in other industries that similarly rely on "cultural matching," social networks, or gender frames to identify "meritous" candidates (Castilla 2008; Rivera 2012).

Personalities and the Organization

In his study of executives in a matrix organization, sociologist Calvin Morrill notes how local cultures of conflict management intersect with gender to produce different identities in the corporation. As individuals confront and resolve conflict at their toy company, they earn monikers such as "the Princess of Power" or "the Iron Man" (Morrill 1991, 1995). Similarly, these two missions' local organizational forms and expectations for interaction can either exacerbate or downplay the effects of identities and affinity networks, producing different biographical trajectories for individual scientists.

For instance, I have described Helen's matrix organization, which combines lateral ties with hierarchical accountabilities. In the extraordinary success of the old girls' network, we witness *both* the importance of women's networks to advancement in lateral scientific organizations (Smith-Doerr 2004) *and* the relative success of minorities in bureaucratic-hierarchical groups (Freeman 1972). After all, at Helen's project office, there is a ladder that can be climbed into positions of authority. The matrix also presents a case where women, especially those placed into subordinate positions at first, can benefit from belonging to an affinity group or from transfer of social capital from higher status networks (Brass 1985; Burt 1998; Ibarra 1993) even as they engage in conflicts over autonomy and the fair allocation of resources that otherwise risk backlash due to role incongruity. Within the matrix too, many women may be employed across the organization but only a few singled out for tokenist effects. For instance, I noted that "personalities" did not typically wear multiple hats, suggesting that matrixed team members perhaps enjoyed the buffering effects of role distance. As for masculine identities, I have already described how the fate of the "cowboy" depends on his organizational position. This was further driven home when I was told that a teammate in Helen's ranks who wore a cowboy hat and lived on a ranch outside of town was "a real personality."

On the other hand, affinity networks on Paris were conspicuous by their *effacement*. In the collective organization, the duality of collectivity and charisma supports a leader's charismatic identity at the same time as it flattens other networks and identity markers within the context of the team, even across institutional borders. Jeremy filled his team roster with high-profile friends as well as a few of his chief rivals in the field; once before and once after launch, this initial group was counterbalanced with independently selected newcomers. Like other social distinctions such as status, existing distinctions between affinity groups such as boys' club members, geeks, counterculturals, and others were locally flattened due to the collectivist interactional principles on the mission. It was only when I moved outside of the team to circulate in planetary science more generally that I noted that Paris members such as Sam, James, Alexa, and Tony—whom I had observed working earnestly and intensively together under high degrees of interpersonal trust—were actually members of entirely distinct affinity networks. I never saw them socialize outside of the mission context, such as at conferences or public events, where networks and stakes became visible and tangible. However, their tone of voice softened when discussing their work on Paris, in contradistinction to their boisterousness, seniority, or other attributes for which they were otherwise well known. Retrospectively, it was strange to be an environment where the field's persistent identity networks and typologies were repeatedly neutralized.

Flattening can create a safe space for team members to interact while giving them a chance to work with a wide range of individuals whom they might not otherwise encounter. This is advantageous for junior scientists, who regularly exclaim at their good fortune of working with those who had "written the book on Mars." However, flattening can also be problematic as it effaces distinctions locally that do matter externally. Paris's initial senior roster was all white men; it took two open calls for participating scientists to include an influx of midcareer and senior women on the team. During my fieldwork, some of the women on the mission organized a TacPlan meeting where all the roles were staffed by women. Described externally as a significant moment in the history of space exploration and sometimes applauded by participants, I also heard some of the women involved downplay its importance, maintaining that it was just for fun.

Effacing distinctions also means that there is no girls' network on Paris. There is no internal hierarchy to climb, of course, and no latent boys' club who actually pull the strings—despite the fact that team members on the mission's original roster enjoy high status in the field more broadly. But with status markers flattened, it is unclear how minorities' local participation translates to promotional opportunities in other projects. This is important because network ties *do* matter when it comes to external validation and continuing projects. It mattered very much for the women on Paris who tried to set up a lunar exploration network but were limited in their ability to leverage their positions on Paris to gain external resources. It also matters when selecting positions of public visibility for the mission. At anniversary parties that I attended and commemorative events at Spacelabs or at the Smithsonian Air and Space Museum, not a single woman was invited to the podium or onto the stage to speak—despite the fact that there are many, many women involved in both the engineering and science sides of the mission who could speak eloquently to the mission's importance. Although it can be a refreshing situation for the minorities who found themselves working in this environment, the flattening of networked identities on a collectivist team can reproduce the very status effects externally that it attempts to eradicate internally. Those who assemble flat teams would do well to consider how similar dynamics might affect minorities in their organizations as well.

Finally, as scientists cycle from one long-term project to another, the biographical outcomes of *organized science* can compound over time. Boys' club members occupied the initial planning roles and responsibilities on Helen: the period of greatest prestige, status, and financial support for the project responsible for laying down organizational pathways, procedures, roles, and rules. As the mission wore on, money and status transferred to new projects, such as the then-nascent Mars Program—and many of these men followed. As Mars became known as the place to be for young upstart engineers and scientists, the men who had played important roles on Helen moved to other projects (including Paris), bringing their continued high status with them. The women beneath them, initially in supporting roles, moved up and promoted their own trusted acolytes as deputies. But as more and more men left for Mars, the organization they left behind became increasingly feminized in terms of its senior staff, which diminished its status on-lab—for instance, when a promising young team member described transferring from a Mars mission to Helen as "moving backwards in your career."

This example recalls patterns of white flight that produce neighborhood effects in urban centers, whereby certain areas maintain higher concentrations of low-income, racially homogenous populations (Massey and Denton 2003). This pattern repeats. Planetary science was dominated by men in the Apollo era, but powerful men left lunar science for flagship opportunities at Mars in the 1970s and then moved to the outer planets via the flagship missions flown through the early 1990s. They then returned to Mars with the fresh funding stream in the Mars Program. In each case, a boys' club follows the funding, sets the agenda, and conducts pioneering work. Only when the missions that the men left behind become less competitive and funding rich (with focus on, say, data analysis instead of gathering or mission building) are they increasingly available for female participation and promotion.

Such field-wide "male flight" produces a systemic power imbalance in planetary science. Despite the apparent gains for women in leadership positions in science due to robust mentorship networks, including those on Helen, their promotions upon men's departure produce increasingly feminized spaces with conspicuously absent power players. As past missions become increasingly feminized, they become the kind of place where young women can find mentors. They commit to the organization—and the mentors—that committed to them. Hence both Rose and Betty informed me that they had received job offers from many different missions but chose to stick with Helen. A concomitant effect of this process along with the mission-specific networks is that such experienced women become stuck in increasingly lower-status organizations, even as young men who are seen as promising are offered opportunities to move to new missions with higher status. Viewed over time, women who are "competent" get to run existing missions but not new ones, while their male peers move to the mission frontier—and up the ladder.

Conclusion

Traweek (1988) asserts that the narratives available for describing a career in physics "could only be men's stories." Planetary science is open to broader participation, but relatively circumscribed identities and narratives remain available as successful possibilities within the field. While these are not just "men's stories," they are primarily white stories: I only met two African Americans in the field, as participants of color were typically immigrants or foreign students (Alegria and Branch 2015; Ray 2019; Lewis 2004).[17] Just as Morrill describes how largely similar groups enforce local ways of differentiating among themselves for the purposes of career advancement (1991), the specific identities available to planetary scientists and their associations with differential merit and status produce different forms of career mobility across their field and in NASA collaborations. Old girls' network and boys' club members experience very different career trajectories than their peers, especially as more senior members of the community act as mentors and assign them to positions of power. Counterculturals and geeks rarely move into powerful situations; cowboys gain accolades from their charismatic successes, while personalities find themselves undermined upon taking on a role they hoped would bring status and respect. Just as each mission's organization displays particular ways of moving across international, interinstitutional, and temporal divides, certain gendered and racial identities and networks can buffer individuals as they make these same moves, shielding them from community outcry. They are simply defending the rules, upholding cultural norms, or doing their jobs well, while other identities or organizational positions produce situations of tokenism or ostracism.

This tour through the people involved in mission work may feel strange coming toward the end of a long book about organizations. Surely, we need to know who these people and their inherent personalities are in advance to understand the scientific collaboration as an organized endeavor. And the scientists I met on mission collaborations were adamant that their success was a question of the personalities in the room. But these personalities were neither individual nor innate. Instead, as I have argued, scientists' ascribed "personalities"—personal characteristics, stylized stories, and narratives about them that I observed in the field—are *produced* through their interactions in these scientific organizations. Taking an organizational lens to the question of "personality" demonstrates how interinstitutional setting, organizational context, and the combination of network ties and self-presentational styles coalesce to reproduce persistent inequalities in science. Interactional dynamics between these groups within the organization and across their institutional contexts explains how and why certain men maintain power across projects, certain women surpass men in the hierarchy, and certain individuals are either celebrated or decried. These inequalities have implications for which scientific observations are accomplished and which are not and for who gets to set up, establish goals for, and command missions. They also have implications for which scientific findings are considered "good," as well as who is seen to be a "good scientist," deserving of future opportunities.

Scientists and engineers alike rely on gendered identities and robust networks as they confront the organizational pressures of remote operations, multiple memberships, and multi-institutional ties over the long *durée*. The ability to stay within an orbital resonance in harmony with one's peers clearly offers an advantage for career placement and opportunity, as individuals navigate the complexities and contradictions of mission work and reputational or scientific success. But those who fall into the fault lines between competing institutional demands, who are unable to accrete toward an affinity network or adopt a rewarded presentational style, risk ostracism and being labeled as bad actors. Should they erupt, like Enceladus, under the pressure of conflicting social forces, their resulting reputations will be attributed not to organizational effects but simply to their "personality."

TEN

The Iterative Loop

The organizational shaping of science—and scientists—that I have described occurs subtly yet consistently over years and even decades of a scientific collaboration's work. How is this possible? It could be simply a result of the routine and ritual character of each mission's interactions, yet sociologists are divided over the obduracy of routines. Routines may provide institutional stability as an organization reproduces itself over time (Zucker 1977), but they can also be opportunities to introduce change (Feldman 2000; Feldman and Pentland 2003). In my observations, ritual meetings such as the PSG, the target teams, and the TacPlan saw limited adjustments to their structure and format despite opportunities for change. Even outside influences, exogenous shocks to the system that could trigger significant change, were incorporated into mission culture (Hannan and Freeman 1984; Haveman, Russo, and Meyer 2001). For instance, when Spacelabs introduced Cisco's WebEx for virtual meetings, both teams used the tool much in the same way they previously ran their telecons, sharing slides or spreadsheets with limited video function. Following severe funding cuts in the aftermath of the 2008 financial crash, Paris personnel were combined into new "super" planning positions, amplifying their motivation to avoid complacency amid overwork. Meanwhile, Helen scientists colored their traceability matrix in red, yellow, and green to demonstrate what science would be eliminated at lower levels of funding. Each mission handled these new hurdles in locally consistent ways. That these sociotechnical systems seemed impervious to significant effects points to a different underlying mechanism for stability on scientific and technical teams.

In factories or in companies, the product of work goes out the door and into circulation in a commercial or public sphere. But in scientific organizations, the products of organizational practices—the knowledge, datasets, people, and instruments involved—*feed back into* each organization. Borrowing a phrase from computer programming, I call this mechanism *iterative looping*: a process by means of which an organization's outputs become its inputs, reproducing and reinforcing organizational norms while ever more tightly coupling process with output. Iterative loops are procedural, stepwise processes that iteratively elaborate upon an object toward a desired end. Unlike infinite loops or the closed circuit of feedback loops, the object of processing becomes more refined with each iteration. As a mechanism, iterative looping describes and explains how mission team members can *both* reinforce organizational stability through periods of change *and* ever more tightly couple their social and natural worlds. Iterative looping also provides a microsociological perspective on—and mechanism for—those situations that economists or political sociologists might identify as due to "path dependence," "lock in," or "increasing returns" (Arthur 1989; David 1985; Krugman 1979): feedback effects whereby decisions taken early on are continually reinforced and become difficult to undo, producing a form of "entrenchment" (Starr 2019).

In this chapter, I use moments of stasis and change to describe the third principle of *organized science*: that scientific outcomes impact scientific organizations. First, I show how what appears to be organizational robustness is due to iterative looping effects, whereby the *products* of group interactional processes loop iteratively back into those same processes once more as their *input*. This loop reifies organizational processes and produces stability over time. Second, I show how moments of radical organizational change create changes in scientific output as well—that is, when the organization does change, scientific results can be seen to change with it. Although these cases emerge from both stability and change, the outcomes under these two conditions ultimately lend support to the third principle of *organized science*.

Modes of Iterative Looping

A key example of this sort of iterative looping concerns the knowledge that mission team members craft through their organizational work. For instance, the framework of *organized science* reveals how the planets themselves are organizationally constituted; our vision of Mars and Saturn differs along with their local processes and social orders of planning as oriented toward consensus or integration. Yet the planets are both the *objects* produced and shaped by organized scientific work and the *subjects* of continuing, iterative scientific questioning based on this ongoing work. *This is the crux of organized science's iterative loop.* With each encounter with the planet, the teams produce an organizationally shaped vision of that world. They then reenroll this perspective on the planet in formulating their next set of questions based on prior findings.

Alongside reinforcing the organizationally shaped qualities of the planets themselves, the iterative loop simultaneously reinforces patterns of social interaction. As each planet seems simply to demand certain types of further questions and instrumental encounters, this naturalizes those Earth-bound social formations that give rise to these types of questions and outcomes. This is not a recursive function but an iterative one. Because successive findings build iteratively upon each other through the context of planning to inform next steps, they suggest new scientific possibilities, results, and discoveries even while reinforcing the same frame of inquiry. In this way, social order—organizational modes of interaction on Earth—and natural order—the resulting knowledge about the planets—are crafted together.

For instance, Nick's experiments under the Paris cobble campaign led to the development of multi-instrumental investigations for understanding the qualities of the cobbles and soil around the spacecraft. Seeing the cobbles did not lead to a moment where team members were convinced that they needed thermal scans of every rock in the area to the exclusion of all other instrumental readings. Instead, it led to a process of taking photos using a range of filters to identify good candidates for getting "the full treatment" of the entire instrument suite, including the thermal instrument as a key player. This happened again and again as scientists explored various aspects of the rocks, craters, and soil the robot encountered. This was also the norm on the Helen PAM collaboration, as I described in the Aurora case. On the remainder of the Helen team, this was not the case, even within its most collegial working groups. For instance, the Titan working group fostered scientists who examined Titan's lakes, seeking variation due to evaporation or precipitation over the seasons. Their questions enrolled the cameras, the infrared instrument, and the radar instrument to answer, among others. These were each published separately as findings, referring to each other to corroborate observations or support past claims. I did not witness a team meeting of any kind where multiple prospective instrumental views—compositional information derived from infrared, shifting cloud cover information from imaging, and terrain data from radar, for instance—were planned collectively as relevant pieces of a shared puzzle. This pattern repeated itself for thermal, radar, or ultraviolet imaging of the rings, stellar occultation tracking, gravity passes, and so on. With few exceptions (PAM among them), the response to questions raised through single instrument views was almost always more single instrument views, carefully integrated into new timelines alongside other instrument requests.

This is *not* because the people involved are bad scientists, are not friendly with each other, or are not open to collaboration. Quite the opposite: for Helenites, the above is *evidence of a good collaboration*, demonstrating mutual respect for individual topics and expertise, as well as early information sharing about results that supported others' premises for observation and publication. Meanwhile, for Parisians, good collaboration means opening a question up to wide engagement. But the sociotechnical systems in which these scientists are embedded do not permit certain types of questions or outcomes. No matter how "good" or promising they are or how many friends they have on different teams, no Helen instrument scientist could credibly speak up at a targeting meeting to request a different instrument's observations, any more than Nick could have successfully pushed an encyclopedic thermal view of regional rocks through Paris's consensus process for weeks on end. Even assuming that such a request made it through Helen's integrative planning process as a top priority for both instrument teams, they might not be able to observe at the same time due to the instruments' placement on the spacecraft. This was especially the case for the radar but also affected other remote sensing instruments, as the Aurora example demonstrates. Follow-up data might only be collected years later, should the spacecraft happen to fly over the same region and use a complementary instrument. This is not impossible: new claims about Titan lake shores, for instance, could rely on fresh radar data as well as a colleague's published imaging or infrared observations, and the cryovolcanos announced at the press conference were observed first by radar, then via infrared. Still, each scientist respectfully speaks first from their own instrument's (and organizational) position.[1] Just as Paris single-instrument work typically takes place within the context of multi-instrument investigations, multi-instrumentalism on Helen follows organizational orderings as well.

How data circulates on the missions also reinforces the way that organizational processes, over time, build up isomorphic organized visions of the planets. I have already described how data about the planets circulate relationally, concordant with mission organizational orders. But as the stuff with which science is done, this data also participates in an iterative loop. The collaboration produces the data through its social processes, which reflect and project particular organizational processes and concerns. Then the collaboration absorbs those same datasets as the means by which its scientific questions are answered, new questions are formulated, and social relations are enacted. This in turn produces more data shaped along the same lines, more visions of the planet in alignment with the group's organizational order. This more deeply entrenches relational ties that underline this orientation and so on. The iterative loop of data circulation therefore repeatedly produces both objects in the world (the planets) and the knowledge-making organization itself (the missions) in the same image. By virtue of this iterative looping, data production becomes a site of epistemic, ontological, and organizational entrenchment.

Even the spacecraft itself is a prime site for iterative looping. The interactional body of the robot demonstrates how participants accounted for their spacecraft's actions in ways that reinforced their local organizational interactions and accountabilities. This held true even during moments that should have provided an exogenous shock to the system, such as when the spacecraft failed or encountered difficulties. When a Paris robot got trapped, I asked a group of scientists over lunch who made the decision to let it go somewhere so dangerous: the answer was hung heads around the table and a solemn, "We all did, Janet; we all did." When the robot's infrared instrument failed following the cobble investigations, the team continued to use it during prime observational hours in the hopes that someday, they might figure out an algorithm to retroactively resolve the problem in the data. Eleven years after the mission began, I observed a meeting that followed up on an operational error. An engineer stood at the front of the room and repeated this phrase printed in italics at the top of his slide: "*We are all responsible.*"

On Helen, similar circumstances had different outcomes, consistent with the organizational imperative of integration. The plasma instrument failure led to one of the ritual meetings between scientists and engineers overseen by the project manager. Although the instrument team PI, Roger, denied that there was sufficient evidence, citing hazards to the other instruments and the spacecraft, the project office turned their device off and their observation time was released to the other teams. Plasma scientists were dismayed at the loss of science observations, but this did not adversely affect other teams' science plans. I also observed several months' worth of meetings and concern over the degradation of the spacecraft's reaction wheels. In each case, the mission's rituals repeated. Procedures were put into place that continued to reinforce distinctions between instruments and between scientists and engineers such as trading time or workforce allocations between them, using software for wheel modeling that was opaque to scientists or instituting rules about the use of hydrazine fuel, the use of reaction wheel flybys, and the use of y-biases. Integration continued to be the appropriate way to think about, use, and respond to the spacecraft as well as the many distinctions among members of the team.

Changes to the spacecraft's body, then, did not serve as an "exogenous shock" to the mission organization. How each team responded to the eventual degradation of their spacecraft over time was consistent with their organizational setting, not a source for significant change. Each spacecraft and its capacities were continually reimagined along existing organizational lines, as described in chapter 7 (Haraway 1997; Mazmanian, Cohn, and Dourish 2014; Orlikowski 2010; Suchman 2011). With each planning cycle, this organizationally consistent vision of the spacecraft was both put into practice and justified by prior interactions, further reinforcing the iterative loop.

An Organizational Exogenous Shock

The iterative loop is a powerful way of ensuring both scientific and organizational stability at the same time. It enrolls processes and products of scientific collaboration in the continual reproduction of natural and social orders. As such, it is very hard to change, or to see how the scientific work could be otherwise. Especially when discovered properties, made visible through *organized science*, appear to be properties of the objects themselves through actors' ontological work in the world, the phenomenon scientists observe seems simply to demand more of the same. Just as economists argue that increasing returns can be a source of monopoly power, looping effects may produce a form of epistemic monopoly, especially if the only available views of that object are through a singular community and their suite of instrumentation. But while the iterative loop increases robustness in the face of change, that does not mean that the missions are entirely impervious to change. One excellent example that occurred on Helen during my time on the project was the removal of Malcolm and replacement with Francis as the new project scientist.

It was not simply that a new leader was put in place.[2] Francis was not previously a member of Helen and espoused quite a different style from the mission's organizational philosophies: a soft-spoken member of the counterculture group committed to its values of collective work and scientific skepticism. Sam reportedly gifted him a copy of Machiavelli's *The Prince* to help him prepare for his new role, saying more about their affinity network's sarcastic view of management than Francis's predilection for manipulation or intrigue. Helen team members were wary. His counterculture friends were eager for news from the big mission, which was otherwise opaque to them as they were excluded from its ranks, but Francis was reticent to gossip given his new relationship with the team. Still, Helen team members took note when Francis and Sam publicly argued against cryovolcanos on Titan in print, in press releases, and at the press conference. Although Francis earnestly insisted that he maintained a sharp distinction between his scientific and managerial work, everyone knew that the lab was grooming him for a competing project. Helenites were used to wearing multiple hats, but such mixed external allegiances made for a difficult case for Francis. He also seemed unwilling to participate in the organizational ostracism of Isabelle, preferring instead to listen to her and try to "bring her into the fold."[3] This managerial style was likely yet more evidence, to Helenites, that Francis simply didn't get it.

When I joined the team, Francis had been in his position for only a few months. But he wanted to change Helen to include ways of working together, across instruments. Convinced that this was the way to do the best science, Francis sought to create institutional-level changes that would enable such work to happen. I have already described the Aurora effort, which was instigated through Francis's conversations with Gabor to provide project support for the meetings. Francis attended the first meeting to reiterate his hope that the venture would succeed. He also sought to create other changes that would facilitate instrumental data sharing across the mission. Yet these had to operate under the existing rules of engagement on Helen in order to be successful. For instance, he pushed for public release of a combined thermal and camera map of the south pole of Enceladus, which required several months to negotiate. Isabelle was unwilling to release camera data to Edward's institution to do the mapping and was wary when he would not share his calibration algorithm with her to do the work at her institution. Of course, she may not have known that there were several competing calibration processes on the thermal team—these and other local pressures upon both instrument teams were invisible to the other. Francis suggested that someone at Spacelabs' image processing facility could do the trick, but this was hardly neutral territory for Isabelle. In the end, Francis and Victoria found someone at Spacelabs associated with the project that both acquiesced to give their data to individually. Thus the norm of group autonomy in integration was upheld.

Francis also attempted to change the rules of the road and asked me to join him, Victoria, Patricia, and a few others on a Data Working Group—just as Jake had charged to Joseph on the prior flagship, twenty years earlier. Our task was to present Helen with a unified rules-of-the-road document that enabled data sharing.[4] Like many advocates of open data, Francis had a vision for openness but failed to account for how data moved in ways concordant with membership norms. These meetings revealed competing goals: Victoria emphasized the autonomy of the PIs and a central coordinating role for the project office, Patricia wished to bring data management practices into line with the goals of the PDS repository, and I attempted to suggest some form of social reciprocity for data exchange. In response, the statement from the PIs and team leaders in an executive summary suggested that "philosophical recommendations" that each subteam could implement as they best saw fit would go further toward facilitating the kinds of investigations that the project office intended than specific rules. The cultural norms of instrumental autonomy were again upheld.

In another effort, Francis and Kenneth, the program representative at Headquarters, conspired to establish a participating scientist program for Helen. Participating scientists are selected by a NASA review panel and join a spacecraft team after it has already launched. Each mission has a slightly different participating scientist's program, tailored for its own needs, and some resist having them at all. On Paris, two groups of participating scientists had already been added to the initial team: one phase before launch and the second two years in to operations to diversify the team's expertise. These were, for the most part, outsiders to the mission, although many knew each other from the Mars community. I observed no distinction in status, form of talk, or authority allocated to these newcomers as opposed to the initial team during my time on the mission. On Helen, Francis and Kenneth also envisioned bringing in scientists with fresh perspectives. Francis, for his part, was keen to expand participation on Helen to newcomers: for instance, Marvin, a counterculture friend who was not a mission member, had recently released a map of certain moons that Helenites had found useful for observation planning. Francis hoped that this was proof that team members had much to gain from welcoming newcomers.

The PIs, team leaders, and interdisciplinary scientists met the initial proposal with some hesitation and convened a working group of PIs and team leaders to study its feasibility. The group's membership was comprised of those who had already expressed resistance to requests from Headquarters to release data more broadly due to the lack of resources that they would be given to satisfy these new goals; they were therefore well placed to address the tension between externally levied requirements and local institutional needs. The group began to meet just as Francis left the project and Victoria was promoted to project scientist. At the next PSG meeting, Victoria reported on the progress of the plan to "create an infusion of new ideas into Helen's science and planning." Proposals would be open to "any of the team associates" who would be expected "to use multiteam investigations" or single datasets to solve problems and might participate in planning the mission for one year at a time.

Kenneth was at the meeting too. He argued that the new scientists would have access to data prerelease and that the program would pay for them to attend PSG meetings. Most of the scientists in the room favored piggybacking the new program onto an existing NASA plan for data analysis, which funds external scientists to work with existing datasets to solve scientific problems. This would keep the new recruits in line with existing regulations such as each instrument's unique rules of the road and local cultures. Helenites also wanted the proposals to be evaluated by existing team members (this was not the case, although several team members sat on the eventual review board) and resisted including newcomers in planning observations. This latter point was essentially moot: recalling how the Titan jumpstart had already divided up all remaining Titan passes in a planning meeting the year before and that even the prior choice of a tour had transformed observational priorities into specific observing opportunities, there would be little that a new team member joining for a year could do to impact the process or get up to speed. Even if they joined operations, what with the new ten-week planning cycle they might not get their requested data back before their contract expired.

The program was also developed to assuage the tensions associated with the long temporal scales of the mission and its overlap with career advancement. Francis and Kenneth had already created a "team associates" roster for existing members who played an active role on the mission such as Edgar, Eleanor, Finn, and Grant. These scientists had started on Helen as graduate students or postdocs and worked concertedly on the mission their whole careers but lacked official status as a coinvestigator. But when Kenneth discussed the participating scientists program at the PSG meeting in early 2011, early-career scientists in the audience wondered if the program could potentially *undermine* their positions with their existing teams.

For instance, Drew's postdoctoral funding at a prestigious university relied on professors in his department who were on different Helen teams. He was already contributing significantly to Helen science and operations but had only recently been nominated to the new team associate position. When he asked if he could apply for a participating scientist position, Kenneth answered, "We're not going to pay you as a team associate to do *x* and pay you as a participating scientist to do *x* because that's called," he paused for dramatic effect, "fraud." Everyone laughed, but Drew nodded earnestly: "Of course, of course," then asked if he could still be involved in operations. Kenneth again described a situation where Drew would have to be responsible for drawing clear lines between his involvements in different roles (perhaps such as Rod once described himself as two different instruments at once). When Kevin, another junior scientist in a similar position to Drew's, explained that his peers were worried about taking faculty jobs and leaving behind the coinvestigators who funded them, therefore having to leave Helen, Kenneth nodded fervently. "That's why we're doing this," he insisted. "This is meant as an opportunity for you guys to step up and mature, to fill a new role, and perhaps have an opportunity to break out on your own. . . . Take the next step: go fly, little bird!" He then switched tones and addressed the crowd: "Hey guys, you've been together for a long time. Let me be blunt: it's normal for people to leave."[5]

The new participating scientists were announced at a PSG later that year. Both Francis and Kenneth looked forward to seeing which fresh faces would join the mission. But as Victoria went through the twelve awardees at the plenary session, only Marvin, Francis's friend, was a true outsider. Drew and Kevin were on the list, as were others who had organized PAM or Titan team meetings and junior scientists I recognized from the Titan group or from my visit to an instrument facility in Rome. Some of them proposed joint instrument work, others single instrument, but the majority were recognizable experts in one instrument. It could have been that the Helen team was so large and distributed, with such limited visibility across teams, that individuals did not recognize these "new" team members and honestly believed that they had selected fresh faces. The pattern repeated itself the next year, and the next.

I recalled an early-career Helenite, in a conversation over a drink at a conference early in my ethnography, once explaining to me why so many of the Helen talks were given by Helen insiders: as he put it, "*You have to be born into Helen.*"[6] Similarly, the new program, once imagined as a way of bringing in fresh faces, became a resource for young Helenites to manage their organizational status within the mission at the same time as managing their career status as scientists at different institutions—where they were expected to move up but, with no other mission to go to, could not afford to move out. The program therefore resolved many of the interinstitutional and temporal tensions described in chapter 5 in ways concordant with the mission's norms. It also became a way to seed the mission with continuing participation from those who already knew its interactional rules.

Gradual Change

It would seem that top-down initiatives were largely unsuccessful in changing Helen's organization. Yet certain things *did* change. At a meeting I attended a year before the mission ended, science planner and thermal scientist George (now Victoria's deputy) put up a slide describing Helen's rules of the road, a version of the one that the data-working group had long ago adapted from PAM, successfully renegotiated by Victoria. The slide began with the following: "The rules of the road offers guidelines for multi-investigation studies only and it applies equally and fairly to all Helen/Probe Teams . . . to data that are not yet in the Planetary Data System." It listed the hierarchical leaders of "teams," such as PI, team lead, or interdisciplinary scientist, requiring associates to be "authenticated" by one of these leaders. The rules reinforced that "each Team has overall responsibility for its investigation," including decisions about analysis and publication but that these teams would be expected to release their processed data for other members of Helen for multi-investigation studies "within a reasonable time" to "identify possible scientifically interesting events." The rules also stated, "Results from single investigations should be published first as much as possible." Thus a model of "equal" and "fair," autonomous, hierarchically sanctioned, multi-instrument collaboration emerged from within the integrated system of Helen organizational structure. It was at this same meeting that Drew's colleague Kevin turned to Gwen in excitement. He had just lined up her recently published ultraviolet data about Enceladus acquired five years prior with data from the infrared instrument he worked with as a participating scientist and noticed some interesting correlations. He presented preliminary findings at a field-wide conference the next spring, a few months before the spacecraft's demise.

Participating scientists, for their part, were largely happy with the program. At an outer planets advisory meeting in 2016 those who came up through Helen's ranks, like a young Titan radar scientist, spoke up fervently to support the program, saying things like "I was able to get a start on Helen," while Daniel reported that he now heard fewer people complain that "Helen was a closed-door thing." Still, even as Marvin thanked the committee and NASA for making it possible for him to join the team, he noted that "something like maybe 50 percent or more of the participating scientists who were brought in were already embedded in the project in some way." As he put it, "it serves a purpose to elevate certain people in ways that the project can't do but should that be a principal result? You had [bullet points on the program announcement] that said bring new people in." Others who were never affiliated with Helen and whose applications had been repeatedly turned down expressed their anger that there was still no external on-ramp onto such a large and important project.

Also in 2016, a group of scientists put together a survey about the participating scientist program to gather data from as many missions as possible. The data gathered indicated that the participating scientist programs on both Paris and Helen had slotted into existing organizational norms. Paris participating scientists extolled being treated "as full team members" and "as equals" and described the mission as "by far the best integrated team I've worked on." One explained that it was "a pleasure to learn about the style of leadership that treated the entire robot as a single entity designed to understand the origin and evolution of the site" and gave "the greatest credit" to the PI for this "intellectual framework." Comments about Jeremy were similarly unrestrained, indicating that "his example was followed by all of the original team members," concordant with the charismatic layer of the organization. Others called it "the gold standard," "one of the main highlights of my career," and "by far the best [participating scientist] experience I could have had" and said it "could not be improved"—and when asked what they should change, most replied, "Nothing." These comments were typical of Paris forms of "happiness"-oriented consensus-talk and effusive, collectivist affect. Helen respondents were also thrilled to be selected, although most noted that they were already affiliated with the teams and were therefore relieved for the "formalization of my position within the mission" and their "secure status as part of the Helen project." One noted a case where the participating scientist had been able to make contributions to operations but the majority indicated that they wished the program had enabled them to get involved in the selection of targets and observations. Despite the effort to create change through the team roster, both examples demonstrate the robust continuation of mission organizational orders.[7]

That does not mean that change did not occur. Working with a graduate student in 2015, I reran the network analyses of copublications. This time we scraped data from the Harvard Astrophysical Database and instituted new tools for visualization over time. The Paris team remained tightly clustered. We detected a latent community of spectroscopists as distinct from scientists using remote sensing data, but these groups overlapped with many dense ties between them. Consistent with Jeremy's commitment to group consensus, a centrality measure found that the PI was not the most central node in the cluster. The consensus format and collectivist structure of sharing "the keys to the spacecraft" were therefore in full force. Centrality was shared between Jeremy, his deputy Tony, and the camera leader on the mission, showing their importance in interrelations across all members of the team. The technique also detected two groupings among the remote sensing cluster, one affiliated with Jeremy and the other with Tony, yet with many interconnections between and across nodes. Rather than showing a split in the collective associated with the two vehicles, upon closer inspection the ties demonstrate instead a widening of Paris's net of synergistic scientific work. At the time, Tony was also a senior member on two other missions, and the members of his publication cluster were busy combining datasets not only across Paris instruments but between Paris and those from other Mars missions as well (figure 15).

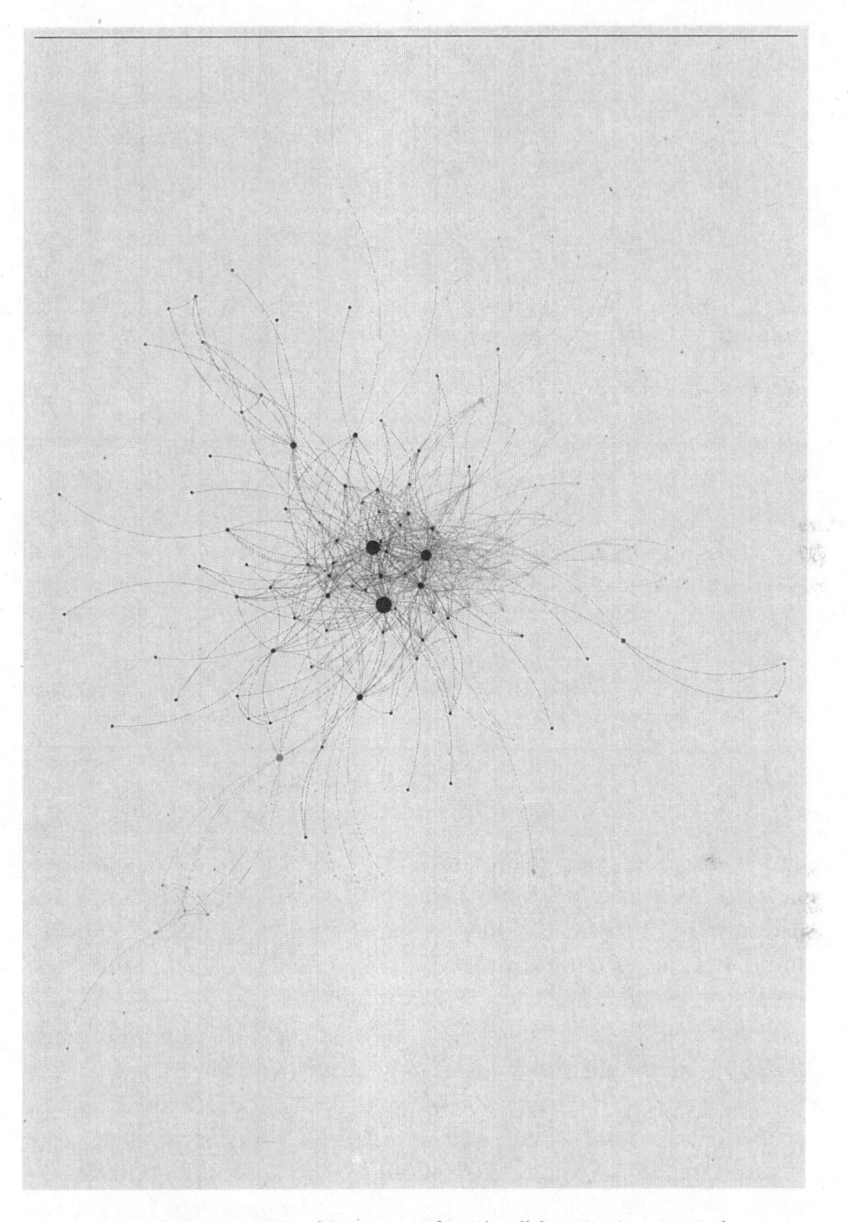

15. Copublication matrix of the core set of Paris's collaboration in 2015. Nodes are individual authors; ties indicate at least five copublications. The computer detects three clusters in this otherwise dense network. A betweenness-centrality measure indicated by node size reveals that the PI, his deputy, and the camera lead play brokerage roles within this overall strongly connected network.

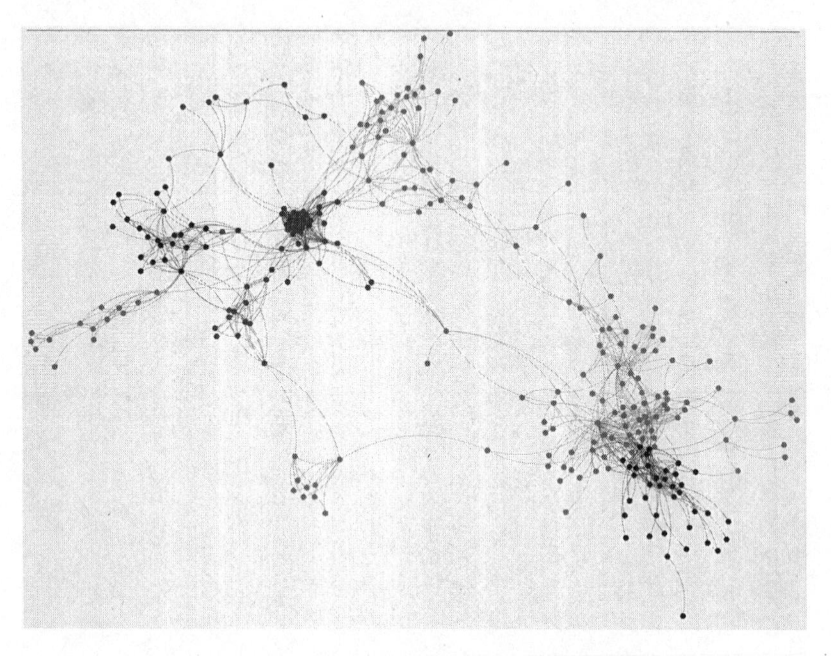

16. Copublication matrix of the core set of Helen's collaboration in 2015, at threshold of fifteen copublications. Nodes are individual authors; the physics component is at bottom right, optical and remote sensing instruments cluster at top left, linked through the dust sensor. Thanks to the aurora investigation, new bridging nodes now robustly connect these groups.

On Helen, mapping publications over time revealed the emergence of new clusters at the tie level of ten and fifteen copublications (figure 16). Each instrument team was visible and relatively distinct, although two PAM instruments are indistinguishable at ten copublications. On the optical side, there were few ties between clusters other than those offered by the interdisciplinary scientists, while the physics side still identified distinct instrument teams but with more interconnections between them, consistent with PAM joint questioning and observational work. Further examinations revealed changes that were consistent with organizational developments that took place during my period on Helen. A new series of bridges emerged between the physics and remote sensing groups, linked by those who had played a role in the Aurora workshop. Two of the young scientists involved in this initiative had begun as postdocs and later been selected as participating scientists on the team, now visible as brokers in the network. The thermal team's publications also revealed a new cluster of multi-instrumental publications tied through interest in Saturn's atmospheres—the outcome of the

CAKEs that piggybacked on the PIE process. The Titan scientists were increasingly tightly clustered around radar scientists' nodes, likely due to the radar team's "affiliate" program that they extended to colleagues in the Titan community. Each of these features was evidence of a robust collaboration of at least ten copublished papers. Clearly, the top-down changes spearheaded by Francis and by Victoria *did* impact mission science. In typical Helen fashion, they did so through the bottom-up context of local collaborations, each with independent, local, and autonomous ways of addressing the challenges of integration together.

Organizational and Scientific Change

Because the processes of *organized science* produce and stabilize the objects and subjects of scientific work, they result in considerable robustness in the face of organizational change. They also produce different visions of the planet—a patchwork quilt that pulls together single instrument and target observations on Helen versus a dense interweaving of instrumental visions on Paris. Those visions in turn affect future iterative interactions with the objects of study and the people on mission teams and extend to the scientific findings of each mission, as well as the data that result as the outcome of the observational requests. It also extends to the people on the missions, whose biographies are shaped by their local interactions. Examining these effects in detail demonstrates how research, data, and people are not only the outcomes of organizational interactions, but as they are shaped by the collaboration, they are also continuing players in those same, ongoing interactions. This thickens our understanding of *organized science* not as a one-way street—from process to product—but rather as an *iterative loop*—from process to product to process again. This is how communities with a shared organizational epistemic culture produce specific planetary knowledge and ontologies, suffused with organizational orders and understandings, and seemingly impervious to change—an imperviousness that is locally attributed to the stable nature of the planet itself.

Much like those electromagnetic field lines that both shape and respond to the surging plasma swirling around a planet, sociotechnical systems produce shaping and orienting effects and are themselves shaped by the interactions within them. These findings therefore hold intriguing possibilities for our understanding of organizational change in the sciences. Periods of change do arise even if their results are slow to materialize. Like in many organizations, I observed that changes were incorporated in ways that were consistent with the organization's local ways of sense-making and decision-making. But even in change, the patterns of consensus versus integration continued to be visible. When there is a change in the production side of the organization, there is also a change in the organization's product—in this case, scientific publications. The principles of *organized science* continue to hold.

There are implications here for those in science studies interested in scientific change. Scientists attribute such change to the progress of knowledge, explaining that we change our theories and tools because we have now uncovered new truths—or at least because our new truths are better than the old ones. Thomas Kuhn transformed our understanding of this process by arguing that the vast majority of scientific work is not revolutionary at all; instead, "normal science" seeks answers to questions that are already laid out within the group's theoretical or explanatory paradigm, and these answers must conform to paradigmatic format or be considered unscientific (as necessary for long-term stability, according to Weber 1968). Social scientists today therefore remain interested in what Kuhn identified as this "essential tension" between tradition and innovation in scientific research (Kuhn 1959). These scholars seek causes for scientific innovation that result from institutional convergences between fields or from the cross-fertilization of techniques and ideas (Foster, Rzhetsky, and Evans 2015; Mody 2014). Others, meanwhile, look at the considerable investments that scientists make in stabilizing their labs, results, and social orders and the arguments they engage in with those who threaten to undermine their social system in moments of controversy. The mechanism of the iterative loop suggests one way in which this "essential tension" is resolved, at least locally. If *organized science* points to ways in which scientific organizations reproduce themselves in the products of their work, reinforced through incorporating those same products as inputs into scientific inquiry, then iterative looping suggests an interactional mechanism for the continued production of "normal science," even in highly heterogeneous collaborative settings.

Coda

I returned to the Division of Planetary Sciences meeting seven years later, this time in Pasadena. Wandering among the posters, Paris results were no longer on display. The project had been funded at the lowest possible level. Participating scientists continued to play an operations role but were not paid for their scientific work. Many of them continued to work for free, their dedication to the mission intact even if their PI was rumored to be working on a new proposal. I recalled Jeremy once telling me that he had set up such a well-oiled machine that even if he were to "get hit by a bus tomorrow," the mission would continue in much the same fashion. Without the full attention of their charismatic leader, the organization was busy bureaucratizing (Weber 1968), adding new deputy PIs from among their ranks to continue their practices and processes. Incidentally, these deputies had originally been added as participating scientists, evidence of the collectively flattened community from which this new layer of management emerged.

Among Helen results, however, I noticed a subtle shift. Where before, I counted one out of ten posters or papers that presented multi-instrumental results, now I counted one in four. On their posters, people credited the PDS and published papers by teammates for the data that they drew together from other parts of the mission. In addition to the accumulation of publicly available data over time, by now, there had been almost two hundred Titan flybys. The data amassed from those passes made up global maps as well as many overlapping squares of observations among radar, infrared, thermal, ultraviolet, and visual data. More and more Titan posters and presentations could draw on these expanded datasets, as well as the tight-knit network of Titan scientists, to produce results. Indeed, a radar scientist's graduate student presented the latest results on Titan's lakes, now able to combine radar with visual and infrared imagery to do so. When I sat with another Titan scientist in October of that year who talked me through his work at his desk, he imported several different instruments' global maps and observations into his geographical information systems software to make claims about dune regions and wind direction. It was a very different experience than the work I had observed among Titan and even radar scientists a few years prior.

Titan was not the only benefactor of this shift. Studies of Enceladus could now draw upon global maps published by thermal team members or by camera team members. When I asked Kevin why this was now the case, he explained that there was so much Helen data out in the public domain that it was now relatively easy to find and work with the datasets. When he used ultraviolet data, for instance, he cited Gwen's published paper and included her as a coauthor. Edgar, for his part, suggested that the PIs were not as wrapped up in the "glory" of the mission by now and so they were not so worried about the data. More than a few scientists pointed out to me that Isabelle no longer participated actively on the teams. These individualist and infrastructural reasons were all credited as reasons for the change. No one mentioned organizational change, the participating scientists, the rules of the road, or the new project scientist, even though these outcomes were precisely the ones that the mission's leadership had hoped to inspire.

Yet even these changes occurred within the existing context of Helen work. Using already-published and archived data retained instrument team

Conclusion

The auditorium that is usually reserved for PSG discussions is flooded with Helen friends, relatives, alumni, and the press. At a back table, a small group of NASA public outreach personnel are logged in to their social media dashboards, sending Twitter updates. There is much affect on display in the room tonight, but no ambiguity. Victoria takes to the podium, and in a personal note, she mentions that her now-married daughter was born when she started working on Helen. Everett had retired several years ago; in his place is a soft-spoken project manager from Helen's early days. Following a rousing speech by a visiting NASA administrator, the Virtual Choir takes to the stage—myself included. We sing a series of classic hits—"The Sound of Science," "The Age of the Proximals," the song listing all the moons of Saturn—this time with lyrics scrubbed of tongue-in-cheek content. We elicit applause and no shortage of tears from team members in the audience. The screen at the front of the room then shows the antechamber in the spacecraft operations center with the flag on the wall, where Connie and her team are assembled to watch for signs of life from the spacecraft. Did it survive its first dip between the rings, its entry into the mission's "Grand Finale"? Cheers erupt as the satellite dish picks up the ping and then starts to receive data. Atmospheres scientists Henry, George, and Max ascend the stage to narrate what they see in the images as they appear on the NASA website, Isabelle's press release is shared widely on social media, and Joseph shakes his head at his screen as his instrument's data appears. "I don't know what I'm looking at!" he exclaims. Getting up from his seat, he walks over to the NASA administrator, also a plasma scientist and seated next to Rod, and they busily discuss the results.

Roger was wrong, I think to myself. Spacecraft do have social lives. That fact is certainly on display in the room tonight.

This book has sought to show how two different scientific collaborations address the question of integrating interdisciplinary perspectives to solve complex problems and allocate resources. Each does so in different ways, as the framework of *organized science* reveals. Helen adopts the matrix model to bridge across divides; Paris works under the interdependent flat model popular among charismatic organizations. One labors to represent all instruments distinctively under a process of integration, while the other aims to bring all instruments together in a process of collectivism that makes them difficult to disentangle. Both are sustained and crafted out of shared practices and narratives, forms of talk and gesture, norms and procedures, even friendships and alliances that reach across considerable divides. Amid different countries, institutions, colleges, laboratories, and disciplines, they assemble a meaningful organizational unit: a mission that members commit to with passion and dedication over years, decades, careers, and even lifetimes.

As I have sought to show throughout this book, the contrasting approaches that emerge from each collaboration do not just represent different local structures or cultures: they produce different kinds of knowledge about the planets themselves emerging from different organizational orders. Under the framework of *organized science*, I have sought to draw attention to how scientific practice as organizational work shapes scientific outcomes, with implications for the robustness and stability of the scientific collaboration. The result is an observable isomorphism between organizational distinctions, practices, and routines and organizational outputs such as scientific papers, data, and personas. It is possible for a discipline to foster several different epistemic cultures and for collaborations within scientific fields to show meaningful differences in practices and outcome as a result.[1] It also turns out to be true that "the organizations for which you work . . . affect your own knowledge processes" (Vaughan 1999, 934–35).

Solutions to the problem of social order, therefore, do present different solutions to the problem of knowledge, even among the self-same community of practitioners (Shapin and Schaffer 1985). This shaping of scientific knowledge occurs through the three principles of *organized science* I have fleshed out throughout this book. First, each collaboration enacts a social form, common practices, and interactional norms to legitimate decision-making and meet the challenges of their field. Second, these organizational elements of a scientific collaboration shape its data, its publications, and its members' reputations. Finally, this organized system of knowledge production is stitched together through looping effects, whereby the outcomes of collaboration become the team's inputs once more. This naturalizes the social, scientific, and technical arrangements peculiar to each organization such that it appears to be simply the way to do science in *this field*, using *that instrument*, with *those people*.

The stability of the laboratory's organizational order or other robust collaboration cultures in well-established fields may lead us to see such differences as disciplinary in nature or perhaps due to the properties of the objects under study. But in fields like planetary science, where NASA's funding crisis shook up the organizational form of the field without injecting new participants or participating disciplines, we witness high degrees of variation in collaboration structures and cultures within, with concomitant effects on scientific outcomes. These ways of working together go far beyond the official "org chart" to become the collaboration's own, unique "form of life" (Wittgenstein, in Latour and Woolgar 1979), embedded in the ways that people talk and act with each other, communicate with, and imagine their instruments and spacecraft. Epistemic cultures (Knorr-Cetina 1999), wherever we may find them, are grounded in organizational principles.[2]

As such, *organized science* suggests new and fruitful possibilities for an organizational sociology of knowledge. Science studies needs sustained attention to scientific organizations and not just for questions of the distribution of scientific resources—that is, how lineages develop among scientists and their students, how fields organize their collaborations, and how minority careers are excluded—but for what those structural considerations mean for knowledge, for knowledge work, and for objects. Because *organized science* suggests a relationship between the ways in which collaborations are organized and how knowledge is produced in the world, these hybrid structural and cultural considerations of inequality and community formation are essential for understanding the manufacture and character of scientific knowledge itself.

There are renewed implications here for the study of the politics of knowledge production. Many scholars have examined the interactions between science and the state, scientific and state institutions and actors, and the politics of knowledge on the national or global stage. But *organized science* shows how knowledge work has *micropolitics* as well. Interactions between individuals within organized frameworks align not just with different philosophies of science but with different politics of decision-making and compromise. These stances on collaboration are also visible in political organizations. The consensus orientation of Paris that forges common ground is reminiscent of the Commonwealth Heads of Government, the parliament in Iceland, or activist groups in the 1960s, all of whom adopt consensus—or collectivist-based decision-making structures. Meanwhile, the perspective of integration on Helen echoes the United Nations, the European Union, or the United States, a form of federalism that allows a patchwork of many different interests to emerge across the group while respecting the autonomy of contributing units to make local decisions that best fit their needs. Behind the scenes on these missions and others, who is in and who is out, who is heard and who is not, and overall members' commitments to various styles of decision-making accord with local ideals of democratic process and good science. These examples are only two varieties that we might expect to find in scientific work. Further studies of other laboratories and collaborations that are attentive to *organized science* will find different arrangements of actors with their own flows and impositions of authority, information, and action, each one a microtechnopolitical stance with implications for the knowledge that results.

Organized science also has implications for our understanding of ontology. In doing mission work, participants build up their knowledge of the planets, spacecraft, and people in their organization's image. They naturalize this organized knowledge too: this is just *how Mars is*, this is just *what data are like*, that's just *her personality*. Science studies scholars call attention to this activity as a form of ontological work in the world: creating boundaries around objects and circumscribing them as natural kinds even though this "natural" identity is suffused with social order through and through (Barad 2007; Haraway 2007; Mol 2002). Others attend to how "the natural order" (i.e., the order of nature) and "the social order" (i.e., ordering among people) are coproduced and intermingled through scientific work (Jasanoff 2004; Knorr-Cetina 1999; Lynch 1994; Martin 1991; Schiebinger 1989). As I show here, the division of labor, modes of interaction, forms of talk, and decision-making techniques that typify the scientific collaboration *as an organization* are written onto the natural world, looping back into the collaboration to stabilize its own epistemic and ontological regime. Those attuned to epistemic, ontological, and ethico-onto-epistemological practices in science studies must consider how the continual production of the organization plays a role in the ordering work that scientists enact in the world (Barad 2007).

For organizational scholars, this book may seem to display yet another set of organizations in action that bring together many different types of experts in technologically enabled, cross-boundary collaboration. Prior studies have examined start-ups in Silicon Valley, production teams in Hollywood, or ties between academics and industry partners to see how groups collaborate across boundaries (Bechky 2006; Neff 2012; Owen-Smith and Powell 2004). But the life span of these collaborations is comparatively short, and it is difficult to build commitment in the light of employee turnover, short delivery dates, or even mobility up or across a chain of command. Paris and Helen, however, allow us to run the clock on the experiment (so to speak) for several decades. On each team, commitment to each local mission is equally strong yet differently expressed. While there are no simple answers for questions of community formation across national, institutional, or temporal boundaries, it is notable that none of these three elements played a defining role in coauthorship networks or other aspects of mission work. These two heterarchical organizations therefore serve as touchstones for thinking about collaborative, integrative work over the long *durée*.

The findings presented here have important applications to the development of new scientific teams. Both missions are astonishing successes, but we might qualify this statement by adding that they each do an outstanding job of producing *different types of knowledge*. Since the Paris structure generates collectivist findings with an interoperable dataset and open data, teams interested in supporting similar types of work or data exchange might consider a similar structure to foster and incentivize the interactions that facilitate these sorts of discoveries. If a scientific community seeks encyclopedic collection of data over a long time, if they need to preserve considerable local autonomy and respect institutional or other boundaries while integrating into a whole, then the Helen model is exemplary. Still, collaborations may not always be in full control of which model to adopt. These two collaborations took shape due to changes in funding policy at a national institution level: changes that aimed to reduce costs but ended up producing variation in how science was done. To that end, even funding agencies considering programmatic changes would do well to consider what models of *organized science* they espouse to promote different knowledge outcomes.

Outside the sciences, technology companies, nonprofits, and industrial firms frequently seek to integrate expertise within their borders. Many of them turn to the organizational styles I have described here. The matrix organization remains popular, and discussion of "multiple hats" and the challenges of project versus line management will sound all too familiar to many of my readers. Technology start-ups and scientific ventures, meanwhile, typically organize around charismatic leaders in flattened groups. As they increase in scale, they must confront the challenge of moving from authority invested in charismatic leadership to that invested in bureaucratic-hierarchical modes—essentially, a shift from Jeremy's leadership style to Everett's. Further, elements like a shared organizational culture that once buffered communication across divides to the organization's success may now splinter under increased division of labor, resulting in misunderstandings or strategic resource pitches—such as those between Helen planning groups. Finally, activist groups or nonprofits may wish to organize themselves in a collective but find difficulty in coming to such consensus without a charismatic leader. The stories of Paris and Helen, then, ring true across many sectors. *Organized science* can point to the normalcy of such challenges in any complex group as well as to how structural changes to these roles and cultural interactions may impact scientific and organizational outcomes in unexpected ways.

But this book would be remiss if it did not remind readers, in the midst of such analysis of organizational orders, division of labor, and other technical jargon, of the experiential importance of being on a "mission." Both teams give a feeling of being part of something greater than oneself. This combines the sense of grandeur and scale that floods into your body while standing at the windows of the spaceflight operations center, watching the blips and pings from distant robots in space, with a more immediate, familiar feeling. It stems from the extraordinary commitment that team members on both missions have to each other and their shared goals. For Paris team members, this was expressed euphorically through the love, care, and commitment to their robots, as a way of caring for the team as well. For Helenites, it was through the persistent metaphor of "family." Both forms of talk speak of emotional labor and lasting bonds.

These social relations endure. It was this form of talk that had Paris engineers reportedly breaking down in tears at their desks as their robot became unmovable and faced death, the cause of considerable personal distress for scientists and planners on the mission team. It was this form of talk that animated conversations on Twitter among Paris engineers, newly visible to the public as their robot faced its demise. It was also this form of talk that dominated the suddenly serious conversations that began to circulate about Helen's impending end. As the spaceflight operations center added a countdown clock to its visual displays of data pings from the distant craft, Betty admitted that she did not know how she was going to feel the day after the spacecraft was due to fly into Saturn's clouds and disintegrate in its gravitational pull. Over coffee at the outdoor plaza at Spacelabs, Victoria described it to me as the loss of a friend, a family member.[3] At Helen's final launch anniversary party, Eugene, a member of the Virtual Choir, sat down at a digital piano at the same thick boardroom table where I had seen so many scientists and engineers gather for target meetings, working groups prioritizations, and PIE resolutions, directly across from where Everett used to adjudicate fair resolutions. Team members became teary-eyed as he performed a

Disney tune with images of Saturn flashing on the screen behind him, singing, "We'll see the universe and dance on Saturn's rings // Fly with me and I will be your wings." It was this sensibility toward the endurance of family that endured as we clustered together and watched, on big outdoor screens, as the final package of data from Helen came down to Earth following the spacecraft's plunge into Saturn's atmosphere. Reporters gathered for comments, while team members hugged, shared champagne, and shed tears.

It was also this form of talk that I witnessed, in a cavernous restaurant in Barcelona, where European and American scientists gathered to celebrate the fifth anniversary of the probe's landing on Titan. The dinner rounded off a long day of science presentations, where everyone discussed their current work on mission data and advanced future mission concepts with their friends. Amid the formal speeches from scientific leaders that commemorated the event were personal stories of the friendship and camaraderie that had brought them through difficult times and successes, the ups and downs of mission work over many long years. Then Edgar stood up. He had been a graduate student when he first started working with the team from Europe in the 1980s. Now married to another Helen scientist and living in the United States, the mission had been a defining element in his career and in his personal biography, which, like for many others, were intimately intertwined. I had previously seen Edgar quip sarcastically with the best of them, but there was no ambiguity in his tone tonight. He spoke from the heart, holding his glass aloft to toast his colleagues:

> When you're leaning science, you learn a lot of things, like how to solve differential equations and how to ask questions. But one thing you don't learn that's surprising is that science is really about the people. Whether it's talking with your colleagues and saying, "I'm right, you're wrong about this," or saying, "Hey, do you want to collaborate with me on this or that," it's about the people; it's a social process. We even have a sociologist studying us. We are family. I think of you guys as my family, I see you as much at least as I see my family. This is the Helen family, and I am grateful for the privilege of working with you.[4]

POSTSCRIPT

Methodological Reflections

Tree branches and foliage tumble over the banks of the shallow river winding outside of Oxford, leaves heavy and green with the flush of spring. Perry is seated in the low boat, the PI of his instrument team standing behind him, pushing a long pole into the water to propel it forward. It's the end of a long day of science team meetings, and the group will grab dinner together at a pub on the outskirts of town once they have tried their hand at this local boating technique. The PI's wife, Gina, is in the boat, as is George, who pulls out a camera and starts his video, focusing on Perry and the PI. "It's gonna be recorded for all time," he says.

"Don't lose it," says Gina, off camera. "You can give it to your sociologist."

George laughs. "Oh, yeah, right."

But Gina asks, "Do people behave differently when she's around? The sociologist?"

"Oh, she's a punter," offers Perry. "I mean, Janet was punting in one of the other boats." (I was at the helm of another boat with a group of French researchers.)

"Spent a year at Cambridge," interjects a recent Oxford graduate behind the camera, explaining my unorthodox technique.

"You'll have to point her out," says Gina.

Perry continues, "Yeah, I just met her this week, but she seems nice. That or perplexed by this weird group, but uh, she's a sociologist trying to figure us out, which is"—he pauses—"difficult."

"If she could just—" Gina starts afresh, speaking carefully. "Once she figures you out, if she could distribute that figuring to all of us who travel with you so that we'll understand as well, that would be nice."[1]

Spacecraft are equipped with various instruments to characterize the distant worlds they visit. They are shaped by the community who outfitted them with what they consider to be the right tools for the job, and their results are discussed and articulated among a community of researchers: elements that, as I have described in this book, shape the way in which knowledge crafting takes place. But where they land, how they traverse, how the properties they study react to them, and how their sensors change over time also play important roles in their data acquisition and analysis.

Like an interplanetary probe, the ethnographer enters an organization at a certain time and place and from there, must forge a path that produces understandings of and visibility into as many phenomena as possible. Over the course of my studies of spacecraft teams, I increasingly began to see myself as like the probes that these teams sent to other worlds. There are many parallels between ethnographic work and spacecraft work, especially in terms of generating trustworthy reports. What functionality—instruments and methods—the probe brings to the site offers one set of questions, but also, What methods did the ethnographer use? What is their methodological heritage? Which tools of the trade were deployed, and why? And just as a spacecraft can create noise in the data that may be mistaken for a signal, do ethnographic techniques introduce artifacts or influence how the people the sociologist follows respond?

Where Did the Probe Go?

As I have laid out in this text, planetary science is a dense life world with many interconnections, distributed across nations and states. Meetings take place linked by virtual and face-to-face encounters; rituals of planning may occur daily or weekly but are also punctuated by other rituals such as team meetings several times a year. In addition, scientists attend several conferences per year, frequently more if they include European and American assemblies and specialist workshops about Titan or Mars. As an ethnographer, trying to be "where the action is" (Goffman 1967) requires being in many places at once.

To accommodate these requirements, my ethnographic work took a two-pronged approach. First, I immersed myself at one of the primary institutional centers for more than a year, from which I attended as many teleconferenced meetings in person as I possibly could—as many as ten per week on Paris and up to five per week on Helen. Because these missions make decisions in such representative assemblies, observation planning and science meetings formed the crux of the mission's interactional environment. I also traveled to other institutional centers, sometimes just for site visits and other times for concurrent meetings or workshops. And I attended key conferences for members of the field where they presented their work to their planetary science peers and heard the official line about current events at

Headquarters. Outside of these events, I interviewed participants, socialized with scientists and science planners at lunch and after-work dinners and drinks, and participated in site-specific activities such as talks, workshops, or events. This gave me a deep appreciation for the forms of talk, work practices, and overarching norms that mission centers infuse into their daily work and that radiate throughout the team in their collaborative activities.

I recorded meetings and interviews, keeping track of the conversation with digital and notebook jottings, which I used later to orient myself within extended conversations and planning points. As many such conversations took place over a long period of time and many meetings—the dust streak campaign or the Enceladus flyby issue, for instance—I used software tags to keep track of issues as they developed, matured, and were resolved, writing memos about them later on. This enabled me to follow the thread later across many times and places. As themes emerged in the site, I spent time recording thoughts and impressions in field notes and memos, all of which were coded alongside my recordings, jottings, interview transcripts, and photographs.

People often wonder if I had to learn the science in order to do this ethnographic work. It certainly helped that I was fascinated by astronomy as a child, as I arrived in the field with more than basic knowledge of the planets—knowing that Titan had a hydrocarbon atmosphere, that all gas giants in our solar system have rings, the names of Mars's two moons, and so on. In line with Harry Collins's argument for "interactional expertise" (Collins and Evans 2002), I developed a broad understanding of as many aspects of planetary science as possible: from surface geology to ring dynamics, from gravity detection to the physics of particles and fields. I cannot *do* this science, but I can follow scientific arguments, read field papers, and understand why scientists think something is awry in their peers' explanations.

The science, it turned out, was not the hard part: it was the many sedimented layers of social relations in the community. For instance, it is not enough to know what a cryovolcano is or how you might know if you had detected one. It is not even enough to know what the detection of a cryovolcano means about the planet's interior (differentiated or undifferentiated core?) and how to follow up on that claim (i.e., gravity experiments). More complex was knowing who the scientists on stage at the press conference were, how or if they knew each other, who they had studied with and who their friends were, if there was long-standing bad blood between their advisors or graduate programs, if they were pitching competing missions, and so on. Like rocks on Mars that are affected by weathering processes, the true story of these social dynamics often required scraping many layers below the scientific surface.

Another element of prior expertise proved important for my research. Both missions involve European members and components, requiring travel to key sites in Europe to understand events outside of the American point of view. Because I attended a French immersion primary school and was fascinated by languages in my youth, I arrived at these meetings able to conduct interviews in French and with varying degrees of conversational ability in Spanish, German, Danish, Italian, and Hungarian. This enabled me to follow hallway conversations, understand informal chats over drinks, and grasp local experiences and framings of events. Even attempting short conversations in their native tongues as well as my Canadian and European citizenship likely contributed to my ability to build rapport with many ESA team members despite being based at Spacelabs for my fieldwork.

How Was the Data Analyzed?

Following a first round of coding after the Paris dataset was collected in 2008, I began to code my notes in earnest from both Paris and Helen in 2010 following my period of intensive primary site immersion. I returned to these notes in 2015, this time using the software tool nVivo. At first, I followed an open coding scheme, then narrowing in on perspicuous topics and common turns of phrase to identify rituals (Goffman 1967), forms of talk (Garfinkel 1967; Goffman 1981; Sacks, Schegloff, and Jefferson 1974), the emergence of social facts (Garfinkel 2002), and shared matters of concern (Latour 2004). While my methods are largely ethnomethodological, following the stabilization of facts over particular transcripts (Garfinkel, Lynch, and Livingston 1981), I also at first deployed a grounded theory approach (Glaser and Strauss 2009) to my overall coding structures, attempting to work from actors' own accounts as much as possible in formulating my analysis. The book went through five drafts as I worked through the rich ethnographic material to formulate the thesis presented here. I circulated near-final drafts to certain members of each science team—selected for their distinct views on the mission and their sociological imaginations—and performed targeted member checks throughout the process to ensure accuracy of both scientific and social events.

The insight that the organization shapes the science first occurred to me at the Puerto Rico meeting, at which point I began to memo about this inductive theme and look for further examples to see if my intuition was correct. This also meant looking for places where the theory was incorrect—a technique Mitch Duneier (2011) calls "inconvenience sampling." For instance, in early conversations about these developing ideas with Victoria, she told me to spend more time with PAM, an example of Helen's multi-instrumental collaboration. While physics was not my best subject, I took a deep breath and dove in. This "inconvenience sample" proved an important test of my hypothesis as I discovered a version of Paris on Helen, a group that turned out to have a similar division of labor and resulting techno-solidarity effects (I also discovered my unexpected enthusiasm for space physics). Returning to Titan meetings and scientific papers over time also demonstrated how this group built upon multiple instruments' published findings, another form of multi-instrumentalism under Helen's premise of team autonomy. In case it was really a question of personality and not social form, I followed many of these scientists to other mission team meetings to observe interactions under different conditions. These engagements did not deny my initial findings but thickened and improved my analysis by, for instance, refining my understanding of core team concepts such as autonomy and collectivism, and making moments like the Aurora meeting revelatory, all of which enabled "the move from how to why" (Katz 2002).

I also put the developing concept of *organized science* to the test by conducting network studies of copublication matrices described in chapters 6 and 10. These were conducted in association with three students, Mauricio Launio, Manish Nag, and Han Zhang, at three distinct moments in 2011, 2013, and 2015. We extended the analysis to other missions, of which I had collected observations and interviews not discussed here. We noted the same patterns recur even when using different software suites, whether we assembled our dataset from internal team lists or from public database scrapes. We were also able to see how multi-instrument publications evolved over time on Helen and Paris and to test for network centrality. In a network analysis omitted from this text, we looked at Helen publications by working group to determine if PAM or Titan were truly unique compared to the other groups and to see if working group or instrument was the primary associative order. Three of the working groups continued to divide their publications by instrument, PAM featured many interconnected nodes, more like Paris, and Titan's interconnected nodes more commonly flowed through its interdisciplinary scientists. As this latter group tends to share and build on instrumental findings rather than craft campaigns, citation analysis may do more to illuminate their collaboration. I do not see these computational investigations as more objective or true than my qualitative work (Daston and Galison 2007; Porter 1995); rather, as investigations of alternative hypotheses, these views both confirmed my intuition and thickened my understanding of the interactional situations I saw play out in the room.

An important moment that ethnographers point to in their work is that of saturation, the point in fieldwork where one stops seeing new things and instead sees and hears the same things over and over again. It took a long time to get to this point on Helen as I was determined to visit as many of the subteams as possible, each of which was unique. Over time, I started to see this distinctiveness as part of the pattern of social life on Helen. I left Spacelabs in 2010 but returned to Helen several times in order to collect more data from different vantage points, test my assumptions, and see if anything had changed. I also returned to Helen a few times while engaged in subsequent fieldwork, sometimes attending a few days of a PSG and eventually rejoining their Virtual Choir as the end of their mission approached. The few changes I observed and was able to track I note in chapter 10. Having kept tabs on the mission for so long, I was able to trace these changes to observed organizational changes that made emerging configurations possible, rather than to shifts in the field or to a spontaneous embrace of new ordering modes. While embedded with Helen too, I visited the Paris team several times in their operations center, including the meeting I cited where the engineers reinforced how "everyone" was responsible for their robot. These missions were truly remarkable for the endurance of their craft and their local social orders.

What Are the Instrument's Constraints?

Like the instruments that scientists send into space, which may pick up artifacts or calibration difficulties along the way, no ethnographer is perfect or able to perfectly capture the organization that she enters or to avoid artifacts that affect her point of view. I was fortunate to be allowed full access to these teams by virtue of coming in under the purview of the leader, whether the Paris PI or Helen project scientist. On Paris, I traveled widely and heard from enough individuals outside of the PI's home base that I am satisfied that I was seen as my own agent on that team. On Helen, however, my arrival shortly after Francis's installation and my simultaneous (though coincidental) departure with him left me inextricably tied to him in ways I could not control. This meant that certain team members associated me with the leader or the central operational node of the collaboration, even though my work was independently funded.

I first began to notice this when I received an email thread in which a mission secretary called me "Francis's person." This instrumental metonymy, common on Helen, continued throughout my ethnography. It was not helped by the fact that a condition of my ethnographic work was that I use insights from studying Helen to help Francis develop his next mission. And because I was introduced to Francis through Sam, whom I knew from Paris but without a sensibility to his own affinity network in planetary science or his role in the cryovolcano debate, I was no doubt even more firmly associated with mission outsiders. I therefore thought often of Gina's question whether people behaved differently because I was around. Occasionally, I was told that people were "behaving themselves" more than usual because I was in the room, but I could not tell if that was a way of describing the discrepancy between the group's actual interactions and their affectively ambiguous forms of talk.

Although I aimed for "psychological mobility" to develop empathy across hierarchical levels (Powdermaker 2000; Turco 2016), I saw no sign that the upper levels of the organization—the PIs, team leaders, and project office—ever lost sight of my initial connection through Francis. People were more welcoming among the ranks. I found more common ground with the counterculturals and geeks than I ever could have among the other affinity groups in planetary science or at Spacelabs. These individuals were always willing to regale me with their often skeptical and detached appraisals of mission events (not unlike Gideon Kunda's "confessional" of attracting more marginalized members of the organization [2006, 244]). I especially took heart from a conversation over dinner in Covent Garden one day, following the PSG in London, when George complimented me on my project. So many people came to study the mission, he said, but only spoke to the upper-ups and never to the people in the trenches who were actually doing the work, as I did. He felt this gave me a much better, and more accurate, picture of the organization. I still consider this to be the highest compliment.

Toward the end of my first wave of fieldwork, early in 2010, I presented a preliminary report of some of my findings to the team, hoping for broader engagement with my developing ideas. People listened politely, but they later bristled about my use of the word "dysfunctional." Although it was their language I was reporting and I was attempting to explain their use of the word as building solidarity among their team, I learned later that to project office members it sounded like I was accusing them of actually *being* dysfunctional. For some, this language identified me as a hostile outsider to whom they should never have opened their doors. Later, when I sent an early draft of a comparative paper to Victoria to look over, she zeroed in on the language difference between Paris and Helen right away, her comments becoming increasingly agitated as the paper progressed. I made the changes she suggested and spent several years attempting to gain the analytic distance necessary to adopt a neutral stance, eventually realizing that I needed to abandon some of the practices of grounded theorizing. But the incident still stung, as it violated several participants' comfort with and support of my work.

I reengaged with the team at their PSG a few months later after an institutional move, following Francis's departure from the team. When I gave a brief presentation about my continuing research, Edgar lobbed a series of ad hominem "questions" at me from the audience. A PI made a comment about my doctorate that emphasized my lower status vis-à-vis the team. Like the traceability matrix, it seemed it was impossible to transition from being "Francis's person" to being something more like "*your* sociologist." James caught up to me after the break at this meeting to declare his colleagues' actions indefensible, while Victoria sent a follow-up email that said after enduring such heckling I was now truly part of the team. For my part, I recalled a PSG meeting when Victoria, discussing the mission's history, said, "Helen got leaner and it got meaner—" and Isabelle interrupted, "It certainly got meaner." Now I knew how she felt.

How Was the Instrument Calibrated?

These experiences on Helen led me to embrace the notion that ethnography of organizations occurs not only in a time, in a place, from an embodied position but from an *organizational position* as well. Of course, an ethnographer's positioning is an essential element of observing social life; as sociologist Seamus Khan argues, "The study of human relations is necessarily an embedded one: to pretend otherwise obscures more than it illuminates" (Khan 2011, 202). And science studies scholars have too long been acolytes of the ethnomethodological prescription for "situated" understanding to pretend otherwise. Organizationally, however, this social fact sits at odds with my initial attempt to gain enough mobility to gather and catalog as many perspectives on the team as possible. In order to calibrate for any instrumental artifacts that my organizational position might introduce, then, I initiated a group project by pulling in four other social scientists—Marisa Cohn, Matthew Bietz, Emma Lawless, and David Reinecke—to conduct ethnography, interviews and site visits on their own terms within the Helen mission.

On the one hand, this assisted with visiting sites I was unable to get to during my immersion period. For this, we deployed a standardized interview script as a typical group project might proceed. On the other, this group assembly also produced five distinctly embodied and uniquely situated perspectives on the same mission. Comparing notes about the same meeting, interviews with the same people, or visits to the same sites proved to be the most illuminating for grappling with our own organizational positioning and the identity boxes into which we were inadvertently put, as well as the positions and identities of those with whom we spoke. To that end, we did not aim to calibrate through standardization but rather benefitted from being very different people, with very different self-presentational styles, gaining access through different organizational inroads and producing different "partial perspectives" on its functioning (Haraway 1988). Mirroring the team we studied, we became a polyvocal, multi-instrumental collaboration as well.

These excellent ethnographers' work informed the findings I report here: examples of the physical versions of Helen on peoples' desks (Lawless), the competing calibration algorithms on the thermal instrument (Bietz), sensibility to "aging with" the spacecraft (Cohn), and the "politics" of spacecraft funding (Reinecke). However, I do not refer extensively to their work or incorporate it into my own ethnography, and I do not present their data as my own. Instead, what was most valuable about this collaboration in the development of our independent accounts of this site is the situated awareness that it generated about our individual "partial perspectives" on Helen. This likely echoes the value that Helenites find in their epistemic endeavor as well.

As mentioned in the introduction, I embrace standpoint theory's notion that objectivity can only arise from combining multiple perspectives on a problem, not from adopting an all-seeing "view from nowhere" (Haraway 1988; Longino 1990). Organizational positioning—or ignorance of it—may present a weakness for an ethnographer if it goes unacknowledged, if one assumes that one's chosen stance within the organization is the legitimate one, or if one is unable to "check" those assumptions and situate them within a matrix of perspectives. If we can acknowledge the partiality of our own ethnographic perspectives as imposed by our field sites and seek out as many alternative viewpoints as possible (including from other ethnographers situated within it), we can enable such complex organizations to speak with the polyvocality they deserve.

List of Sites Visited

AMERICAN SITES: NASA Ames Research Laboratory in Menlo Park, CA; Arizona State University in Tempe, AZ; Ohio State University in Columbus, OH; Space Science Institute in Boulder, CO; SETI Institute in Mountain View, CA; Honeybee Robotics in New York; Cornell University in Ithaca, NY; Washington University in St. Louis in Missouri; Smithsonian Institute in Washington, DC; US Geological Survey in Flagstaff, AZ; California Technical

Institute in Pasadena, CA; the Jet Propulsion Laboratory in La Cañada Flintridge, CA; Johns Hopkins Applied Physics Laboratory in Columbia, MD; Goddard Spaceflight Center in Annapolis, MD; Southwest Research Institute in Boulder, CO; Johnson Spaceflight Center in Houston, TX; Lunar and Planetary Institute in Houston, TX; Space Sciences Institute in Boulder, CO; Laboratory for Atmospheric and Space Physics in Boulder, CO; US Geological Survey in Federal Center, CO; and Brown University in Providence, RI.

EUROPEAN SITES: University of Rome Sapenzia, Rome; National Institute for Astrophysics' Institute for Space Astrophysics and Planetology, Rome; The Italian Space Agency, Rome; European Space Agency Archives, Florence; King's College, London; Oxford University, Oxford; The Paris Observatory, Paris; ESA Headquarters in Nordwijk, Netherlands; Copenhagen University, Denmark; Deep Space Network Ground Station, Madrid; and ESA Ground Station, Madrid.

CONFERENCES AND WORKSHOPS: American Geophysical Union Conference, San Francisco, 2008, 2009, 2016; Seventh International Conference on Mars, Pasadena, CA, 2007; Landing Site Workshop for the Mars Science Laboratory Rover Project, Pasadena, CA, 2008; Division of Planetary Science Meeting, Puerto Rico, 2009; Lunar and Planetary Science Conference, Houston, 2009, 2017; Titan Surface Workshop, 2009; Fifth Anniversary of ESA Probe Landing, Barcelona, 2010; MEPAG (Mars Exploration Program Advisory Group) Meetings, 2007, 2008, 2009, 2010; OPAG (Outer Planet Advisory Group) Meetings, 2009, 2010, 2013, 2015, 2016.

MISSION TEAM MEETINGS: Paris team meetings, January 2007 (Pasadena, CA), July 2007, January (Washington, DC), May 2010 (Pasadena, CA), July 2011 (Pasadena, CA); Helen PSG meetings, January 2009 (Pasadena, CA), June 2009 (London), October 2009 (Pasadena, CA), January 2010 (Pasadena, CA), June 2010 (Munich), October 2010 (Pasadena, CA), January 2011 (Pasadena, CA), January 2012 (Pasadena, CA), June 2012 (Sardinia, Italy); PSG-collocated workshops (Magnetospheres-Satellites and Aurora); Titan Jumpstart, April 2009 (Columbia, MD); PAM team meeting, April 2011 (Annapolis, MD); Thermal instrument team meeting, June 2009 (Oxford); Radio team meeting, June 2009 (London); Radar team meeting, August 2009 (Pasadena); Infrared team meeting, September 2009 (Denver); Imaging team meeting, January 2010 (Pasadena); meetings for all working groups (Titan, Atmospheres, atmospheric investigation subteams, cross-discipline, Satellites, Rings, and magnetospheres), 2009–10; spacecraft funerals, 2011 and 2017.

ACKNOWLEDGMENTS

Any project that has been a decade in the making has accumulated too many people to thank. My gratitude first and foremost goes to the mission teams and the people that welcomed me into their lives for so many years, especially Steve Squyres, Jim Bell, Robert Pappalardo, Bob Mitchell, and Linda Spilker for permission to study their teams and their invaluable feedback on the project at its various stages. I am also grateful for the friendship of Emily Dean, Jeff Moore, Scott Maxwell, Kim Lichtenberg, and Nora Kelly Alonge and for repeated conversations with Daniel Gautier, Scott Edgington, Tamas Gombosi, Matt Hedman, Jeff Johnson, Sarah Milkovich, Rob Lange, Ralph Lorenz, Eldar Noë, Zibi Turtle, Jani Radebaugh, and Trina Ray, among many others. To the indefatigable members of Helen's Virtual Choir, my admiration, appreciation, and love always.

Several institutions also housed me while working on this project: namely, Cornell University Science & Technology Studies Department, the University of California Irvine Informatics Department, Princeton University Sociology Department and Society of Fellows, and NASA Ames Intelligent Systems Division. I am grateful for the National Science Foundation's support of this project at all its stages, including the Science and Society Doctoral Dissertation Research Grant #0645945, Virtual Organizations as Sociotechnical Systems Grant #0968608, Office of Cyberinfrastructures Grant #0968616, and Cyber-Human Systems Grant #1552469, and to the NASA History Office—History of Science Society Fellowship in Space History for facilitating archival research. The folks at Kahve Coffee Lounge, Literature and Latte, and Data and Society enabled good spirits while writing.

Scholars in science studies and sociology were incredibly generous with their comments on earlier drafts. Special thanks to Paul DiMaggio, Betsy Armstrong, Miguel Centeno, Mitchell Duneier, Matt Salganik, Kim Lane

Scheppele, Paul Starr, and Viviana Zelizer; to danah boyd, Joseph Kaye, Karen Knorr, Michael Lynch, Trevor Pinch, David Stark, Sharon Traweek, Fred Turner, Diane Vaughan, and Judy Wajcman; and to Clark Bernier, Leah Reisman, David Schwartz, and Samantha Jaroszewski. Paul Dourish and members of the "Spaceteams" research group Melissa Mazmanian, Matthew Bietz, David Reinecke, Emma Lawless, and Marisa Cohn provided feedback every step of the way. Chapters and excerpts underwent substantial changes following generative comments from Sarah Thébaud, C. J. Pascoe, Laurel Smith-Doerr, Enobong Branch, the students in the Princeton Center for the Study of Social Organization Workshop, the digitalSTS community (especially Steve Sawyer, Ingrid Ericson, David Nemer, and Steve Jackson), Angela Creager, and Erika Milam. Matt Hedman, Scott Maxwell, Curt Niebur, Bob Pappalardo, and Jadi Radebaugh provided feedback and checked for accuracy. Jordan Bimm, Alex Kindel, Mauricio Lanio, Manish Nag, Brandon Stewart, Tobiah Waldron, Meghan Wheeler, and Han Zhang, as well as Jill, Merrill, Kathleen, and Cathy, provided invaluable assistance along the way. Many thanks to my outstanding editors at the University of Chicago Press, Doug Mitchell and Karen Darling, and to my anonymous reviewers for making this book stronger.

Feedback and discussion at talks and presentations helped shape the book's content. Conversations were especially generative at Brigham Young University; SpaceX technologies; the University of British Columbia; the University of California Irvine, Santa Barbara, Los Angeles, and San Diego; York University; Georgia Tech; California's Organizations Research Workshop; Yale University; and Cornell University and also at research centers such as the Data and Society Research Institute in New York and the Institute of Advanced Study in Princeton; and at NASA meetings like the Cassini Project Science Group Meeting, the Europa Jupiter System Mission Science Definition Team Meeting, the Outer Planets Advisory Group, and the NASA Science Mission Directorate.

I think often of Jake's statement that outer planets missions involve "births, deaths, marriages." While working on this project, I celebrated several marriages of team members, mourned the loss of their spacecraft, and witnessed the passing of individuals who played a central role in these projects. I also progressed in my own career, got married, and had children—which I too identify by period of planetary investigation as well as calendar date. I dedicate this project to Craig Sylvester, whose support and engagement have been with me without fail through these milestones and beyond.

APPENDIX

Acronym and Technical Dictionary

"One of the fun parts of a science planning job, and this helps to alleviate some of the stress, is to come up with acronyms. (*laughs*) In science planning, it's notorious."

—George

Like all professional or technical organizations, life at NASA is suffused with specialist vocabulary, with many collaborations featuring an "acronym dictionary" for teammates. Rolling easily off members' tongues, they are a significant hurdle for newcomers and outsiders, who get lost in the flow of conversation from the start. For the sake of clarity, I have changed names of organizational subunits and spelled out many acronyms to limit confusion, but some remain unavoidable. The reader may wish to bookmark this page and return to it for ease of reference.

Acronym	Meaning
ALH000841	A meteorite from Mars, found in 1984. Analysis in the early 1990s indicated possible microbial origins for its structures, prompting the start of the Mars Program.
ASI	The Italian Space Agency. ASI built the radio antenna on the Helen spacecraft as an independent contribution.
CAKE	A campaign by the Atmospheres working group that requested cross-targeting team observational time, leveraging the PIE process to secure it.
Conduct!	The Paris planning software.
Discovery mission	NASA's small mission category—flying projects costing under $1 billion, typically led by a single PI.

ESA	The European Space Agency. ESA built the Titan probe and staffs each Helen team with European scientists under the original memorandum of agreement between the agencies.
Facility instrument	An instrument team that was staffed by NASA, not by a PI. Individuals apply to join at the outset of the mission with individual proposals and contracts. Helen and the prior outer planets flagship both had facility instruments.
Flagship	NASA's largest category of science missions, typically featuring nine to thirteen distinct instrument teams who coordinate science observations between them.
IAU	International Astronomical Union, the body of elected scientists who make decisions about, among other things, what to name different planetary features and moons.
IDS	Interdisciplinary scientist. Selected at the outset of the mission in 1989, Helen's ten interdisciplinary scientists are equivalent in social status to instrument PIs. They are allowed to combine instrument data and requests from more than one instrument, typically two.
ITAR	The International Traffic in Armaments Regulations, United States law that restricts circulation of information or technical materials to noncitizens.
PAM	The physics group of magnetospheric and particle scientists, operates more collectively, like the "postcommunitarian" communities identified by Knorr-Cetina and Shrum.
NASA	National Aeronautics and Space Agency, the United States' space agency.
PDS	Planetary Data System, a NASA-wide open database and searchable library with all data from all spacecraft missions. Mission teams are responsible for submitting data to the PDS on a regular schedule.
PI	Principal investigator, the leader of a team of scientists. On Helen, there are eight PIs that lead instrument teams that they assembled to solve specific problems ("investigation"). On Paris, there is only one PI and one science team that hosts many instruments and investigations.
PIE	Preintegrated events, observational requests levied by one discipline working group's targeting team over another's.
PSG	Project Science Group meeting, a meeting of all affiliated Helen scientists held thrice yearly, twice hosted by NASA and once by ESA.
Segment/Sequence	A section of Helen's tour, planned by targeting teams. Segments are composed of sequences of activities that the spacecraft executes.
TacPlan	A Paris meeting of all relevant science and engineering ("tactical operations") personnel for planning a day's activities on Mars, contrast with Helen's sequential process, where scientists met with science planners in targeting teams who then hand off plans to the engineers.
Target Team	A Helen group staffed with representatives from each instrument team who meet to plan sequences of activities for a specific spacecraft sequence as the craft flies past their object of interest.

Team Lead	NASA-selected leaders of "facility instrument" teams, at the same hierarchical level as PIs but with no power or authority over their team members, as each one is individually contracted by NASA.
Tour	The preestablished path that the spacecraft takes around the planet and its moons, uses gravity assists as it flies past these targets in order to change its attitude, heading, or inclination.
WebEx	Commercial videoconferencing software by Cisco Systems.

NOTES

INTRODUCTION

1. In short, I take a poststructuralist stance on organizational structure. This approach might unduly entangle methods and phenomena thanks to the rise of poststructuralist-inspired investigations that blur the tired structure-agency divide in organizational theory (Barley 1996; Giddens 1986; Powell and Colyvas 2013). In any case, such a consideration would not impact my choice to observe the active structuring of an organization in action through ritualized decision-making meetings.
2. In this formulation, I offer dual apologies to John Law (1994) for the similarities in expression with his "organizational ordering" and for transforming a verb into a noun. I hope, however, to retain something of order's "verb-ness" through my ethnomethodological commitment to *order* as a situated, practical, continual accomplishment.
3. During my fieldwork, Helen staff initiated a census to account for all active mission personnel, indicating the mission's broad reach and porous boundaries.
4. Principally, the European nations included but were not limited to the UK, Spain, Germany, Italy, and France.
5. During my ethnographic work, I attended meetings and interviewed scientists and engineers involved in other, contrasting missions, which aided tremendously in formulating the central thesis of this text. I am grateful to the leaders and members of these spacecraft teams for their gracious invitations and time spent in discussion with me.
6. Other narratives about these missions are told in Buratti (2017), Clancey (2012), Colwell (2017), Conway (2015), Groen and Hampden-Turner (2005), Lorenz (2017), Meltzer (2014), Mirmalek (2009, 2020), Rosental (2007), Squyres (2005), and Vertesi (2015).
7. For ethnomethodologists too, using members' categories allows the analyst to stay true to the interactions that compose local sense-making instead of disappearing into lofty formal analysis (what Harald Garfinkel abbreviates as "FA").

CHAPTER ONE

1. Stacia Zabusky describes her study as concerned with epistemology, not collaborative work (Zabusky 1995, 37). However, I observed no ready distinction between the science I was interested in tracing and the micropolitics of collaboration in the room.

2. For this historical account, I am grateful to many conversations with Alain, one of Helen's founding fathers; to archival documents provided by Cliff, an interdisciplinary scientist on the original science definition team; as well as to European Space Agency archives accessed thanks to a NASA History Office-History of Science Society Fellowship.
3. Contrasting Helen to its precursor mission decades earlier, a science planner explained the clarity and visibility provided by the prior mission's bureaucratic-hierarchical structure as follows: "Everybody knew everybody. Everybody knew what their job was. Everybody knew what everybody else's job was. The lines of communication were crystal clear. The operational interface agreements probably weren't even written down. If they were written down, I certainly didn't read them. It was just *wonderful*" (Interview, November 12, 2009).
4. This was consistent with the lab's managerial structure in the early aughts, as it supplemented its matrix organization with a turn to Total Quality Management. Under the new managerial principles espoused by the lab director—himself the project scientist on a prior flagship who had once had to adjudicate debates among PIs—and his colleague trained at the Harvard Business School, decision-making of import was now encouraged at lower levels of the hierarchy (Westwick 2007).
5. This is a members' term, not the analytical definition offered by Balakrishnan et al. (2011).
6. It was also because of these losses that the NASA administrator in charge determined that the Paris mission would fly two identical vehicles to mitigate potential losses—and public relations—in case one failed upon landing. An outstanding history of the Mars Program is relayed by Erik Conway (2015); here, I relay only those details relevant to crafting Paris as a sociotechnical organization.
7. Operational engineers have another perspective again. It is well known when a mission is announced, money starts to flow from NASA, providing jobs and bodies on lab to support the development phase. But mission costs that come during the operations phase tend to be less prestigious and dwindle the longer the spacecraft is in operation. For this reason, an operations engineer explained to me that development-phase engineers were likely to "throw [problems] over the wall" to operations—that is, make decisions that solve problems and save money up front in development but require complexity or navigating impossibilities in operations. The loss of the scan platform is one such case of a cost-savings decision during development that balloons complexity during operations.
8. Interview, August 3, 2009.
9. In their studies of classification systems, Geof Bowker and Susan Leigh Star apply the concept of torque to social experience, showing how classification régimes such as racial categories in apartheid South Africa can produce "tremendous strain on individual biographies" when a person and their assigned category are not in alignment. The authors identify this as a form of social and categorical *torque* (Bowker and Star 1999).
10. Even the congressional hearing and protests at Helen's launch due to its planned radioactive power source paled in comparison to the scan platform's prominence in Helenites' sense of their history (for analysis of prior clashes over NASA priorities and environmental concerns, see Maher 2017).

CHAPTER TWO

1. The "virtual organization" fad of the late 1990s produced many studies of the challenges of distance work, including limited common ground between sites, the main-

tenance of bureaucratic control over distance workers, and the construction of e-laboratories (see especially Jarvenpaa and Leidner 1999; Nohria and Berkley 1994; Olson and Olson 2000; Olson, Zimmerman, and Bos 2008; Ramayah et al. 2003).

2. There is an occasional sixth targeting team for when the spacecraft is as far from the planet as possible in its tour. This period is open season for all disciplines to observe, but the spacecraft is so far away from the planet and its moons that targeted observations are rarely feasible.
3. Some science planners, such as George, hold doctorates in a scientific topic relevant to planetary science; others, such as Rose, hold a dual doctorate in engineering as well.
4. Rings targeting team, July 8, 2009.
5. Satellites targeting team, July 10, 2009.
6. Saturn targeting team, August 8, 2009.
7. Compare with collaborative temporal work and the fixation of space and place described in Jackson et al. (2011), Schegloff (1972), and Steinhardt and Jackson (2015).
8. Reviewing PSG agendas since the 1980s, the sequence of events is remarkably similar. In the final two years of the mission, however, this was held up to new scrutiny and changes were made to the structure. The turnover in project scientist leadership also produced a few changes over the years. For instance, when Francis came into this role, he had heard that people sat together congregated in cliques, from whence they launched coordinated contestations of events. He therefore created a U-shaped set of tables at the front of the room for the top level of the hierarchy: PIs, Team leaders, senior personnel in the project, NASA and ESA representatives, and interdisciplinary scientists. With Facebook, email, and SMS systems at fingertips, much of the backchannel conversation soon moved online, occurring simultaneously to events in the room (Vertesi 2014). After I left, the new project scientist, Victoria, reportedly introduced laptop-free meetings, locally referred to as "topless" meetings.
9. This is a ritual form that Gideon Kunda calls "talking down" (Kunda 2006): a method of information exchange from management to membership, although "talking across" between members within matrixed groups is also common at the PSG.
10. The problem does not arise from the science team but is established by the project office. I did not witness a PSG address questions that might concern the rank and file on the mission, such as getting more data, integrating junior scholars into the process, organizing to petition politicians for funds, or coordinating communication of science findings.
11. Plenary session PSG, January 27, 2009.
12. Titan working group meeting, January 26, 2009.
13. Although not articulated, the suggestion would make for a challenge in allocating observational time fairly among instruments if the camera and infrared received extra passes to accomplish a radar goal.
14. Personal communication, October 27, 2009.
15. For another example, see the discussion of PIEs in the next chapter.
16. Interview, November 14, 2009.
17. PSG meeting, January 27, 2009.
18. Meeting observation, January 27, 2009. In a discussion over a difficult observation, Karen asked, "Is this gonna be one where you'll have to make cookies to spend [fuel]?" Isabelle questioned, "Is this the standard way of doing things on this team, baking cookies?" "How do you think we got all that stuff on [the prior mission]? I baked cookies like crazy," stated Gwen, to much laughter. Isabelle rejoined, "I need to hire a cookie baker!"

19. A few examples are as follows: a scientist, Grant, during a Titan meeting suggested splitting a flyby between two instruments who were competing for it: "That would at least make everybody equally unhappy." Or science planner Tim in an interview described the difficulty of interfacing between scientists and engineers: "You have to keep them equally unhappy."
20. Interview, June 29, 2009.

CHAPTER THREE

1. Atmospheres meeting, November 16, 2009.
2. Rings meeting, July 29, 2009.
3. Atmospheres meeting, August 3, 2009.
4. Nichols (2018) argues that such comments, offered when speakers admittedly "know better," manage status in a masculine cultural environment. Despite—or perhaps because of—Helen's many women in leadership positions, I witnessed several exchanges that could be cataloged in this way, from the moniker of "topless meetings" for laptop-free assemblies to a comment about adding "no procreation" to the mission contract when an expert was missing due to maternity leave to my own harassment by a member of the team in a public forum, described in the methods appendix.
5. This also conformed to the interactional norm of managing in-group status by asserting one group's processes as better than another.
6. PIE Meeting, November 17, 2009.
7. PIE Meeting, November 17, 2009. Satellites resolved the problem by using intergroup negotiation, convincing colleagues on the Rings targeting team that *it was in the Rings group's best interest* to look at moonlets to better understand the dynamics of the ring system. Thus several Satellite observations went ahead under the cover of Rings priorities. This is one of the few examples I observed on Helen where a team achieved an observation by making their priority a shared one across another team.
8. Selections here taken from field notes and meeting recordings on July 24, 2009.
9. One interviewee expressed this classic divide in planetary science with a common turn of phrase: "Left to their own devices, the scientists would do insane things with the spacecraft, and left to their own devices, the engineers would be happy to fly a brick." The distinction between science and engineering was organizationally robust on Helen, reinforced through meetings such as this one. As I will describe, on Paris, the distinction was acknowledged in terms of expertise, but organizationally, individuals worked constantly across the divide so that no such "versus" could take place.
10. In addition to project manager (i.e. head of engineering staff), Everett also held the title of program manager overseeing the mission, including its ESA component. One scientist called him "the daddy of the mission who people went to," while my colleague Marisa Cohn interviewed an engineer in the ranks who described Everett as "the man at the top who sees everything." Everett himself graciously allowed me and my colleagues to study the mission but stated several times that he did not think we would uncover anything that he did not already know—evidence of his position at the top of a well-functioning chain of command. Everett also controlled the budgets, requiring PIs and team leaders to go to him to request resources. Many of them thought he was an outstanding manager who handled budgets and decision-making well, while others expressed concern that he was often in charge of scientific decision-making. In a later interview, a scientist recalled seeing Everett present about Helen for a congressional visitor despite Everett not being a scientist; he realized

then and there that "there could never be a parallel relationship between the project scientist and the project manager position; it was because he [the project manager] was in charge."

11. This required asking Isabelle, the camera PI, about cutting camera observations. Her response, reported in a later meeting, was "if radio can deliver, then she doesn't give a damn." This was relayed to Antonio, who responded with considerable detail about what exactly they would observe—perhaps to indicate that they could, indeed, deliver.
12. I use the male pronoun because there are no women appointed to these roles. I was unable to observe the mission in depth during Victoria's tenure as project scientist.
13. The choir is also a valuable site for ethnographic analysis. As a member, I had a front-line view of backstage organizational talk about the life of working on the mission. Their lyrics wittily combined joy, irreverence, celebration, self-deprecation, and wry humor in a way that spoke from the heart of Helen culture.
14. PSG meeting lunch, January 30, 2009.
15. Interview, July 7, 2009. In another interview, her colleague similarly described Helen as "a family that in it has its own little dysfunctional elements and we love them anyway."
16. Interview, July 27, 2009. As I have mentioned, many sought to correct me once I began to present my findings, arguing against my use of the word "dysfunctional"—despite the fact that it came from years of observing their own talk. To be clear, this does not mean that I argue the mission *is*, in a measurable sense, dysfunctional.

CHAPTER FOUR

1. TacPlan, October 4, 2006.
2. TacPlan, March 5, 2007.
3. On adjacency pairs, see Sacks, Schegloff, and Jefferson (1974) and Schegloff (2007). So common and obvious was this adjacency pair that it was frequently subject to joking, such as stating "I'm ecstatic!" or "I'm the happiest!" Consistent with jokes on the Paris team—and in stark contrast to the jokes using affective ambiguity on Helen—these restatements maintained positive affect and earnestness in demeanor.
4. Interview, November 14, 2009.
5. Several times, I heard from team members that even though they didn't know exactly what it was I did, they were sure that I was making a valuable contribution. This echoed how individuals felt about or perceived other members of their team, given the clear separation of roles and respect for expertise.
6. To paraphrase Turco's analysis of TechCo, this change in communication environment was locally equated with changes in decision-making structure (Turco 2016, 169), again demonstrating the inseparable hybridity of culture and structure.
7. Paris science team meeting, July 18, 2011.

CHAPTER FIVE

1. Future work will examine the perennial problems with personal, institutional, and mission funding excluded from detailed analysis here.
2. Spacelabs itself was also beset with interinstitutional complexities as it is a subsidiary of an engineering school, not an official NASA center. During my fieldwork on Helen, the lab implemented government background checks for badge approval. Consistent with the techniques that DiMaggio and Powell describe by means of which organizations imitate each other in order to gain legitimacy ("institutional

isomorphism"; DiMaggio and Powell 1983), this would bring Spacelabs into line with other government centers and NASA facilities and help with the seamless implementation of contracts. But the lab had a dual identity as university research institute and a contract center. A group of scientists and engineers—including Mark from Paris and Mickey from Helen—brought a legal case against Spacelabs to protect their right to information privacy. The case went to the Supreme Court, where it was dismissed, and the plaintiffs left the lab for different research centers and technical positions altogether. Years later, notices from the National Labor Relations Board remained posted around Spacelabs indicating that the lab had retracted the "unlawful written warnings" issued to these individuals during and after their lawsuit—evidence of the cost of unresolved interinstitutional tensions in the field.

3. Ad hoc navigation and science planning meeting, November 11, 2009.
4. On constraining surface interpretation, see chapter 7 of Vertesi (2015).
5. It was also the case with a new space telescope, which ballooned to many billion dollars over budget and was years behind schedule. Since this was planned in an adjacent scientific community, its impact on planetary scientists was experienced as an exogenous pressure as opposed to the internal jostle for funding between missions from the small planetary science budget.
6. In many ways, planetary scientists were responding to a crisis in public funding caused by the global financial crash, which compounded with Spacelabs' local emphasis on the period of mission development at the expense of operations. On the ground, local resource prioritization was visible as a series of decisions that drew ire from Paris and Helen team members and the community at large. It could also be seen as an effective pitch for continued funding by Spacelabs and its new rover mission in a period of financial strain: a strategic move that kept the lab competitive and active, and maintained internal continuity in knowledge management despite the collapse of public funds.
7. While it is not my focus here, not all international collaborations were locally considered virtuous. Euro-American collaborations, American-Japanese collaborations, and even launches from India's space agency during my period of study were usually discussed with this kind of praise, while plaudits for the emergent space program in China were more reserved.
8. Translations reported here are paraphrased or rely on field notes made in English following group social events.
9. My interviews with the ESA project manager, François, with founding scientist Alain and with the director of the Italian Space Agency continually highlighted this point.
10. This was certainly a benefit during funding crises in the United States: when NASA cannot afford to produce the whole spacecraft by itself, international instruments or other contributions can fill in the gaps. Historian Patrick McCray describes how this strategy proved useful in the astronomy community with a US-Canada partnership for an eight-meter telescope (2006, 189–90).
11. Field notes, January 17, 2010.
12. Even as US-Russia relations soured toward the close of my project, scientists spoke amicably of relationships with Russian colleagues, even when Roscosmos took responsibility for launching American astronauts at the close of the shuttle program.
13. Field notes, April 28, 2011; translated from French.
14. Personal conversation, October 28, 2008.
15. Described in personal conversations in French and in German at a Paris team meeting in July 2007.

16. My own national status was a hallmark of my experience on Paris. At one stage, I was invited to officially join the team, but my citizenship was reportedly too complicated. When I began work on Helen, I expected a similar set of hurdles and was astonished when my paperwork was processed easily, and I received a facility badge with no delay. I noted this discrepancy in passing to the associate who assisted me with lab access. Her explanation was institutional: everyone was used to international participants on Helen, and the associate who worked for Paris was not widely liked, so her superiors were more likely to kick back her applications with problems. The friction that I had once associated with transnational and intergovernmental tensions was actually due to local social relations, producing a clash with the Paris value of collectivism.
17. Meeting observation, March 25, 2009.
18. Field notes, January 30, 2009.
19. Personal conversation, August 25, 2011.
20. Mirmalek (2020) argues that the experience of Mars time was bound up with organizational concerns at the lab, where individuals were asked to work without human needs, like machines themselves.

CHAPTER SIX

1. Paris weekly science tagup meeting, April 26, 2007.
2. Paris weekly science tagup meeting November 19, 2007; targets requested in TacPlan meeting, August 23, 2007.
3. Like Neff's studies of Silicon Alley, in conversation with qualitative analysis, the graphs explore and attempt to either validate or falsify claims that arise from interpretivist work (Howard 2002; Neff 2012) and can illuminate the visible distinctions in scientific "cycles of credit" (Latour and Woolgar 1979).
4. It is beyond the scope of this book to articulate where these datasets travel outside of the mission once they are released to the broader community of scientists.
5. Outside of PAM, Helen's components are held together by interdisciplinary scientists and a few individuals, like postdocs Kevin and Drew, who draw funding from multiple teams' coinvestigators. This is in line with Burt's (2000) argument that both high and low status individuals may hold brokerage positions. Paris relies less on brokerage: its higher clustering coefficient and network density and its lower average path length than Helen's speaks to individual members' greater propensity toward cross-team publications.
6. Compare with Olson and Olson (2000). We did not conduct a citation network analysis, which, based on junior scholars' cited experiences, might be well placed to detect institutional or national ties.
7. Neither David nor Theodore were born in the US, nor were they native English speakers. Given that they were already working in a new cultural environment, they were perhaps more likely to miss subtle distinctions in coded practices on each team.
8. Helen PSG PAM meeting, October 21, 2009; the first workshop was held on October 19. Gabor later called this "really . . . a new science." PAM had already been the locus for collaboration between instruments in the early days of the mission when the magnetometer PI flew out to the lab to argue that they had just seen a strange feature at the south pole of Enceladus. Describing this discovery, later legendary on the team, she explained, "What we were able to do was—the only way to understand an observation like that is to pull all the data centers together, and so it was, I think, the first time that all the teams worked closely together. . . . It's the way that science should be done,

and we had been doing that, and I think the Enceladus data then showed the other teams that you could actually do this with all the datasets" (Interview, June 7, 2010).

9. Helen PSG, Aurora working group meeting, February 1, 2010.
10. Helen PSG, Aurora working group meeting, June 3, 2010. As Max explained, "We knew about this back in 1991, we had a discussion about it, and they [ultraviolet] wanted us [infrared] to change our orientation to fit theirs, and we wanted them to change their orientation to fit ours . . . and in the end we were like, well, if you're not gonna help us, then we're not gonna help you. . . . I think if we'd known there was some science we could have done together we might have fought a little harder . . . but this is a very small part of infrared observations." Note the use of instrument metonymy, battle talk, and resisting an unfair trade ("If you're not gonna help us, then . . ."), which are Helen interactional norms.
11. A later paper coauthored by Kevin and Gwen in 2017 again combined results from ultraviolet and infrared but not due to planning. In a later chapter, I will discuss the impact of organizational change and time on the Helen dataset that facilitated such "collaborations" to emerge across instrument teams on the flagship mission.
12. Infrared team meeting, September 1, 2009.

CHAPTER SEVEN

1. Compare with Ziewitz (2011), wherein screens and screen work are enacted in practice as either perspicuous or absent; the present analysis emphasizes the *organization's* enactment through screen work.
2. On spreadsheets' incorporation into sociotechnical practices, see Dourish (2017).
3. Based on the practices of geomorphology, planetary scientists are taught to draw geological contact points onto maps in order to identify land features on photographs taken from orbit or from the air. On transferring Earthly geological practices to the planets, see Messeri (2014), Shindell (2010), and Vertesi (2015).
4. This joke played on the excitement about this mission milestone but also on the importance of accurate annotations; see chapter 5.
5. Titan jumpstart meeting, April 15, 2009.
6. On "saving the phenomena," see Kuhn (1958).
7. Scientists could email their requests for observations to the software secretary in advance, but these were all subject to the same discussion and trimming as those offered after the meeting officially began. Certain software secretaries were especially adept at making quick changes or suggesting effective ways to synergize across observations. In one meeting, the PI remarked that watching his colleague expertly make these changes in Conduct! in real time was "like watching a piano play" (TacPlan meeting, March 27, 2007).
8. Interview, June 29, 2009.
9. Despite the importance of digital distance collaboration tools, Walter's paper printouts remind us of the continuing, essential role for paper in knowledge work (Sellen and Harper 2003).
10. I learned how important the *just so* aspect of these movements was when, publishing an early piece about this aspect of the team, a professional illustrator depicted their interpretation along the margins of the text. But this was not the dance, I protested! They would never move their arms *that way*, only precisely *like this*! I was overruled in favor of artistic license.
11. Rod attended this meeting via teleconference, so I could not see whether he used his body or other tools to work this problem through.

12. I am grateful to Emma Lawless for her observations in this location in 2013.
13. This explains the conclusion of the Titan jumpstart exchange above, when Eleanor and Jerome worked to decide how to point their remote sensing instruments. Their right-hand rule gesture helps them articulate where Helen is pointing and coordinates between their instruments, but there is some confusion between them because each does not fully know what the other's instrument requires. In their gestures, it is also not clear which part of the hand corresponds to which axis: Edgar and Jerome both use their index finger to indicate primary pointing for their instrument; in Jerome's case, negative-y; and in Edgar's case, positive-z. It is two PAM scientists, Howard and Daniel, who solve the problem. They identify where the spacecraft will be and how it will rotate across the sun-Titan plane by bringing the gestures closer into their bodies. For Daniel, the spacecraft's orientation relative to the sun-Titan plane is as tangible a feature for planning and as sensible to humans as Titan is a targeted object for Eleanor or Jerome. It is not that Eleanor and Jerome do not know what these elements *are*; rather, Daniel has an *embodied* sense of where they are and what they mean for the spacecraft as it tumbles through space.
14. Many science studies scholars follow Bruno Latour in denying a hard distinction between "the social" and "the technical." Here, I disentangle the distant craft from the team on Earth only to show how their many entanglements are organizationally specific and reinforcing, as team members interactionally produce spacecrafts' bodies and their possibilities on Earth. As this enactment is organizationally situated and accomplished, this produces a hybrid spacecraft shot through with organizational order.

CHAPTER EIGHT

1. As work with planetary data transitioned to digital systems and everyday computing, this may have facilitated some crossover with broader norms in computing culture. More research is necessary to trace whether the shift in attitudes also emerged in response to the growth of the planetary sciences and the unavailability of funds and faculty positions to support newcomers in the wake of the "lost decade" of missions in the 1980s.
2. I am indebted in this discussion to ideas developed in earlier papers with fellow scholars (Vertesi and Dourish 2011; Vertesi et al. 2016).
3. The growth of large-scale scientific collaborations under "cyberinfrastructure" initiatives in the late 1990s and early 2000s produced a range of studies aiming to better understand the principles and tensions of data creation, exchange, and management in scientific communities (Baker and Yarmey 2009; Bietz and Lee 2009; Birnholtz and Bietz 2003; Borgman 2015; Borgman et al. 2016; Bowker 2008; Cummings and Kiesler 2005; Edwards 2010; Edwards et al. 2011; Lee, Dourish, and Mark 2006; Millerand et al. 2013; Olson, Zimmerman, and Bos 2008; Ribes and Bowker 2008; Ribes and Finholt 2009; Wynholds et al. 2012).
4. Language from economic sociology illuminates how such exchanges become socially meaningful. In Viviana Zelizer's account, the circulation of money both establishes boundaries and enacts intimate relationships. Children's allowances, marital joint credit cards, and even sexual transactions demonstrate that how much and to whom currency is given is a way of managing or violating relationship norms. As money becomes the material with which relationships are constituted and maintained, financial exchange (or in my case, data exchange) becomes a form of what Zelizer calls "relational work" (2005, 2012).

5. File sharing, open data, and talk of "the knowledge economy" dominates popular, academic, and policy discussions about the exchange and mobility of large datasets and infuses such practices with moral overtones. How much and how rapidly a scientific project shares its data are now metrics of success, with agencies such as the US National Science Foundation and journals such as *Science* requiring researchers to make their data publicly available. From this point of view, scientific data sharing is a simple question of reinforcing well-known Mertonian norms, such as communalism and openness, and *not* sharing is simply being a bad scientist. This chapter sets aside such moral positions to focus instead on how local data circulation practices are organizationally consistent and must be upheld to support an organization's knowledge production aims.
6. The tools of trace ethnography (Geiger and Ribes 2011) reveal the heated debates that took place over these Wikipedia entries. This representative quote is from the discussion on the article about the moon Methone.
7. Personal correspondence, August 19, 2010. Isabelle's correspondence with the Wikipedia page editors requests credit be given only to the instrument team, removing her name (inserted by another user) and those of other scientists. In our interview, Isabelle also related going to the IAU to request standardization of discovery announcements by adding credit to the instrument team to resolve this issue in future. Their circulars reporting these discoveries list her name, her institution (responsible for image planning and processing), and the Helen imaging team. Isabelle told me that her goal was "to share the credit" with her teammates. But in a private conversation, two Helenites told the story differently, explaining that Isabelle had "twisted arms" at the IAU to remove attribution to instrument operators or low-level scientists and to state that discoveries had to include the team leader's name alongside the instrument team. The latter assertion circulated among Helenites, although I could not find support for this interpretation in the IAU's documentation. I relate this competing tale of events to show how such stories circulated on the team as a way of elucidating Isabelle's "personality"—an element I will return to in the next chapter.
8. The line between "proprietary" and "validation" periods was murky, as I observed during the PAM Aurora meeting when Joseph—who had once edited the prior flagship's rules of the road for data management—corrected Gabor: "Nobody has a proprietary period on Helen. We have a *validation* period. We were not allowed to have a proprietary period." Gabor agreed but explained that the validation period still worked the same way. "In effect, they [the instruments] don't have to share data" (October 21, 2009).
9. This period was variously reported to me with certainty as six, nine, or twelve months, speaking less to the contractual particulars and more to the polyvocality, autonomy, and lack of visibility between units in Helen's organizational form.
10. Of course, team members could come to consensus that the PI should have privilege as first author, but in practice, this was only the case on select reports summarizing mission findings—or on the paper reporting the cobble investigation, which Jeremy reportedly argued had enrolled the entire team.
11. Historical documentation of these rules of the road and their discussion were retrieved from the NASA Ames History Office's personal archives of scientists Elliott Leventhall and Al Seiff, team members on relevant prior missions from the 1970s through to the 1990s. I quote here from a 1978 version.
12. Helen PSG, January 27, 2009.
13. Interview, June 18, 2007.

14. Because the Italian Space Agency provided the radio antenna instrument with no further exchange of funds and because scientists in Italy were fully funded in their research, they may not have felt the pressure to "publish or perish" that their American counterparts on soft money and ongoing contracts did. Further, radio scientists often published in astrophysics or astronomy, where norms of data release may differ.
15. Radio team meeting, June 22, 2009.
16. Members of Isabelle's team I spoke to were divided over whether this constituted a true problem. Many of the scientists I spoke to attributed the primary concern to a few individuals, whom they described as "personalities" or "older" scientists; they claimed that the majority of them supported public release.
17. Interviews, December 3, 2009, June 13, 2012, January 6, 2020. Individuals who were there recalled that it was likely two different meetings and that one dealt more with budgetary concerns. The stories that circulated decades afterward, however, chiefly conflated these meetings into one. Unfortunately minutes, recordings, or notes were not made available for research.
18. Battles over Isabelle's spending and her control over team members' budgets was another common thread in oral history interviews. Isabelle was also the subject of an anonymous tip regarding misuse of travel funds, which she described as retaliatory. She reported resorting to a Freedom of Information Act request to gain access to relevant documents, meeting the NASA administrator, and threatening legal action to clear her name.
19. Everett reportedly exerted considerable oversight over instrument budgets, especially Isabelle's. In my time on the mission, I was never permitted to view a mission budget; Francis, despite being project scientist, reported finally seeing a partial budget just before leaving the team. At one of the Headquarters meetings involving Isabelle, Malcolm, and Everett, Malcolm recalled a presentation that Isabelle made to outline several objections to the image release policy, her first point indicating that it was an "onerous budget obligation" on her team. When the administrator asked for budgetary particulars, Everett provided a number for her yearly budget in her stead, prompting an official to nudge Malcolm and ask, "Is she for real?"
20. Interview, December 3, 2009.
21. After I left the mission, I heard that Victoria as the new project scientist had successfully negotiated the ability to use the PAM rules of the road across the teams, lowering barriers to data sharing across the mission. This was hailed as exciting news for multi-instrumental science. It also curtailed Isabelle's ability to control her instrument's data circulation and use in addition to existing restrictions upon monitoring team members' contracts and her instrument's budget.

CHAPTER NINE

1. Interview, date redacted.
2. Interview, August 29, 2007.
3. Interview, September 19, 2009.
4. On Helen, other instrument scientists and engineers often requested images to put their own data—thermal, ultraviolet, navigation, and so on into context. Because these requests could fill up the data cap before the camera team's own observations could be added, several team members told me that Isabelle deleted others' images to save space for her own team's needs and priorities, leading to accusations of egotism and obstructionism. For her part, Isabelle insists that she never refused anyone access to an image that they required. The confusion may center around which

images are essential for alignment or pointing ("context") and which are part of a scientific investigation. Scientists using nonvisual instruments sought "context" images to ground their scientific claims (in the style of "co-registration" or "constraints" discussed in Vertesi 2015), but Isabelle asserted that this constituted a scientific use of an image and thus required a preexisting collaborative agreement with her team.

5. The Discretionary Instrument is a reference to a phenomenon in allocating telescope time, wherein the chief scientist may use the telescope for their own projects instead of partitioning it out to participating scientists. Isabelle does not recall this instance, but she did describe a moment during her negotiations with individual scientists for data sharing agreements when the Helen project office announced to everyone that they would take charge of all support imaging for other instruments, overriding her own negotiations to that effect and her team's drafted data-sharing plan.
6. See chapter 3, note 4, on off-color humor in organizational context and Kanter (1977, 221–24) on "testing" as a form of boundary work in a dominant environment.
7. The literature on gender is vast, as is the literature on masculinities and growing studies of femininities. My analysis of "personalities" and affinity networks is indebted to Raewyn Connell's concept of masculinities (1995; Connell and Messerschmidt 2005) and its instantiation among scientists (Milam and Nye 2015), computer engineers (Cooper 2000; Ensmenger 2015), and high school groups (Pascoe 2012), with implications for hegemony (Schippers 2007).
8. On competence as a virtue for women but not for men, see Fletcher (1999). Women's networks and mentorship relationships have long been associated with better outcomes in STEM careers, although recent studies have qualified which types of networks and relationships matter most for success (Gaughan, Melkers, and Welch 2018).
9. Trautner and Kwan (2010) show that workplace dress plays a role in perceived competence, while Dellinger and Williams (1997) argue that we should consider women as active agents in their bodily self-presentation while navigating workplace power structures. I agree that women's—and men's—choices of clothing, hair styling, and jewelry enacts local notions of workplace competence, but here, I argue that such choices also signal membership in latent homosocial groups.
10. For connections between Burning Man and technology companies, see Turner (2009). See also Chen (2009).
11. Historians of technology describe the role of countercultural visionaries in the development of quantum physics, computing, space policy, and engineering (Kaiser 2012; Kaiser and McCray 2016; Maher 2017; McCray 2012; Turner 2006) in contrast with military-industrial developments (Edwards 1997).
12. These men see themselves as allies to improving the status of women at NASA and foster friendships and mentoring relationships with certain women in the field. Some of them even initiated a "men's auxiliary" for those who wished to help improve the status of women in planetary science. Unlike boys' club members, whose wives and partners typically do not work outside the home, their spouses were employed, albeit not in high-status scientific or technical careers. Although they profess a liberal and open attitude toward women in the field, an interviewee also described them as yet another network of high-status men that was difficult to break into.
13. This collusion of judgments of "excellence" as a question of "fit" within relatively tight or otherwise exclusionary networks is also described in studies of groups as varied as particle physicists (Traweek 1988), consultants (Rivera 2012), Web 2.0 workers (Marwick 2013), and academic review panel members (Lamont 2010). The

combination of locally judged principles of "merit" combined with rigid social networks serve a gatekeeping role that maintains homophily within power groups, as in Castilla and Benard (2010).

14. In line with prior findings about women's and men's networks (Brass 1985; Burt 1998), I observed limited capacity for individual women who were mentored or seen as "competent" in one environment to move up to another mission or be promoted to a position of more authority, as was more frequently the case with boys' club acolytes. While they were praised locally for their "competence," girls' club members were not perceived as major players. It may be because, starting out as deputies to powerful men, they were able to climb the organizational ladder without threatening existing powerful communities. But they are also not entrusted with founding positions on new projects precisely because they were not founders before.
15. Castilla's "paradox of meritocracy" describes how members of a dominant group are promoted and given higher bonuses than their minority counterparts in "meritocratic" organizations (Castilla 2008; Castilla and Benard 2010).
16. This resonates with John Law's (1994) study of a hierarchical government laboratory, where he describes such charismatic individuals as the result of a "mode of ordering" that retroactively makes sense of certain individuals' ability to wield agency in the face of structure. Such individuals were also labeled "cowboys."
17. While Americans were subject to these gendered networks and requirements, non-American participants on these missions could evade them and participate with limited ill effects. For instance, Sarah, a deputy PI eventually promoted to lead a European instrument, was celebrated for her discovery of the plumes and treated differently from Isabelle, even though both kept a tight watch on who could publish about their instrument's full quality data and how to note instrument leader credit. A French scientist I interviewed reported being baffled by the gender dynamics she encountered when she took a few months to visit a participating institution in the US affiliated with her instrument. The European men I met were either blind to these networks or, in a few cases, frustrated by their exclusion from them and by their resulting inability to move up in status or position in the community. I cannot describe the African Americans' experiences without de-anonymization, but they did not contravene my findings.

CHAPTER TEN

1. DiMaggio discusses the role of network and organizational positioning in cognition, which may also affect self-reporting of collaborative ties and orientation (1997; Srivastava and Banaji 2011).
2. Discussions of CEO importance to organizational change also dominate the literature on corporations: two sociological perspectives are Hannan and Freeman (1984) and Haveman, Russo, and Meyer (2001).
3. Perhaps by way of local explanation for this otherwise inexplicable behavior, a false rumor circulated at this time that Isabelle and Francis had once briefly dated.
4. Accounts on Helen differ as to whether there was ever a team-wide rules of the road in the early days of the mission. Some spoke of rules crafted under Malcolm's tenure that met with controversy; others pointed only to their local team's rules. This is consistent with the independent and autonomous orientation of the mission's teams, where a one-size-fits-all approach imposed from the top down could be problematic and accused of unfairness, as it would restrict each team from managing their boundaries and membership as they deemed fit.

5. PSG meeting, January 2011.
6. Field notes, March 23–27, 2009. On this topic, an affectively ambiguous comment about the Helen team retweeted on Twitter stated, "I could be their loo cleaner & after 15 years be promoted to assistant head loo cleaner :) :)" (@Mercury. 2010. Twitter, April 9, 2010, 19:30 p.m., [original account deleted]).
7. I am grateful to Louise Procktor for the opportunity to review this data.

CONCLUSION

1. This does not replace the concept of epistemic cultures but instead demonstrates how organizational knowledge is transmitted between generations of scholars in a laboratory. Further, many aspects of collaborative work are clearly associated with a discipline: note that the physics team on Helen works like many other particle physics collaborations. The present analysis offers a possibility for fine-grained attribution of epistemic outcomes.
2. Knorr-Cetina (1999) also refers to Durkheim's division of labor as part of the reason particle physicists feel so strongly associated with their detectors (123–36). Structural aspects of the collaboration's organization suffuse the descriptions of the field sites in *Epistemic Cultures*, but perhaps because the axis of comparison is established along disciplinary lines, these features may be submerged from the reader's view.
3. As the spacecraft was suddenly personified in members' talk, coming to life just before its fiery demise, I wondered if there was a new social order afoot, perhaps one more akin to Paris's. However, a few weeks later, I walked onto the lawn of the local technical university to join the Helenites for the conclusion of their mission. Hundreds of chairs were set up on the lawn in front of screens attuned to that small antechamber at Spacelabs with the American flag on the wall that I had glimpsed through the window on my way to Helen's project office. The chairs were clustered into sections labeled with signs—reserving them for each of the twelve instrument teams. These organizational orders endured to the very end of the mission.
4. Probe workshop, January 14, 2010.

POSTSCRIPT

1. Thermal team meeting, Oxford, June 18, 2009. My sincere thanks to George for, indeed, giving the video to the sociologist.

REFERENCES

Acker, J. 1990. "Hierarchies, Jobs, Bodies: A Theory of Gendered Organizations." *Gender & Society* 4 (2): 139–58.

Alač, Morana. 2009. "Moving Android: On Social Robots and Body-in-Interaction." *Social Studies of Science* 39 (4): 491–528.

———. 2011. *Handling Digital Brains: A Laboratory Study of Multimodal Semiotic Interaction in the Age of Computers*. Inside Technology. Cambridge, MA: MIT Press.

Alegria, Sharla N., and Enobong Hannah Branch. 2015. "Causes and Consequences of Inequality in the STEM: Diversity and Its Discontents." *International Journal of Gender, Science and Technology* 7 (3): 321–42.

Ames, Morgan. 2019. *The Charisma Machine: The Life, Death and Legacy of One Laptop per Child*. Cambridge, MA: MIT Press.

Anderson, Ryan B., and James F. Bell III. 2013. "Correlating Multispectral Imaging and Compositional Data from the Mars Exploration Rovers and Implications for Mars Science Laboratory." *Icarus* 223 (1): 157–80.

Andreas, Joel. 2007. "The Structure of Charismatic Mobilization: A Case Study of Rebellion during the Chinese Cultural Revolution." *American Sociological Review* 72 (3): 434–58.

Anicich, Eric M., Nathanael J. Fast, Nir Halevy, and Adam D. Galinsky. 2015. "When the Bases of Social Hierarchy Collide: Power without Status Drives Interpersonal Conflict." *Organization Science* 27 (1): 123–40.

Appelbaum, Steven H., David Nadeau, and Michael Cyr. 2008. "Performance Evaluation in a Matrix Organization: A Case Study (Part One)." *Industrial and Commercial Training* 40 (5): 236–41.

Arthur, W. Brian. 1989. "Competing Technologies, Increasing Returns, and Lock-In by Historical Events." *Economic Journal* 99 (394): 116.

Baker, Karen S., and Lynn Yarmey. 2009. "Data Stewardship: Environmental Data Curation and a Web-of-Repositories." *International Journal of Digital Curation* 4 (2): 12–27.

Balakrishnan, Aruna D., Sara Kiesler, Jonathon N. Cummings, and Reza Zadeh. 2011. "Research Team Integration: What It Is and Why It Matters." In *Proceedings of the ACM 2011 Conference on Computer Supported Cooperative Work*, 523–32. New York: ACM.

Barad, Karen Michelle. 2007. *Meeting the Universe Halfway: Quantum Physics and the Entanglement of Matter and Meaning*. Durham, NC: Duke University Press.

Barley, Stephen R. 1996. "Technology as an Occasion for Structuring: Evidence from Observation of CT Scanners and the Social Order of Radiology Departments." *Administrative Science Quarterly* 31 (1): 78–108.

Bechky, Beth A. 2006. "Gaffers, Gofers, and Grips: Role-Based Coordination in Temporary Organizations." *Organization Science* 17 (1): 3–21.

Bezrukova, K., K. Jehn, and E. Zanutto. 2009. "Do Workgroup Faultlines Help or Hurt? A Moderated Model of Faultlines, Team Identification, and Group Performance." *Organization Science* 20 (1): 35–50.

Biagioli, Mario. 1993. *Galileo, Courtier: The Practice of Science in the Culture of Absolutism*. Chicago: University of Chicago Press.

Bietz, Matthew J., and Charlotte P. Lee. 2009. "Collaboration in Metagenomics: Sequence Databases and the Organization of Scientific Work." In *Proceedings of the 11th European Conference on Computer Supported Cooperative Work, 7–11 September 2009, Vienna, Austria*, 243–62. London: Springer Verlag.

Birnholtz, Jeremy P., and Matthew J. Bietz. 2003. "Data at Work: Supporting Sharing in Science and Engineering." In *GROUP '03 Proceedings of the 2003 International ACM SIGGROUP Conference on Supporting Group Work*, 339–48. New York: ACM.

Blau, Peter M. 1980. *The Dynamics of Bureaucracy: A Study of Interpersonal Relations in Two Government Agencies*. Rev. ed. Chicago: University of Chicago Press.

Bloor, David. 1991. *Knowledge and Social Imagery*. 2nd ed. Chicago: University of Chicago Press.

Borgman, Christine L. 2015. *Big Data, Little Data, No Data: Scholarship in the Networked World*. Cambridge, MA: MIT Press.

Borgman, Christine L., Peter T. Darch, Ashley E. Sands, Jillian C. Wallis, and Sharon Traweek. 2014. "The Ups and Downs of Knowledge Infrastructures in Science: Implications for Data Management." In *Proceedings of the 14th ACM/IEEE-CS Joint Conference on Digital Libraries*, 257–66. Piscataway, NJ: IEEE.

Borgman, Christine L., Peter T. Darch, Ashley S. Sands, and Milena S. Golshan. 2016. "The Durability and Fragility of Knowledge Infrastructures: Lessons Learned from Astronomy." In *Proceedings of the 79th ASIS&T Annual Meeting: Creating Knowledge, Enhancing Lives through Information & Technology*, 57:1–57:10. Silver Springs, MD: American Society for Information Science.

Bowker, Geoffrey C. 2008. *Memory Practices in the Sciences*. Paperback ed. Inside Technology. Cambridge, MA: MIT.

Bowker, Geoffrey C., and Susan Leigh Star. 1999. *Sorting Things Out: Classification and Its Consequences*. Cambridge, MA: MIT Press.

Brass, D. J. 1985. "Men's and Women's Networks: A Study of Interaction Patterns and Influence in an Organization." *Academy of Management Journal* 28 (2): 327–43.

Buratti, Bonnie Jean. 2017. *Worlds Fantastic, Worlds Familiar: A Guided Tour of the Solar System*. Cambridge, MA: Cambridge University Press.

Burawoy, Michael. 1979. *Manufacturing Consent: Changes in the Labor Process under Monopoly Capitalism*. Chicago: University of Chicago Press.

———. 1985. *The Politics of Production: Factory Regimes under Capitalism and Socialism*. New York: Verso.

Burns, Lawton R. 1989. "Matrix Management in Hospitals: Testing Theories of Matrix Structure and Development." *Administrative Science Quarterly* 34 (3): 349–68.

Burt, Ron S. 1998. "The Gender of Social Capital." *Rationality and Society* 10 (1): 5–46.

———. 2000. "The Network Structure of Social Capital." *Research in Organizational Behavior* 22:345–423.

———. 2004. "Structural Holes and Good Ideas." *American Journal of Sociology* 110 (2): 349–99.

Burton, M. E., M. K. Dougherty, and C. T. Russell. 2010. "Saturn's Internal Planetary Magnetic Field." *Geophysical Research Letters* 37 (24): L24105, 5pp.

Canales, Jimena. 2011. *A Tenth of a Second: A History*. Paperback ed. Chicago: University of Chicago Press.

Castilla, Emilio J. 2008. "Gender, Race, and Meritocracy in Organizational Careers." *American Journal of Sociology* 113 (6): 1479–526.

Castilla, Emilio J., George J. Lan, and Ben A. Rissing. 2013a. "Social Networks and Employment: Mechanisms (Part 1)." *Sociology Compass* 7 (12): 999–1012.

———. 2013b. "Social Networks and Employment: Outcomes (Part 2)." *Sociology Compass* 7 (12): 1013–26.

Castilla, Emilio J., and Stephen Benard. 2010. "The Paradox of Meritocracy in Organizations." *Administrative Science Quarterly* 55 (4): 543–76.

Centellas, K. M., R. E. Smardon, and S. Fifield. 2014. "Calibrating Translational Cancer Research: Collaboration without Consensus in Interdisciplinary Laboratory Meetings." *Science, Technology & Human Values* 39 (3): 311–35.

Chen, Katherine Kang-Ning. 2009. *Enabling Creative Chaos: The Organization behind the Burning Man Event*. Chicago: University of Chicago Press.

Cheng, Leo, Nicole Spanovich, Alicia Vaughan, and Robert Lange. 2008. "Opposite Ends of the Spectrum: Cassini and Mars Exploration Rover Science Operations." Paper presented at the SpaceOps 2008 Conference, Heidelberg, Germany, May 12–16.

Clancey, William. 2012. *Working on Mars: Voyages of Scientific Discovery with the Mars Exploration Rovers*. Cambridge, MA: MIT Press.

Clark, B. C., R. V. Morris, S. M. McLennan, R. Gellert, B. Jolliff, A. H. Knoll, S. W. Squyres, et al. 2005. "Chemistry and Mineralogy of Outcrops at Meridiani Planum." *Earth and Planetary Science Letters* 240 (1): 73–94.

Cleland, David I. 1981. "The Cultural Ambience of the Matrix Organization." *Management Review* 70 (11): 24–31.

Coates, A. J., G. H. Jones, G. R. Lewis, A. Wellbrock, D. T. Young, F. J. Crary, R. E. Johnson, T. A. Cassidy, and T. W. Hill. 2010. "Negative Ions in the Enceladus Plume." *Icarus* 206 (2): 618–22.

Cohn, Marisa. 2016. "Convivial Decay: Entangled Lifetimes in a Geriatric Infrastructure." In *Proceedings of the ACM 2016 Conference on Computer Supported Cooperative Work*, 1509–21. New York: ACM.

Collins, Harry M. 1985. *Changing Order: Replication and Induction in Scientific Practice*. London: Sage.

———. 1998. "The Meaning of Data: Open and Closed Evidential Cultures in the Search for Gravitational Waves." *American Journal of Sociology* 104 (2): 293–338.

———. 2004. *Gravity's Shadow: The Search for Gravitational Waves*. Chicago: University of Chicago Press.

Collins, Harry M., and Robert Evans. 2002. "The Third Wave of Science Studies: Studies of Expertise and Experience." *Social Studies of Science* 32 (2): 235–96.

Collins, Randall. 2004. *Interaction Ritual Chains*. Princeton Studies in Cultural Sociology. Princeton, NJ: Princeton University Press.

Colwell, Joshua. 2017. *The Ringed Planet: Cassini's Voyage of Discovery at Saturn*. San Rafael, CA: Morgan & Claypool.

Connell, R. W. 1995. *Masculinities*. Crows Nest, New South Wales: Allen & Unwin.

Connell, R. W., and James W. Messerschmidt. 2005. "Hegemonic Masculinity: Rethinking the Concept." *Gender and Society* 19 (6): 829–59.

Conway, Erik M. 2015. *Exploration and Engineering: The Jet Propulsion Laboratory and the Quest for Mars*. Baltimore, MD: Johns Hopkins University Press.

Conway, Melvin E. 1968. "How Do Committees Invent?" *Datamation* 14 (5): 28–31.

Cooper, Marianne. 2000. "Being the 'Go-To Guy': Fatherhood, Masculinity, and the Organization of Work in Silicon Valley." *Qualitative Sociology* 23 (4): 379–405.

Correll, Shelley J. 2004. "Constraints into Preferences: Gender, Status, and Emerging Career Aspirations." *American Sociological Review* 69 (1): 93–113.

Correll, Shelley J., Cecilia L. Ridgeway, Ezra W. Zuckerman, Sharon Jank, Sara Jordan-Bloch, and Sandra Nakagawa. 2017. "It's the Conventional Thought That Counts: How Third-Order Inference Produces Status Advantage." *American Sociological Review* 82 (2): 297–327.

Cravens, T. E., R. L. McNutt Jr., J. H. Waite Jr., I. P. Robertson, J. G. Luhmann, W. Kasprzak, and W.-H. Ip. 2009. "Plume Ionosphere of Enceladus as Seen by the Cassini Ion and Neutral Mass Spectrometer." *Geophysical Research Letters* 36 (8): L08106, 4pp.

Cummings, Jonathon N., and Sara Kiesler. 2005. "Collaborative Research across Disciplinary and Organizational Boundaries." *Social Studies of Science* 35 (5): 703–22.

Cuzzi, J. N., J. A. Burns, S. Charnoz, R. N. Clark, J. E. Colwell, L. Dones, L. W. Esposito, et al. 2010. "An Evolving View of Saturn's Dynamic Rings." *Science* 327 (5972): 1470–75.

Daston, Lorraine. 1995. "The Moral Economy of Science." *Osiris* 10:2–24.

Daston, Lorraine, and Peter Galison. 2007. *Objectivity*. New York: Zone Books.

David, Paul A. 1985. "Clio and the Economics of QWERTY." *American Economic Review* 75 (2): 332–37.

Davis, Stanley M., and Paul R. Lawrence. 1977. *Matrix*. Addison-Wesley Series on Organization Development. Reading, MA: Addison-Wesley.

———. 1979. "Problems of Matrix Organizations." In *Matrix Organization & Project Management*, edited by Raymond Hill and Bernard J. White, 134–51. Michigan Business Papers, no. 64. Ann Arbor: Division of Research, Graduate School of Business Administration, University of Michigan.

De Vaan, Mathijs, Balazs Vedres, and David Stark. 2015. "Game Changer: The Topology of Creativity." *American Journal of Sociology* 120 (4): 1144–94.

Dellinger, Kirsten, and Christine Williams. 1997. "Makeup at Work: Negotiating Appearance Rules in the Workplace." *Gender and Society* 11 (2): 151–77.

DiMaggio, Paul J. 1997. "Culture and Cognition." *Annual Review of Sociology* 23:263–87.

DiMaggio, Paul J., and Walter W. Powell. 1983. "The Iron Cage Revisited: Institutional Isomorphism and Collective Rationality in Organizational Fields." *American Sociological Review* 48 (2): 147–60.

Doing, Park. 2004. "'Lab Hands' and the 'Scarlet O': Epistemic Politics and (Scientific) Labor." *Social Studies of Science* 34 (3): 299–323.

Dourish, Paul. 2017. *The Stuff of Bits: An Essay on the Materialities of Information*. Cambridge, MA: MIT Press.

Dove, Edward S., David Townend, Eric M. Meslin, Martin Bobrow, Katherine Littler, Dianne Nicol, Jantina de Vries, et al. 2016. "Ethics Review for International Data-Intensive Research." *Science* 351 (6280): 1399–1400.

Dunbar-Hester, Christina. 2008. "Geeks, Meta-Geeks, and Gender Trouble: Activism, Identity, and Low-Power FM Radio." *Social Studies of Science* 38 (2): 201–32.

Duneier, Mitchell. 2011. "How Not to Lie with Ethnography." *Sociological Methodology* 41 (1): 1–11.

Durkheim, Émile. 1893. *The Division of Labor in Society*. 1997 ed. New York: Free Press.

———. 1912. *The Elementary Forms of Religious Life*. Translated by Carol Cosman and Mark Sydney Cladis. 2008 ed. Oxford World's Classics. Oxford: Oxford University Press.

Durkheim, Émile, Steven Lukes, and W. D. Halls. 2013. *The Rules of Sociological Method and Selected Texts on Sociology and Its Method*. New York: Free Press.

Dyudina, U. A., A. P. Ingersoll, S. P. Ewald, C. C. Porco, G. Fischer, W. S. Kurth, and R. A. West. 2010. "Detection of Visible Lightning on Saturn." *Geophysical Research Letters* 37 (9): L09205, 5pp.

Eagly, Alice H., and Steven J. Karau. 2002. "Role Congruity Theory of Prejudice toward Female Leaders." *Psychological Review* 109 (3): 573–98.

Edwards, Paul N. 1997. *The Closed World: Computers and the Politics of Discourse in Cold War America*. 1st MIT Press paperback ed. Inside Technology. Cambridge, MA: MIT.

Edwards, Paul N. 2010. *A Vast Machine: Computer Models, Climate Data, and the Politics of Global Warming*. Cambridge, MA: MIT Press.

Edwards, Paul N., M. S. Mayernik, Archer L. Batcheller, Geoffrey C. Bowker, and Christine L. Borgman. 2011. "Science Friction: Data, Metadata, and Collaboration." *Social Studies of Science* 41 (5): 667–90.

Ensmenger, Nathan. 2015. "'Beards, Sandals, and Other Signs of Rugged Individualism': Masculine Culture within the Computing Professions." *Osiris* 30 (1): 38–65.

Ezell, Edward Clinton, and Linda Neuman Ezell. 2009. *On Mars: Exploration of the Red Planet, 1958–1978—the NASA History*. Newburyport: Dover.

Fara, Patricia. 2002. *Newton: The Making of Genius*. London: Macmillan.

Farrand, Bill, James F. Bell III, Jeff R. Johnson, B. L. Jolliff, A. H. Knoll, S. M. McLennan, S. W. Squyres, et al. 2006. "Visible and Near Infrared Multispectral Analysis of Rocks at Meridiani Planum, Mars by the Mars Exploration Rover Opportunity." *Journal of Geophysical Research* 12:E06S02, 28pp.

Farrand, William H., Jeffrey R. Johnson, Melissa S. Rice, Alian Wang, and James F. Bell III. 2016. "VNIR Multispectral Observations of Aqueous Alteration Materials by the Pancams on the Spirit and Opportunity Mars Exploration Rovers." *American Mineralogist* 101 (9): 2005–19.

Feldman, Martha S. 2000. "Organizational Routines as a Source of Continuous Change." *Organization Science* 11 (6): 611–29.

Feldman, Martha S., and Brian T. Pentland. 2003. "Reconceptualizing Organizational Routines as a Source of Flexibility and Change." *Administrative Science Quarterly* 48 (1): 94.

Ferguson, Trish, ed. 2013. *Victorian Time: Technologies, Standardizations, Catastrophes*. Palgrave Studies in Nineteenth-Century Writing and Culture. New York: Palgrave Macmillan.

Fernandez, Roberto M., and M. Lourdes Sosa. 2005. "Gendering the Job: Networks and Recruitment at a Call Center." *American Journal of Sociology* 111 (3): 859–904.

Fine, Gary Alan. 2010. *Authors of the Storm: Meteorologists and the Culture of Prediction*. Paperback ed. Chicago: University of Chicago Press.

———. 2012. *Tiny Publics: A Theory of Group Action and Culture*. Russell Sage Foundation Series on Trust. New York: Russell Sage Foundation.

Fiore, S. M. 2008. "Interdisciplinarity as Teamwork: How the Science of Teams Can Inform Team Science." *Small Group Research* 39 (3): 251–77.

Fletcher, Joyce K. 1999. *Disappearing Acts: Gender, Power, and Relational Practice at Work*. 1st MIT Press paperback ed. Cambridge, MA: MIT Press.

Fletcher, L. N., G. S. Orton, N. A. Teanby, P. G. J. Irwin, and G. L. Bjoraker. 2009. "Methane and Its Isotopologues on Saturn from Cassini/CIRS Observations." *Icarus* 199 (2): 351–67.

Ford, Robert C., and W. Alan Randolph. 1992. "Cross-Functional Structures: A Review and Integration of Matrix Organization and Project Management." *Journal of Management* 18 (2): 267–94.

Foster, J. G., A. Rzhetsky, and J. A. Evans. 2015. "Tradition and Innovation in Scientists' Research Strategies." *American Sociological Review* 80 (5): 875–908.

Foucault, Michel. 1977. *Discipline and Punish: The Birth of the Prison*. New York: Random House.

Fox, M. F. 2001. "Women, Science, and Academia: Graduate Education and Careers." *Gender & Society* 15 (5): 654–66.

———. 2010. "Women and Men Faculty in Academic Science and Engineering: Social-Organizational Indicators and Implications." *American Behavioral Scientist* 53 (7): 997–1012.

Freeland, Robert, and Ezra Zuckerman Sivan. 2018. "The Problems and Promise of Hierarchy: Voice Rights and the Firm." *Sociological Science* 5:143–81.

Freeman, Jo. 1972. "The Tyranny of Structurelessness." *Berkeley Journal of Sociology* 17: 151–64.

French, Richard G., Essam A. Marouf, Nicole J. Rappaport, and Colleen A. McGhee. 2010. "Occultation Observations of Saturn's B-Ring and Cassini Division." *The Astronomical Journal* 139 (4): 1649–67.

Friedland, William H. 1964. "For a Sociological Concept of Charisma." *Social Forces* 43 (1): 18–26.

Galbraith, Jay R. 2009. *Designing Matrix Organizations That Actually Work: How IBM, Procter & Gamble, and Others Design for Success*. 1st ed. Jossey-Bass Business & Management Series. San Francisco: Jossey-Bass.

Galison, Peter. 1997. *Image and Logic*. Chicago: University of Chicago Press.

———. 1998. "Trading Zone: Coordinating Action and Belief." In *The Science Studies Reader*, edited by Mario Biagioli, 137–60. New York: Routledge.

———. 2004. *Einstein's Clocks, Poincaré's Maps: Empires of Time*. New York: Norton.

Garfinkel, Harald. 1967. *Studies in Ethnomethodology*. Upper Saddle River, NJ: Prentice Hall.

———. 2002. *Ethnomethodology's Program: Working out Durkheim's Aphorism*. Lanham, MD: Rowman & Littlefield.

Garfinkel, Harald, Michael Lynch, and E. Livingston. 1981. "The Work of a Discovering Science Construed with Materials from the Optically Discovered Pulsar." *Philosophy of Social Science* 11:121–58.

Gaughan, Monica, Julia Melkers, and Eric Welch. 2018. "Differential Social Network Effects on Scholarly Productivity: An Intersectional Analysis." *Science, Technology & Human Values* 43 (3): 570–99.

Geiger, R. S., and D. Ribes. 2011. "Trace Ethnography: Following Coordination through Documentary Practices." In *2011 44th Hawaii International Conference on System Sciences (HICSS)*, 1480–89. Kauai, HI: IEEE.

Giddens, Anthony. 1986. *Central Problems in Social Theory: Action, Structure and Contradiction in Social Analysis*. Contemporary Social Theory. London: Macmillan.

Girard, Monique, and David Stark. 2003. "Heterarchies of Value in Manhattan-Based New Media Firms." *Theory, Culture & Society* 20 (3): 77–105.

Glaser, Barney G., and Anselm L. Strauss. 2009. *The Discovery of Grounded Theory: Strategies for Qualitative Research*. New Brunswick, NJ: Aldine.

Goffman, Erving. 1959. *The Presentation of Self in Everyday Life*. New York: Doubleday.

———. 1961a. *Asylums: Essays on the Social Situation of Mental Patients and Other Inmates*. New York: Doubleday Anchor.

———. 1961b. *Encounters: Two Studies in the Sociology of Interaction*. Hammondsworth, UK: Penguin Press.
———. 1967. *Interaction Ritual*. Chicago: Aldine.
———. 1981. *Forms of Talk*. University of Pennsylvania Publications in Conduct and Communication. Philadelphia: University of Pennsylvania Press.
Gore, Rick. 1981. "Voyager 1 at Saturn: Riddles of the Rings." *National Geographic Magazine* 160 (3): 3–31.
Groen, Bram, and Charles Hampden-Turner. 2005. *The Titans of Saturn: Leadership and Performance Lessons from the Cassini—Huygens Mission*. London: Cyan Books.
Gurnett, D. A., J. B. Groene, A. M. Persoon, J. D. Menietti, S.-Y. Ye, W. S. Kurth, R. J. MacDowall, and A. Lecacheux. 2010. "The Reversal of the Rotational Modulation Rates of the North and South Components of Saturn Kilometric Radiation near Equinox." *Geophysical Research Letters* 37 (24): L24101, 4pp.
Gusterson, Hugh. 1996. *Nuclear Rites: A Weapons Laboratory at the End of the Cold War*. Berkeley: University of California Press.
Hahn, Joseph M., Joseph N. Spitale, and Carolyn C. Porco. 2009. "Dynamics of the Sharp Edges of Broad Planetary Rings." *Astrophysical Journal* 699 (1): 686–710.
Hannan, Michael, and John Freeman. 1984. "Structural Inertia and Organizational Change." *American Sociological Review* 49 (2):149–64.
Hanson, Norwood Russell. 1958. *Patterns of Discovery: An Inquiry into the Conceptual Foundations of Science*. Cambridge, MA: Cambridge University Press.
Haraway, Donna J. 1988. "Situated Knowledges: The Science Question in Feminism and the Privilege of Partial Perspective." *Feminist Studies* 14 (3): 575–99.
———. 1991. *Simians, Cyborgs, and Women*. New York: Routledge.
———. 1997. *Modest_Witness@Second_Millenium. FemaleMan Meets OncoMouse: Feminism and Technoscience*. 1st ed. New York: Routledge.
———. 2007. *When Species Meet*. Minneapolis: University of Minnesota Press.
Hart, C., and M. Van Vugt. 2006. "From Fault Line to Group Fission: Understanding Membership Changes in Small Groups." *Personality and Social Psychology Łdots* 32 (3): 392–404.
Haveman, Heather A., Michael V. Russo, and Alan D. Meyer. 2001. "Organizational Environments in Flux: The Impact of Regulatory Punctuations on Organizational Domains, CEO Succession, and Performance." *Organization Science* 12 (3): 253–73.
Hecht, Gabrielle. 1998. *The Radiance of France: Nuclear Power and National Identity after World War II*. Inside Technology. Cambridge, MA: MIT Press.
Heckscher, Charles C. 1994. "Defining the Post-bureaucratic Type." In *The Post-bureaucratic Organization: New Perspectives on Organizational Change*, edited by Anne Donnellon and Charles C. Heckscher, 14–62. Thousand Oaks, CA: Sage.
Heckscher, Charles C., and Anne Donnellon, eds. 1994. *The Post-bureaucratic Organization: New Perspectives on Organizational Change*. Thousand Oaks, CA: Sage.
Heckscher, Charles C., and Paul S. Adler, eds. 2006. *The Firm as a Collaborative Community: Reconstructing Trust in the Knowledge Economy*. Oxford: Oxford University Press.
Hedman, M. M., J. A. Burns, M. S. Tiscareno, and C. C. Porco. 2009. "Organizing Some Very Tenuous Things: Resonant Structures in Saturn's Faint Rings." *Icarus* 202 (1): 260–79.
Heilman, Madeline E. 2001. "Description and Prescription: How Gender Stereotypes Prevent Women's Ascent Up the Organizational Ladder." *Journal of Social Issues* 57 (4): 657–74.
Helmreich, Stefan. 2007. "An Anthropologist Underwater: Immersive Soundscapes, Submarine Cyborgs, and Transductive Ethnography." *American Ethnologist* 34 (4): 621–41.

———. 2009. *Alien Ocean*. Cambridge, MA: MIT Press.

———. 2010. "Listening against Soundscapes." *Anthropology News* 51 (9): 10.

Helmreich, Stefan, Sophia Roosth, and Michele Ilana Friedner. 2015. *Sounding the Limits of Life: Essays in the Anthropology of Biology and Beyond*. Princeton Studies in Culture and Technology. Princeton, NJ: Princeton University Press.

Hendrix, Amanda R., Candice J. Hansen, and Greg M. Holsclaw. 2010. "The Ultraviolet Reflectance of Enceladus: Implications for Surface Composition." *Icarus* 206 (2): 608–17.

Herbert, Melissa S. 1998. *Camouflage Isn't Only for Combat: Gender, Sexuality and Women in the Military*. New York: New York University Press.

Hersch, Matthew H. 2012. *Inventing the American Astronaut*. 1st ed. Palgrave Studies in the History of Science and Technology. New York: Palgrave Macmillan.

Hinds, Pamela J., and Diane E. Bailey. 2003. "Out of Sight, Out of Sync: Understanding Conflict in Distributed Teams." *Organization Science* 14 (6): 615–32.

Hinds, Pamela J., and M. Mortensen. 2005. "Understanding Conflict in Geographically Distributed Teams: The Moderating Effects of Shared Identity, Shared Context, and Spontaneous Communication." *Organization Science* 16 (3): 290–307.

Hochschild, Arlie Russell. 1979. *The Managed Heart: Commercialization of Human Feeling*. 2012 ed. Berkeley: University of California Press.

———. 2000. *The Time Bind: When Work Becomes Home and Home Becomes Work*. New York: Holt.

Howard, Philip N. 2002. "Network Ethnography and the Hypermedia Organization: New Media, New Organizations, New Methods." *New Media & Society* 4 (4): 550–74.

Hughes, Thomas P. 1999. "The Evolution of Large Technological Systems." In *The Science Studies Reader*, edited by Mario Biagioli, 202–23. New York: Routledge.

Hurowitz, J. A., S. M. McLennan, N. J. Tosca, R. E. Arvidson, J. R. Michalski, D. W. Ming, C. Schröder, and S. W. Squyres. 2006. "In Situ and Experimental Evidence for Acidic Weathering of Rocks and Soils on Mars: Evidence for Acidic Weathering on Mars." *Journal of Geophysical Research: Planets* 111 (E2): E02S19, 16pp.

Ibarra, Herminia. 1993. "Personal Networks of Women and Minorities in Management: A Conceptual Framework." *Academy of Management Review* 18 (1): 56–87.

———. 1997. "Paving an Alternative Route: Gender Differences in Managerial Networks." *Social Psychology Quarterly* 60 (1): 91–102.

Isaacson, Walter. 2011. *Steve Jobs*. New York: Simon and Schuster.

Jackson, Steven J., David Ribes, Ayse Buyuktur, and Geoffrey C. Bowker. 2011. "Collaborative Rhythm: Temporal Dissonance and Alignment in Collaborative Scientific Work." In *CSCW '11 Proceedings of the ACM 2011 Conference on Computer Supported Cooperative Work*, 245–54. New York: ACM.

Janssen, M. A., R. D. Lorenz, R. West, F. Paganelli, R. M. Lopes, R. L. Kirk, C. Elachi, et al. 2009. "Titan's Surface at 2.2-Cm Wavelength Imaged by the Cassini RADAR Radiometer: Calibration and First Results." *Icarus* 200 (1): 222–39.

Jarvenpaa, Sirkka L., and Dorothy E. Leidner. 1999. "Communication and Trust in Global Virtual Teams." *Organization Science* 10 (6): 791–815.

Jasanoff, Sheila, ed. 2004. *States of Knowledge: The Co-Production of Science and Social Order*. International Library of Sociology. London: Routledge.

Jemielniak, Dariusz. 2014. *Common Knowledge? An Ethnography of Wikipedia*. Stanford: Stanford University Press.

Jones, G. H., C. S. Arridge, A. J. Coates, G. R. Lewis, S. Kanani, A. Wellbrock, D. T. Young, et al. 2009. "Fine Jet Structure of Electrically Charged Grains in Enceladus' Plume." *Geophysical Research Letters* 36 (16): L16204, 6pp.

Junker, Andrew. 2014. "Follower Agency and Charismatic Mobilization in Falun Gong." *Sociology of Religion* 75 (3): 418–41.

Kaiser, David. 2012. *How the Hippies Saved Physics: Science, Counterculture, and the Quantum Revival*. New York: W. W. Norton.

Kaiser, David, and W. Patrick McCray, eds. 2016. *Groovy Science: Knowledge, Innovation, and American Counterculture*. Chicago: University of Chicago Press.

Kanter, Rosabeth Moss. 1993. *Men and Women of the Corporation*. 2nd ed. New York: Basic Books.

Kaplan, Sarah. 2011. "Strategy and PowerPoint: An Inquiry into the Epistemic Culture and Machinery of Strategy Making." *Organization Science* 22 (2): 320–46.

Katz, Jack. 2002. "From How to Why: On Luminous Description and Causal Inference in Ethnography (Part 2)." *Ethnography* 3 (1): 63–90.

Kellogg, Katherine C. 2009. "Operating Room: Relational Spaces and Microinstitutional Change in Surgery." *American Journal of Sociology* 115 (3): 657–711.

Kellogg, Katherine C., Wanda J. Orlikowski, and JoAnne Yates. 2006. "Life in the Trading Zone: Structuring Coordination across Boundaries in Postbureaucratic Organizations." *Organization Science* 17 (1): 22–44.

Khan, Shamus Rahman. 2011. *Privilege: The Making of an Adolescent Elite at St. Paul's School*. Princeton Studies in Cultural Sociology. Princeton, NJ: Princeton University Press.

Kinch, Kjartan, J. N. Sohl-Dickstein, James F. Bell III, Jeff R. Johnson, Walter Goetz, and Geoff Landis. 2007. "Dust Deposition on the Mars Exploration Rover Panoramic Camera (Pancam) Calibration Targets." *Journal of Geophysical Research* 112 (E6): E06S03, 21pp.

Kling, Rob. 1991. "Computerization and Social Transformations." *Science, Technology, & Human Values* 16 (3): 342–67.

Knoblauch, Hubert. 2013. *PowerPoint, Communication, and the Knowledge Society*. Learning in Doing. Cambridge, MA: Cambridge University Press.

Knorr-Cetina, Karin. 1999. *Epistemic Cultures: How the Sciences Make Knowledge*. Cambridge, MA: Harvard University Press.

Knorr-Cetina, Karin, and Klaus Amann. 1990. "Image Dissection in Natural Scientific Inquiry." *Science, Technology & Human Values* 15 (3): 259–83.

Kohler, Robert E. 1994. *Lords of the Fly: Drosophila Genetics and the Experimental Life*. Chicago: University of Chicago Press.

Krugman, Paul R. 1979. "Increasing Returns, Monopolistic Competition, and International Trade." *Journal of International Economics* 9 (4): 469–79.

Kuhn, Thomas S. 1958. *The Copernican Revolution: Planetary Astronomy in the Development of Western Thought*. Cambridge, MA: Harvard University Press.

———. 1959. *The Essential Tension: Selected Studies in Scientific Tradition and Change*. Chicago: University of Chicago Press.

———. 1962. "The Structure of Scientific Revolutions." In *Foundations of the Unity of Science*, vol. 2. Chicago: University of Chicago Press.

Kunda, Gideon. 2006. *Engineering Culture: Control and Commitment in a High-Tech Corporation*. Philadelphia: Temple University Press.

Kunda, Gideon, and Stephen R. Barley. 2006. *Gurus, Hired Guns, and Warm Bodies Itinerant Experts in a Knowledge Economy*. Princeton, NJ: Princeton University Press.

Kvande, E. 1999. "'In the Belly of the Beast': Constructing Femininities in Engineering Organizations." *European Journal of Women's Studies* 6 (3): 305–28.

Lamont, Michèle. 2010. *How Professors Think: Inside the Curious World of Academic Judgment*. 1st paperback ed. Cambridge, MA: Harvard University Press.

Lamont, Michèle, and Virag Molnár. 2002. "The Study of Boundaries in the Social Sciences." *Annual Review of Sociology* 28:167–95.

Latour, Bruno. 1990. "Drawing Things Together." In *Representation in Scientific Practice*, edited by Michael Lynch and Steve Woolgar, 19–68. Cambridge, MA: MIT Press.

———. 1991. "Technology Is Society Made Durable." In *A Sociology of Monsters: Essays on Power, Technology and Domination*, 103–32. Sociological Review Monograph 38. New York: Routledge.

———. 2004. "Why Has Critique Run Out of Steam? From Matters of Fact to Matters of Concern." *Critical Inquiry* 30 (2): 25–248.

Latour, Bruno, and Steve Woolgar. 1979. *Laboratory Life: The Construction of Scientific Facts*. 1st ed. Princeton, NJ: Princeton University Press.

Law, John. 1994. *Organizing Modernity*. Oxford: Blackwell.

Law, John, and Annemarie Mol, eds. 2002. *Complexities: Social Studies of Knowledge Practices*. Science and Cultural Theory. Durham, NC: Duke University Press.

Lee, Charlotte P., Paul Dourish, and Gloria Mark. 2006. "The Human Infrastructure of Cyberinfrastructure." In *CSCW '06 Proceedings of the 2006 20th Anniversary Conference on Computer Supported Cooperative Work*, 483–92. New York: ACM.

Lee, Ching Kwan. 1999. "From Organized Dependence to Disorganized Despotism: Changing Labour Regimes in Chinese Factories." *China Quarterly* 157 (157): 44–71.

Leer, Kristoffer, Walter Goetz, Marjorie A. Chan, Steven Gorevan, Mikkel Fougt Hansen, Christian Lundmand Jensen, Gunther Kletetschka, Alastair Kusack, and Morten Bo Madsen. 2011. "RAT Magnet Experiment on the Mars Exploration Rovers: Spirit and Opportunity beyond Sol 500." *Journal of Geophysical Research* 116 (E7): E00F18, 8pp.

Lemmon, M. T., M. J. Wolff, M. D. Smith, R. T. Clancy, D. Banfield, G. A. Landis, A. Ghosh, et al. 2004. "Atmospheric Imaging Results from the Mars Exploration Rovers: Spirit and Opportunity." *Science* 306 (5702): 1753–56.

Leonardi, Paul M. 2012a. *Car Crashes without Cars: Lessons about Simulation Technology and Organizational Change from Automotive Design*. Acting with Technology. Cambridge, MA: MIT Press.

———. 2012b. "Materiality, Sociomateriality, and Socio-technical Systems: What Do These Terms Mean? How Are They Different? Do We Need Them?" In *Materiality and Organizing: Social Interaction in a Technological World*, edited by Paul M. Leonardi, Bonnie A. Nardi, and Jannis Kallinikos, 25–48. 1st ed. Oxford: Oxford University Press.

Leonardi, Paul M., and Stephen R. Barley. 2010. "What's under Construction Here? Social Action, Materiality, and Power in Constructivist Studies of Technology and Organizing." *Academy of Management Annals* 4 (1): 1–51.

Leonardi, Paul M., Bonnie A. Nardi, and Jannis Kallinikos, eds. 2012. *Materiality and Organizing: Social Interaction in a Technological World*. 1st ed. Oxford: Oxford University Press.

Lewis, Amanda E. 2004. "'What Group?' Studying Whites and Whiteness in the Era of 'Color-Blindness.'" *Sociological Theory* 22 (4): 623–46.

Leyrat, Cedric, Linda J. Spilker, Nicolas Altobelli, Stuart Pilorz, and Cecile Ferrari. 2008. "Infrared Observations of Saturn's Rings by Cassini CIRS Phase Angle and Local Time Dependence." *Planetary and Space Science* 6 (1): 117–33.

Li, Liming, Barney J. Conrath, Peter J. Gierasch, Richard K. Achterberg, Conor A. Nixon, Amy A. Simon-Miller, F. Michael Flasar, et al. 2010. "Saturn's Emitted Power." *Journal of Geophysical Research: Planets* 115 (E11): E11002, 16pp.

Lincoln, Anne E., Stephanie Pincus, Janet Bandows Koster, and Phoebe S. Leboy. 2012. "The Matilda Effect in Science: Awards and Prizes in the US, 1990s and 2000s." *Social Studies of Science* 42 (2): 307–20.

Linde, C. 2009. *Working the Past: Narrative and Institutional Memory*. Oxford: Oxford University Press.

Lipsky, Michael. 2010. *Street-Level Bureaucracy: Dilemmas of the Individual in Public Services*. 30th anniversary expanded ed. New York: Russell Sage Foundation.

Longino, Helen E. 1990. "Values and Objectivity." In *Science as Social Knowledge: Values and Objectivity in Scientific Inquiry*, 62–82. Princeton, NJ: Princeton University Press.

Lorenz, Ralph. 2017. *NASA Cassini-Huygens Manual*. Newbury Park, CA: Haynes North America.

Lynch, Michael, ed. 1985a. *Art and Artifact in Laboratory Science: A Study of Shop Work and Shop Talk in a Research Laboratory*. Studies in Ethnomethodology. London: Routledge.

———. 1985b. "Discipline and the Material Form of Images: An Analysis of Scientific Visibility." *Social Studies of Science* 15:37–66.

———. 1991. "Laboratory Space and the Technological Complex: An Investigation of Topical Contextures." *Science in Context* 4 (1): 51–78.

———. 1994. *Scientific Practice and Ordinary Action*. Cambridge, MA: Cambridge University Press.

Mackenzie, Adrian. 2006. *Transductions: Bodies and Machines at Speed*. Technologies. London: Continuum.

MacKenzie, Donald A. 2006. *An Engine, Not a Camera: How Financial Models Shape Markets*. Inside Technology. Cambridge, MA: MIT Press.

Maher, Neil M. 2017. *Apollo in the Age of Aquarius*. Cambridge, MA: Harvard University Press.

Mahler, Julianne. 2016. "NASA Contracting and the Direction of Space Science." *Administration & Society* 48 (6): 711–35.

Marcus, George E. 1995. "Ethnography in/of the World System: The Emergence of Multi-sited Ethnography." *Annual Review of Anthropology* 24:95–117.

Martin, Emily. 1991. "The Egg and the Sperm: How Science Has Constructed a Romance Based on Stereotypical Male-Female Roles." *Signs* 16 (3): 485–501.

Marwick, Alice Emily. 2013. *Status Update: Celebrity, Publicity, and Branding in the Social Media Age*. New Haven, CT: Yale University Press.

Massey, Douglas S., and Nancy A. Denton. 2003. *American Apartheid: Segregation and the Making of the Underclass*. Cambridge, MA: Harvard University Press.

Mazmanian, Melissa, Marisa Cohn, and Paul Dourish. 2014. "Dynamic Reconfiguration in Planetary Exploration: A Sociomaterial Ethnography." *MIS Quarterly* 38 (3): 831–48.

Mazmanian, Melissa, Wanda J. Orlikowski, and JoAnne Yates. 2013. "The Autonomy Paradox: The Implications of Mobile Email Devices for Knowledge Professionals." *Organization Science* 24 (5): 1337–57.

McCray, W. Patrick. 2000. "Large Telescopes and the Moral Economy of Recent Astronomy." *Social Studies of Science* 30 (5): 685–711.

———. 2006. *Giant Telescopes: Astronomical Ambition and the Promise of Technology*. 1st Harvard University Press paperback ed. Cambridge, MA: Harvard University Press.

———. 2012. *The Visioneers: How a Group of Elite Scientists Pursued Space Colonies, Nanotechnologies, and a Limitless Future*. Princeton, NJ: Princeton University Press.

McCurdy, Howard E. 1994. *Inside NASA: High Technology and Organizational Change in the U.S. Space Program*. Baltimore, MD: Johns Hopkins University Press.

McDonald, Steve. 2011. "What's in the 'Old Boys' Network? Accessing Social Capital in Gendered and Racialized Networks." *Social Networks* 33 (4): 317–30.

Medina, Eden. 2014. *Cybernetic Revolutionaries: Technology and Politics in Allende's Chile*. 1st paperback ed. Cambridge, MA: MIT Press.

Meltzer, Michael. 2014. *The Cassini-Huygens Visit to Saturn*. New York: Springer Verlag.

Merton, Robert K. 1938. "Science and the Social Order." In *The Sociology of Science: Theoretical and Empirical Investigations*, edited by Norman W. Storer, 321–37. Sociology of Science in Europe. Chicago: University of Chicago Press.

———. 1942. "The Normative Structure of Science." In *The Sociology of Science: Theoretical and Empirical Investigations*, edited by Norman W. Storer, 267–78. Sociology of Science in Europe. Chicago: University of Chicago Press.

———. 1965. *On the Shoulders of Giants*. New York: Harcourt, Brace & World.

———. 1968. "The Matthew Effect in Science." *Science* 159 (3810): 56–63.

Messeri, Lisa. 2014. "Earth as Analog: The Disciplinary Debate and Astronaut Training That Took Geology to the Moon." *Astropolitics* 12 (2–3): 196–209.

———. 2016. *Placing Outer Space: An Earthly Ethnography of Other Worlds*. Experimental Futures: Technological Lives, Scientific Arts, Anthropological Voices. Durham, NC: Duke University Press.

Messeri, Lisa, and Janet Vertesi. 2015. "The Greatest Missions Never Flown: Anticipatory Discourse and the 'Projectory' in Technological Communities." *Technology and Culture* 56 (1): 54–85.

Michel, Alexandra, and Stanton Emerson Fisher Wortham. 2009. *Bullish on Uncertainty: How Organizational Cultures Transform Participants*. New York: Cambridge University Press.

Milam, Erika, and Robert Nye, eds. 2015. *Scientific Masculinities*. Chicago: University of Chicago Press.

Miles, Raymond E., and Charles C. Snow. 1992. "Causes of Failure in Network Organizations." *California Management Review* 34 (4): 53–72.

Millerand, Florence, David Ribes, Karen S. Baker, and Geoffrey C. Bowker. 2013. "Making an Issue out of a Standard: Storytelling Practices in a Scientific Community." *Science, Technology & Human Values* 38 (1): 7–43.

Mirmalek, Z. 2020. *Making Time on Mars*. Cambridge, MA: MIT Press.

Mitroff, Ian. 1974. "Norms and Counter-Norms in a Select Group of the Apollo Moon Scientists." *American Sociological Review* 39:579–95.

Mody, Cyrus C. M. 2014. "Essential Tensions and Representational Strategies." In *Representation in Scientific Practice Revisited*, edited by Catelijne Coopmans, Michael Lynch, Janet Vertesi, and Steve Woolgar, 223–48. Cambridge, MA: MIT Press.

Mol, Annemarie. 2002. *The Body Multiple: Ontology in Medical Practice*. Durham, NC: Duke University Press.

Morrill, Calvin. 1991. "Conflict Management, Honor, and Organizational Change." *American Journal of Sociology* 97 (3): 585–621.

———. 1995. *The Executive Way: Conflict Management in Corporations*. Chicago: University of Chicago Press.

Myers, Natasha. 2008. "Molecular Embodiments and the Body-Work of Modeling in Protein Crystallography." *Social Studies of Science* 38:163–99.

Nader, Laura. 1972. "Up the Anthropologist: Perspectives Gained from Studying Up." In *Reinventing Anthropology*, edited by Dell Hymes, 284–311. Ann Arbor: University of Michigan Press.

Neff, Gina. 2012. *Venture Labor: Work and the Burden of Risk in Innovative Industries*. Acting with Technology. Cambridge, MA: MIT Press.

Nell Trautner, Mary, and Kwan, Samantha. 2010, "Gendered appearance norms: An analysis of employment discrimination lawsuits, 1970–2008." In *Research in the Sociology of Work*, Vol. 20, *Gender and Sexuality in the Workplace*, edited by Christine Williams and Kirsten Dellinger, 127–50. Bingley: Emerald Group.

Nelson, Robert M., Lucas W. Kamp, Rosaly M. C. Lopes, Dennis L. Matson, Randolph L. Kirk, Bruce W. Hapke, Stephen D. Wall, et al. 2009. "Photometric Changes on Saturn's Titan: Evidence for Active Cryovolcanism." *Geophysical Research Letters* 36 (4): L04202, 5pp.

Nichols, Kitty. 2018. "Moving beyond Ideas of Laddism: Conceptualising 'Mischievous Masculinities' as a New Way of Understanding Everyday Sexism and Gender Relations." *Journal of Gender Studies* 27 (1): 73–85.

Nixon, C. A., D. E. Jennings, J.-M. Flaud, B. Bézard, N. A. Teanby, P. G. J. Irwin, T. M. Ansty, A. Coustenis, S. Vinatier, and F. M. Flasar. 2009. "Titan's Prolific Propane: The Cassini CIRS Perspective." *Planetary and Space Science* 57 (13): 1573–85.

Nohria, Nitin, and James D. Berkley. 1994. "The Virtual Organization: Bureaucracy, Technology, and the Implosion of Control." In *The Post-bureaucratic Organization: New Perspectives on Organizational Change*, edited by Anne Donnellon and Charles C. Heckscher, 108–28. Thousand Oaks, CA: Sage.

O'Connell, Joseph. 1993. "Metrology: The Creation of Universality by the Circulation of Particulars." *Social Studies of Science* 23 (1): 129–73.

Olson, Gary M., Ann Zimmerman, and Nathan Bos, eds. 2008. *Scientific Collaboration on the Internet*. Cambridge, MA: MIT Press.

Olson, Gary M., and Judy Olson. 2000. "Distance Matters." *Human Computer Interaction* 15 (2): 139–78.

O'Reilly, Jessica. 2017. *The Technocratic Antarctic: An Ethnography of Scientific Expertise and Environmental Governance*. Expertise. Ithaca, NY: Cornell University Press.

Orlikowski, Wanda J. 2010. "The Sociomateriality of Organisational Life: Considering Technology in Management Research." *Cambridge Journal of Economics* 34 (1): 125–41.

Orlikowski, Wanda J., and JoAnne Yates. 1994. "Genre Repertoire: The Structuring of Communicative Practices in Organizations." *Administrative Science Quarterly* 39 (4): 541–74.

Owen-Smith, Jason, and Walter W. Powell. 2004. "Knowledge Networks as Channels and Conduits: The Effects of Spillovers in the Boston Biotechnology Community." *Organization Science* 15 (1): 5–21.

Paradis, Elise, and Mathieu Albert. 2013. "Separate and Unequal: Epistemic Habitus and Interdisciplinary Hierarchies in Canadian Medical Research." Paper presented at the Annual Meeting of the American Sociological Association, New York, August 10, 2013.

Pascoe, C. J. 2012. *Dude, You're a Fag: Masculinity and Sexuality in High School*. Berkeley: University of California Press.

Pedulla, David S. 2014. "The Positive Consequences of Negative Stereotypes: Race, Sexual Orientation, and the Job Application Process." *Social Psychology Quarterly* 77 (1): 75–94.

Peters, Benjamin. 2016. *How Not to Network a Nation: The Uneasy History of the Soviet Internet*. Information Policy. Cambridge, MA: MIT Press.

Peterson, David. 2015. "All That Is Solid: Bench-Building at the Frontiers of Two Experimental Sciences." *American Sociological Review* 80 (6): 1201–25.

Pinch, Trevor. 1986. *Confronting Nature: The Sociology of Solar-Neutrino Detection*. Dordrecht, Netherlands: D. Reidal.

Pinch, Trevor, and Wiebe E. Bijker. 1987. "The Social Construction of Facts and Artifacts: Or How the Sociology of Science and the Sociology of Technology Might Benefit Each Other." In *The Social Construction of Technological Systems: New Directions in the Sociology and History of Technology*, edited by Trevor Pinch, Wiebe E. Bijker, and Thomas P. Hughes. Cambridge, MA: MIT Press.

Polletta, Francesca. 2002. *Freedom Is an Endless Meeting: Democracy in American Social Movements*. Chicago: University of Chicago Press.

Popper, Karl R. 1962. *Conjectures and Refutations; the Growth of Scientific Knowledge*. New York: Basic Books.

Porter, Theodore. 1995. *Trust In Numbers: The Pursuit of Objectivity in Science and Public Life*. Princeton, NJ: Princeton University Press.

Powdermaker, Hortense. 2000. *Stranger and Friend: The Way of an Anthropologist*. New York: Norton.

Powell, Walter W., and Jeannette Colyvas. 2013. "The Micro-foundations of Institutional Theory." In *The SAGE Handbook of Organizational Institutionalism*, edited by Royston Greenwood, Christine Oliver, Kerstin Sahlin, and Roy Suddaby, 276–98. New York: Sage.

Powell, Walter W., Kenneth W. Koput, and Laurel Smith-Doerr. 1996. "Interorganizational Collaboration and the Locus of Innovation: Networks of Learning in Biotechnology." *Administrative Science Quarterly* 41 (1): 116.

Radin, Joanna. 2017. *Life on Ice: A History of New Uses for Cold Blood*. Chicago: University of Chicago Press.

Ramayah, T., J. Muhamad, M. N. Aizzat, and P. L. Koay. 2003. "Internal Group Dynamics, Team Characteristics and Team Effectiveness: A Preliminary Study of Virtual Teams." *International Journal of Knowledge, Culture, and Change Management* 3:415–35.

Rappaport, Nicole J., Pierre-Yves Longaretti, Richard G. French, Essam A. Marouf, and Colleen A. McGhee. 2009. "A Procedure to Analyze Nonlinear Density Waves in Saturn's Rings Using Several Occultation Profiles." *Icarus* 199 (1): 154–73.

Ray, Victor. 2019. "A Theory of Racialized Organizations." *American Sociological Review* 84 (1): 26–53.

Reagle, Joseph Michael. 2010. *Good Faith Collaboration: The Culture of Wikipedia*. History and Foundations of Information Science. Cambridge, MA: MIT Press.

Ribes, David, and Geoffrey C. Bowker. 2008. "Organizing for Multidisciplinary Collaboration: The Case of the Geosciences Network." In *Scientific Collaboration on the Internet*, edited by Gary M. Olson, Ann Zimmerman, and Nathan Bos, 310–30. Cambridge, MA: MIT Press.

Ribes, David, and Steven J. Jackson. 2013. "Data Bite Man: The Work of Sustaining a Long-Term Study." In *Raw Data Is an Oxymoron*, edited by Lisa Gitelman, 147–66. Cambridge, MA: MIT Press.

Ribes, David, and Thomas A. Finholt. 2009. "The Long Now of Technology Infrastructure: Articulating Tensions in Development." *Journal of the Association for Information Systems* 10 (5): 375–98.

Rico, R., E. Molleman, M. Sanchez-Manzanares, and G. S. Van der Vegt. 2007. "The Effects of Diversity Faultlines and Team Task Autonomy on Decision Quality and Social Integration." *Journal of Management* 33 (1): 111–32.

Ridgeway, Cecilia L. 2011. *Framed by Gender: How Gender Inequality Persists in the Modern World*. New York: Oxford University Press.

Ridgeway, Cecilia L., and Shelley J. Correll. 2004. "Unpacking the Gender System: A Theoretical Perspective on Gender Beliefs and Social Relations." *Gender and Society* 18 (4): 510–31.

Rivera, Lauren A. 2012. "Hiring as Cultural Matching: The Case of Elite Professional Service Firms." *American Sociological Review* 77 (6): 999–1022.

———. 2015. *Pedigree: How Elite Students Get Elite Jobs*. Princeton, NJ: Princeton University Press.

Rivera, Lauren A., and András Tilcsik. 2016. "Class Advantage, Commitment Penalty: The Gendered Effect of Social Class Signals in an Elite Labor Market." *American Sociological Review* 81 (6): 1097–131.

Robertson, I. P., T. E. Cravens, J. H. Waite, R. V. Yelle, V. Vuitton, A. J. Coates, J. E. Wahlund, et al. 2009. "Structure of Titan's Ionosphere: Model Comparisons with Cassini Data." *Planetary and Space Science* 57 (14): 1834–46.

Robinson, S. L., J. O'Reilly, and W. Wang. 2013. "Invisible at Work: An Integrated Model of Workplace Ostracism." *Journal of Management* 39 (1): 203–31.

Rosental, Claude. 2007. *Les capitalistes de la science: Enquête sur les démonstrateurs de la Silicon Valley et de la NASA*. Paris: CNRS.

Rossiter, M. W. 1993. "The Matthew Matilda Effect in Science." *Social Studies of Science* 23 (2): 325–41.

Rothschild-Whitt, Joyce. 1979. "The Collectivist Organization: An Alternative to Rational-Bureaucratic Models." *American Sociological Review* 44 (4): 509–27.

Rudman, Laurie A., and Kimberly Fairchild. 2004. "Reactions to Counterstereotypic Behavior: The Role of Backlash in Cultural Stereotype Maintenance." *Journal of Personality and Social Psychology* 87 (2): 157–76.

Rudman, Laurie A., and Peter Glick. 2001. "Prescriptive Gender Stereotypes and Backlash toward Agentic Women." *Journal of Social Issues* 57 (4): 743–62.

Šabanović, Selma. 2010. "Robots in Society, Society in Robots: Mutual Shaping of Society and Technology as a Framework for Social Robot Design." *International Journal of Social Robotics* 2 (4): 439–50.

Sacks, Harvey, E. A. Schegloff, and Gail Jefferson. 1974. "A Simplest Systematics for the Organisation of Turn-Taking for Conversation." *Language* 50:696–735.

Salonius, Annalisa. 2008. "Working in the Lab: Social Organization of Research and Training in Biomedical Research Labs in Canada and its Relationship to Research Funding." PhD diss., McGill University, Montreal, Canada.

Schaffer, S. 1988. "Astronomers Mark Time: Discipline and the Personal Equation." *Science in Context* 2 (1): 115–45.

Schairer, Cynthia. 2006. "Diffused Embodiment, Extended Visions: The Prosthetics of Martian Geology." Paper presented at the Society for Social Studies of Science, Vancouver, BC, Canada, November 2.

Schegloff, Emanuel. 1972. "Notes on a Conversational Practice: Formulating Place." In *Studies in Social Interaction*, edited by David Sudnow, 75–119. New York: Free Press.

———. 2007. *Sequence Organization in Interaction: A Primer in Conversation Analysis*. Cambridge, MA: Cambridge University Press.

Schiebinger, Londa. 1989. *The Mind Has No Sex? Women in the Origins of Modern Science*. Cambridge, MA: Harvard University Press.

Schippers, Mimi. 2007. "Recovering the Feminine Other: Masculinity, Femininity, and Gender Hegemony." *Theory and Society* 36 (1): 85–102.

Schivelbusch, Wolfgang. 1986. *The Railway Journey: The Industrialization of Time and Space in the 19th Century*. Berkeley: University of California Press.

Schug, Joanna, Nicholas P. Alt, and Karl Christoph Klauer. 2015. "Gendered Race Prototypes: Evidence for the Non-prototypicality of Asian Men and Black Women." *Journal of Experimental Social Psychology* 56:121–25.

Sellen, Abigail J., and Richard H. R. Harper. 2003. *The Myth of the Paperless Office*. 1st paperback ed. Cambridge, MA: MIT Press.

Shapin, Steven. 1989. "The Invisible Technician." 77 (6): 554–63.

———. 1994. *A Social History of Truth*. Chicago: University of Chicago Press.

Shapin, Steven, and Simon Schaffer. 1985. *Leviathan and the Air Pump: Hobbes, Boyle and the Experimental Life*. Princeton, NJ: Princeton University Press.

Sheehan, Michael. 2007. *The International Politics of Space*. 1st ed. New York: Routledge.

Shih, Johanna. 2004. "Project Time in Silicon Valley." *Qualitative Sociology* 27 (2): 223–45.

Shils, Edward. 1965. "Charisma, Order, and Status." *American Sociological Review* 30 (2): 199–213.

Shindell, Matthew Benjamin. 2010. "Domesticating the Planets: Instruments and Practices in the Development of Planetary Geology." *Spontaneous Generations: A Journal for the History and Philosophy of Science* 4 (1): 191–230.

Shrum, Wesley, Joel Genuth, and Ivan Chompalov. 2007. *Structures of Scientific Collaboration*. Inside Technology. Cambridge, MA: MIT Press.

Slaughter, S., T. Campbell, M. Holleman, and E. Morgan. 2002. "The 'Traffic' in Graduate Students: Graduate Students as Tokens of Exchange between Academe and Industry." *Science, Technology & Human Values* 27 (2): 282–312.

Smith, H. T., M. Shappirio, R. E. Johnson, D. Reisenfeld, E. C. Sittler, F. J. Crary, D. J. McComas, and D. T. Young. 2008. "Enceladus: A Potential Source of Ammonia Products and Molecular Nitrogen for Saturn's Magnetosphere." *Journal of Geophysical Research: Space Physics* 113 (A11): A112086, 11pp.

Smith-Doerr, Laurel. 2004. *Women's Work: Gender Equality vs. Hierarchy in the Life Sciences*. Boulder, CO: Lynne Reinner.

Squyres, S. W. 2005. *Roving Mars: Spirit, Opportunity, and the Exploration of the Red Planet*. New York: Hyperion.

Squyres, S. W., Raymond E. Arvidson, Diana L. Blaney, Benton C. Clark, Larry Crumpler, William H. Farrand, Stephen Gorevan, et al. 2006. "Rocks of the Columbia Hills." *Journal of Geophysical Research* 111 (E2): E02S11, 19pp.

Srivastava, Sameer B., and Mahzarin R. Banaji. 2011. "Culture, Cognition, and Collaborative Networks in Organizations." *American Sociological Review* 76 (2): 207–33.

Star, Susan Leigh, and James R. Griesemer. 1989. "Institutional Ecology, 'Translations,' and Boundary Objects: Amateurs and Professionals in Berkeley's Museum of Vertebrate Zoology, 1907–39." *Social Studies of Science* 19 (3): 387–420.

Stark, David. 2009. *The Sense of Dissonance: Accounts of Worth in Economic Life*. Princeton, NJ: Princeton University Press.

Starr, Paul. 2019. *Entrenchment: Wealth, Power, and the Constitution of Democratic Societies*. New Haven, CT: Yale University Press.

Steinhardt, Stephanie B., and Steven J. Jackson. 2014. "Reconciling Rhythms: Plans and Temporal Alignment in Collaborative Scientific Work." In *CSCW '14 Proceedings of the 17th ACM Conference on Computer Supported Cooperative Work & Social Computing*, 134–45. New York: ACM.

———. 2015. "Anticipation Work: Cultivating Vision in Collective Practice." In *Proceedings of the 18th ACM Conference on Computer Supported Cooperative Work & Social Computing*, 443–53. New York: ACM.

Stinchcomb, A. L. 1965. "Social Structures and Organizations." In *Handbook of Organizations*, edited by James G. March, 142–93. Chicago: Rand McNally.

Suchman, Lucy. 2011. "Subject Objects." *Feminist Theory* 12 (2): 119–45.

Sutton, Robert I. 2010. *The No Asshole Rule: Building a Civilized Workplace and Surviving One That Isn't*. New York: Business Plus.

Tajfel, Henri. 1982. "Social Psychology of Intergroup Relations." *Annual Review of Psychology* 33:1–39.

Taylor, T. L. 2012. *Raising the Stakes: E-Sports and the Professionalization of Computer Gaming.* Cambridge, MA: MIT Press.

Thompson, Charis. 2005. *Making Parents: The Ontological Choreography of Reproductive Technologies.* Inside Technology. Cambridge, MA: MIT Press.

Thornton, Patricia H., William Ocasio, and Michael Lounsbury. 2012. *The Institutional Logics Perspective a New Approach to Culture, Structure, and Process.* Oxford: Oxford University Press.

Thorpe, Charles, and Steven Shapin. 2000. "Who Was J. Robert Oppenheimer? Charisma and Complex Organization." *Social Studies of Science* 30 (4): 545–90.

Tiscareno, Matthew S., Randall P. Perrine, Derek C. Richardson, Matthew M. Hedman, John W. Weiss, Carolyn C. Porco, and Joseph A. Burns. 2010. "An Analytic Parameterization of Self-Gravity Wakes in Saturn's Rings, with Application to Occultations and Propellers." *The Astronomical Journal* 139 (2): 492–503.

Tosi, F., R. Orosei, R. Seu, A. Coradini, J. I. Lunine, G. Filacchione, A. I. Gavrishin, et al. 2010. "Correlations between VIMS and RADAR Data over the Surface of Titan: Implications for Titan's Surface Properties." *Icarus* 208 (1): 366–84.

Traweek, Sharon. 1988. *Beamtimes and Lifetimes: The World of High Energy Physicists.* Cambridge, MA: Harvard University Press.

Tsing, Anna Lowenhaupt. 2005. *Friction: An Ethnography of Global Connection.* Princeton: Princeton University Press.

Turco, Catherine. 2016. *The Conversational Firm.* New York: Columbia University Press.

Turner, Fred. 2006. *From Counterculture to Cyberculture: Stewart Brand, the Whole Earth Network, and the Rise of Digital Utopianism.* Chicago: University of Chicago Press.

———. 2009. "Burning Man at Google: A Cultural Infrastructure for New Media Production." *New Media & Society* 11 (1–2): 73–94.

Van Maanen, John, and Gideon Kunda. 1989. "'Real Feelings': Emotional Expression and Organization Culture." In *Research in Organizational Behavior*, vol. 11, edited by Barry M. Staw and L. L. Cummings, 43–103. Greenwich, CT: JAI.

Van Maanen, John, and Stephen R. Barley. 1985a. "Cultural Organization: Fragments of a Theory." In *Organizational Culture*, edited by P. J. Frost, L. F. Moore, M. R. Louis, C. C. Lundberg, and J. Martin, 31–53. Beverly Hills: Sage.

———. 1985b. "Occupational Communities: Culture and Control in Organizations." In *Research in Organizational Behavior*, vol. 6, edited by Barry M. Staw and L. L. Cummings. Greenwich, CT: JAI.

Vaughan, Diane. 1996. *The Challenger Launch Decision: Risky Technology, Culture, and Deviance at NASA.* Chicago: University of Chicago Press.

———. 1999. "The Rôle of the Organization in the Production of Techno-scientific Knowledge." *Social Studies of Science* 29 (6): 913–43.

Vedres, Balázs, and David Stark. 2010. "Structural Folds: Generative Disruption in Overlapping Groups." *American Journal of Sociology* 115 (4): 1150–90.

Vertesi, Janet. 2014. "Seamful Spaces: Heterogeneous Infrastructures in Interaction." *Science, Technology & Human Values* 39 (2): 264–84.

———. 2015. *Seeing like a Rover: How Robots, Teams, and Images Craft Knowledge of Mars.* Chicago: University of Chicago Press.

Vertesi, Janet, and Paul Dourish. 2011. "The Value of Data: Considering the Context of Production in Data Economies." In *CSCW '11 Proceedings of the ACM 2011 Conference on Computer Supported Cooperative Work*, 533. New York: ACM Press.

Vertesi, Janet, Jofish Kaye, Samantha N. Jaroszewski, Vera D. Khovanskaya, and Jenna Song. 2016. "Data Narratives: Uncovering Tensions in Personal Data Management."

In *CSCW '11 Proceedings of the ACM 2011 Conference on Computer Supported Cooperative Work*, 477–89. New York: ACM Press.

Volkoff, Olga, Diane M. Strong, and Michael B. Elmes. 2007. "Technological Embeddedness and Organizational Change." *Organization Science* 18 (5): 832–48.

Waite, J. H., Jr., W. S. Lewis, B. A. Magee, J. I. Lunine, W. B. McKinnon, C. R. Glein, O. Mousis, et al. 2009. "Liquid Water on Enceladus from Observations of Ammonia and ^{40}Ar in the Plume." *Nature* 460 (7254): 487–90.

Wajcman, Judy. 2015. *Pressed for Time: The Acceleration of Life in Digital Capitalism*. Chicago: University of Chicago Press.

Watson, Warren E., Kamalesh Kumar, and Larry K. Michaelsen. 1993. "Cultural Diversity's Impact on Interaction Process and Performance: Comparing Homogeneous and Diverse Task Groups." *Academy of Management Journal* 36 (3): 590–602.

Weber, Max. 1968. *Economy and Society: An Outline of Interpretive Sociology*. New York: Bedminster Press.

———. 2004. *The Vocation Lectures*. Edited by David S. Owen and Tracy B. Strong. Translated by Rodney Livingstone. Indianapolis: Hackett.

Weeks, John. 2004. *Unpopular Culture: The Ritual of Complaint in a British Bank*. Chicago: University of Chicago Press.

Weitz, C. M., R. C. Anderson, J. F. Bell III, W. H. Farrand, K. E. Herkenhoff, J. R. Johnson, B. L. Jolliff, R. V. Morris, S. W. Squyres, and R. J. Sullivan. 2006. "Soil Grain Analyses at Meridiani Planum, Mars." *Journal of Geophysical Research: Planets* 111 (E12): E12S04, 26pp.

Wessen, Randii R., and David Porter. 1998. "Market-Based Approaches for Controlling Space Mission Costs: The Cassini Resource Exchange." *Journal of Reducing Space Mission Cost* 1 (1): 9–25.

Westwick, Peter. 2007. *Into the Black: JPL and the American Space Program, 1976–2004*. New Haven, CT: Yale University Press.

Whittington, Kjersten Bunker, and Laurel Smith-Doerr. 2005. "Gender and Commercial Science: Women's Patenting in the Life Sciences." *Journal of Technology Transfer* 30 (4): 355–70.

Whyte, William H. 1956. *The Organization Man*. Garden City, NJ: Doubleday.

Winner, Langdon. 1993. "Upon Opening the Black Box and Finding It Empty: Social Constructivism and the Philosophy of Technology." *Science, Technology & Human Values* 18 (3): 362–78.

Wolfe, Tom. 1980. *The Right Stuff*. London: Bantam.

Woolgar, Steve, and Javier Lezaun. 2013. "The Wrong Bin Bag: A Turn to Ontology in Science and Technology Studies?" *Social Studies of Science* 43 (3): 321–40.

Wynholds, Laura A., Jillian C. Wallis, Christine L. Borgman, Ashley Sands, and Sharon Traweek. 2012. "Data, Data Use, and Scientific Inquiry: Two Case Studies of Data Practices." In *JCDL '12 Proceedings of the 12th ACM/IEEE-CS Joint Conference on Digital Libraries*, 19. New York: ACM.

Xie, Yu, and Kimberlee A. Shauman. 2003. *Women in Science: Career Processes and Outcomes*. Cambridge, MA: Harvard University Press.

Yates, JoAnne, and Wanda J. Orlikowski. 2007. "The PowerPoint Presentation and Its Corollaries: How Genres Shape Communicative Action in Organizations." In *Communicative Practices in Workplaces and the Professions: Cultural Perspectives on the Regulation of Discourse and Organizations*, edited by Mark Zachry and Charlotte Thralls, 67–92. Amityville, NY: Baywood.

Zablocki, Benjamin David. 1980. *Alienation and Charisma: A Study of Contemporary American Communes*. New York: Free Press.

Zabusky, Stacia E. 1995. *Launching Europe: An Ethnography of European Cooperation in Space Science*. Princeton, NJ: Princeton University Press.

Zelizer, Viviana A. 2005. *The Purchase of Intimacy*. Princeton, NJ: Princeton University Press.

———. 2012. "How I Became a Relational Economic Sociologist and What Does That Mean?" *Politics & Society* 40 (2): 145–74.

Zerubavel, Eviatar. 1982. "The Standardization of Time: A Sociohistorical Perspective." *American Journal of Sociology* 88 (1): 1–23.

———. 2004. *Time Maps: Collective Memory and the Social Shape of the Past*. Chicago: University of Chicago Press.

Ziewitz, Malte. 2011. "How to Attend to Screens? Technology, Ontology and Precarious Enactments." *Encounters* 4 (2): 203–28.

Zucker, Lynne G. 1977. "The Role of Institutionalization in Cultural Persistence." *American Sociological Review* 42 (5): 726–43.

Zuckerman, Ezra W. 2010. "Speaking with One Voice: A 'Stanford School' Approach to Organizational Hierarchy." In *Stanford's Organization Theory Renaissance, 1970–2000*, 289–307. Research in the Sociology of Organizations 28. Bingley, UK: Emerald Group.

INDEX